AF541036

The Birds of the Delhi Area

Sudhir Vyas

Principal Photographer: Amit Sharma

juggernaut

JUGGERNAUT BOOKS
C-I-128, First Floor, Sangam Vihar, Near Holi Chowk,
New Delhi 110080, India

First published by Juggernaut Books 2023

10 9 8 7 6 5 4 3 2 1

P-ISBN: 9789353451646
E-ISBN: 9789353451653

Book design by Mugdha Sethi

Additional typesetting support by R. Ajith Kumar

Maps prepared by Anis Khan
The international boundaries on the maps of India are neither purported to be correct nor authentic by Survey of India directives.

Printed at Thomson Press India Ltd

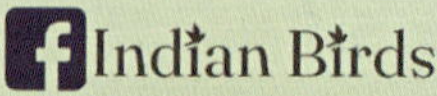

This publication was conceptualized and supported by Kanwar Bir Singh, founder of the Indian Birds community on Facebook, the largest and most active online community of bird enthusiasts in India.

Contents

Introduction

The Delhi neighbourhood is remarkable for the wealth and diversity of its avifauna, and the city itself is one of the most bird-rich national capitals in the world. With its variety of habitats and landscapes, more than 470 bird species have been recorded in areas that are within a day's trip from the city. However, the Delhi area has also witnessed large-scale changes over recent decades. Population growth and burgeoning urban sprawl have, and continue to, exert formidable pressures on the land and critical bird habitats.

Over the last century and more, Delhi has been privileged to have had a small but active group of individuals interested in ornithology call the city home, their notes illustrating in vivid detail how Delhi's birdlife has evolved with the city itself. More recently, and especially since the late 1990s, a greater awareness, the increased availability of quality field guides and optical equipment and the spread of internet-based groups and social media channels have fostered the growth of a vibrant community of bird enthusiasts; drawn from all walks of life, of different age groups and working in varied professions, many are refreshingly competent and knowledgeable birdwatchers and photographers today.

The diversity of the region's avifauna has been chronicled over the years, starting with Maj Gen. Hutson's *The Birds about Delhi* (1954), *Usha Ganguli's A Guide to the Birds of the Delhi Area* (1975) and the *Bird Atlas of Delhi and Haryana* compiled by Bill Harvey et al. (2006).

More than fifteen years of residence in Delhi, home postings during a foreign service career straddling four decades since the 1970s, offered opportunities to watch and enjoy the birds of India's capital city and its neighbourhood. This experience has given me a perspective of the profound changes that have impacted its environment and its avifauna over the years and continue to do so. This book builds upon my monograph on the birds of the Delhi area, *'The Birds of the Delhi Area: An Annotated Checklist'*, published in the online journal *Indian Birds* (2019), now extensively redone in the light of the new material and data accumulated since then. This is a compendium of up-to-date information on Delhi's birdlife for anyone interested in the ornithology of the city.

This is not a field guide as several excellent resources are already available. It has a dual objective: first, for the growing community of Delhi's birdwatchers, it attempts to provide insights into some deeper and broader aspects of the birds they see around them – beyond simple identification – to touch upon other facets and to contextualize the presence of a bird species in the Delhi area within its overall range and movements in the hope that it will encourage them to explore some of these fascinating aspects in as much detail as may catch their fancy.

As a second objective, it establishes an updated, thoroughly researched and annotated checklist of the birds of the Delhi area based upon a collation of reliable and verified information and data from multiple sources. This should benefit and be of use to Delhi's birdwatching community and meet the specific needs of policy-makers and those engaged in conservation fields as well.

At the time of writing, the Delhi checklist stands at 471 species, with another 22 that have not been re-recorded since 1975, with all records antedating that year being considered historical. I have used 31 March 2023 as the last date for the inclusion of new records. Any omissions and errors in this work remain my own.

Finally, I wish to acknowledge the encouragement, support and advice generously given by Anita Mani and her Indian Pitta imprint, and Kanwar Bir Singh, in bringing this project to fruition. I remain indebted to Amit Sharma whose outstanding bird and habitat images illuminate the text, along with those by Abhishek Gulshan, Anand Arya, Ashish Loya, Debashish Chakravarti, Janardhan Barthwal, K.B. Singh, M. Ramgopal, Rohit Sharma, Vijay Dhasmana and Vijay Kumar Sethi, who have all contributed their photographs that add to the effort. I should mention that all these images are from the Delhi area itself, except for a very few which are flagged as such in their captions. And my grateful thanks for the constant attention and the contributions of the members of the Juggernaut Books Pvt. Ltd team, Arani Sinha and Devangana Ojha, during the preparation of this book, and the immense creative effort put in by Mugdha Sethi, who designed it. My thanks as well to Anis Khan for designing the various maps, and to the growing community of Delhi birders, whose bird observations and records, both in print and online, have provided so much of the source material for this work.

1 The Delhi Area

Take a satellite map of Delhi and its neighbourhood – within, say, a radius of about 100 km, roughly the area that can be covered on a day's outing from the city – and then, in your mind's eye, if you peel away the layers upon layers of human impact on this historic city and the urban sprawl that marks it today, the topography of the land would be thrown up in bold relief.

The Yamuna River would be the most prominent feature, marking the western fringes of the Gangetic Plain. To its west, a range of low, eroded hills, comprising the Delhi Ridge, would appear as a crooked Y shaped landform; these are the northernmost extensions of the Aravalli Hills that extend diagonally for 700 km across eastern Rajasthan to southern Haryana and Delhi, the oldest mountain range in India. The Aravallis enter the National Capital Territory of Delhi (NCT) from Gurugram (formerly Gurgaon) district of Haryana, where the two arms of the 'Y' diverge, one carrying on northwards past Mehrauli and New Delhi to reach the right bank of the Yamuna north of Shahjahanabad, while the second branches to the northeast as a broad belt of rocky, scrubby ravines that extends from Sohna, in Haryana, to terminate at the picturesque ruins of Tughlaqabad. These features separate two drainages, the floodplains of the upper Ganga–Yamuna basin on their east from the drier, sandier tracts of Rajasthan and western Haryana.

The area we are looking at covers the political extent of the NCT with its satellite towns and rural landscapes on both sides of the Yamuna; thus, broadly, this includes the districts of Sonipat and Panipat (but short of Karnal in Haryana) towards the north; Shamli, Baghpat, Ghaziabad, GautamBuddha Nagar, Hapur, Meerut and Bulandshahr (but short of Muzaffarnagar and Aligarh in UP) towards the east; Gurugram, Faridabad, Palwal and Nuh (up to Kosi Kalan) towards the south; and Rohtak, Jhajjar, Dadri and parts of Rewari district (but short of Mahendragarh in Haryana) to the west.

Southern Ridge, south of Gurugram

The area has a continental climate with hot, dry summers with temperatures sometimes rising to 46°C in some years, and brilliant, bracing winters with cool, crisp air and skies of unblemished blue, of which Delhi used to be proud before these pleasures were lost to pollution and smog. It receives about 60 cm of rain annually, mainly between July and September, when the monsoon raises the humidity to uncomfortable levels. The natural vegetation is xerophytic in character and can withstand these fluctuations in temperature and humidity. It has features in common with the flora of eastern Rajasthan, the Punjab plains and the drier Upper Gangetic Plain, dry monsoon forest types adapted to a semi-arid environment.

Delhi's topography has anchored the many city incarnations in the past, from historic Indraprastha, and Lal Kot and Qila Rai Pithora of the eighth and twelfth centuries respectively to the seventeenth-century Shahjahanabad and twentieth-century New Delhi. These geographical features have impacted the composition of its birdlife as well.

A Changing City

As the saying goes, the only constant is change, and so it is in Delhi, as few parts of the world have been subjected to such intensive human impact over the centuries. The area has witnessed exponential growth on limited natural resources.

Contemporary photographs from around the time King George V announced the transfer of the seat of Indian government from Calcutta to Delhi in 1911 show much of the country southwards of Shahjahanabad as a bare wasteland, where 'the eye swept "a vast manworn plain" littered with the crumbling debris of earlier, abandoned Delhis'.[1] North of the old city, the land was lowlying, subject to inundation by monsoon floods in the wet season.

The making of the new capital, inaugurated in 1931, was a carefully planned affair. Raisina Hill, an outcrop of the Delhi Ridge south of Shahjahanabad, was chosen as its focal point. Sydney Basil-Edwardes, one of the earliest ornithologists to study the birds of the area in the winter of 1924–25, when the city was actually being built, describes the landscape thus:[2]

> ... Raisina, or New Delhi as it is also called, is a gently undulating, open tract–a city and its environs in embryo, which is developing rapidly. The country round about consists of open spaces, covered either with short grass or low, scrubby bushes, with the babool *Acacia arabica* growing freely... Deserted towns and tombs lie in every direction... The banks of the River Jumna [sic] are, in winter, bare stretches of sand on which the tamarix or jhow *Tamarix dioica* grows in great profusion. At Okla is situated a weir and sluice-gates where the waters of the Jumna are controlled and diverted by canals. Some sort of a park and garden exists here – a pleasing contrast to the bare and rather monotonous aspect of the banks which stretch out on either side beyond the grounds at Okla. The historic Ridge of Delhi, which is an extension of the Mewat branch of the Aravalli Hills, forms a conspicuous and characteristic physical feature ... Off the Karnal Road lies the comparatively large 'Horse-shoe Jheel', so-called on account of its shape.

What a contrast to the current scenario! The new capital, planned on the level ground between the Ridge and the Yamuna River, was designed to be a garden city, 'in the main a sea of foliage', with the tree species planted to line its main avenues being selected for their shade and evergreen quality.[3] Most were not native to the area but they, along with the ornamentals and other trees planted after Independence, define New Delhi's environment today.

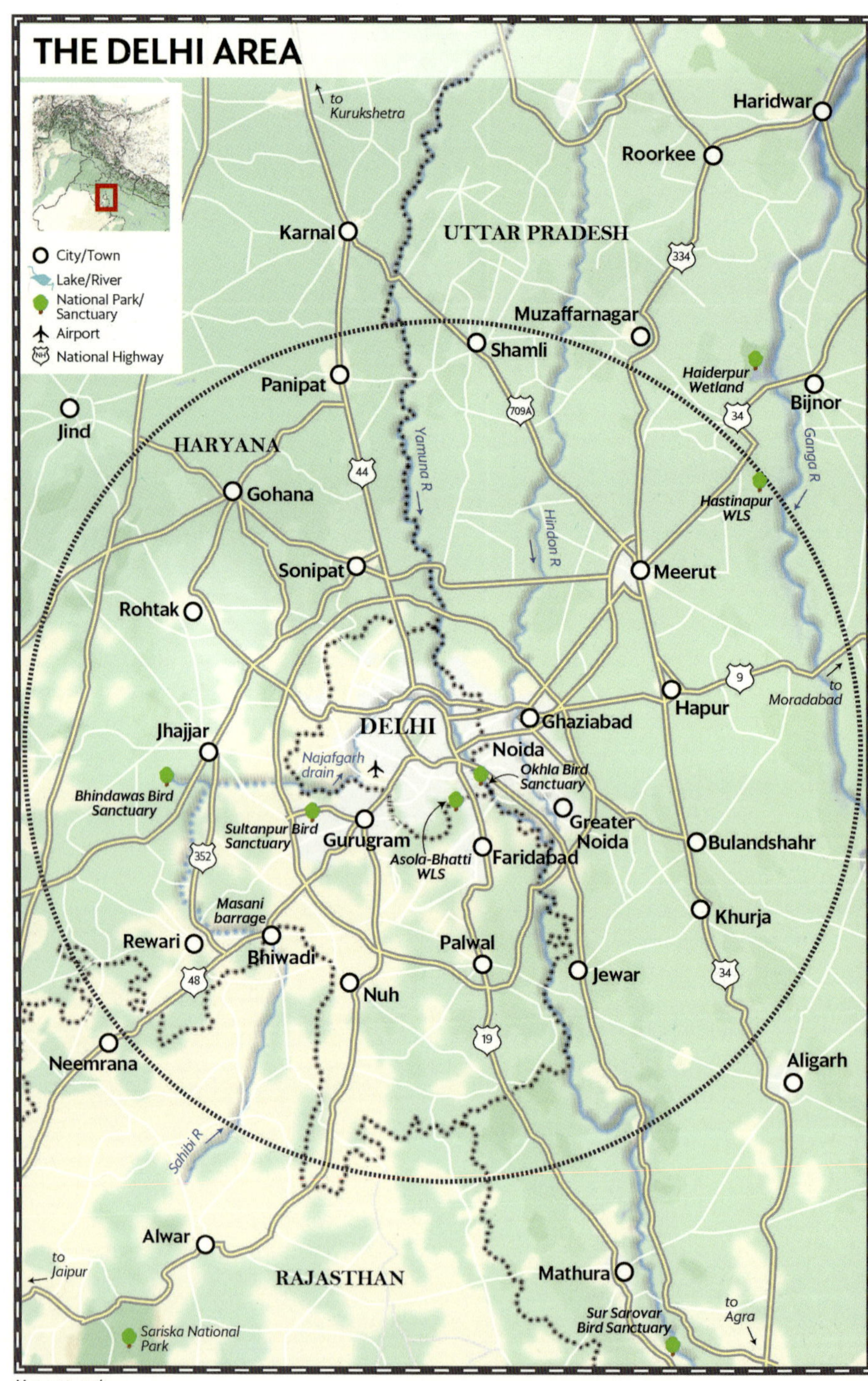

Map not to scale

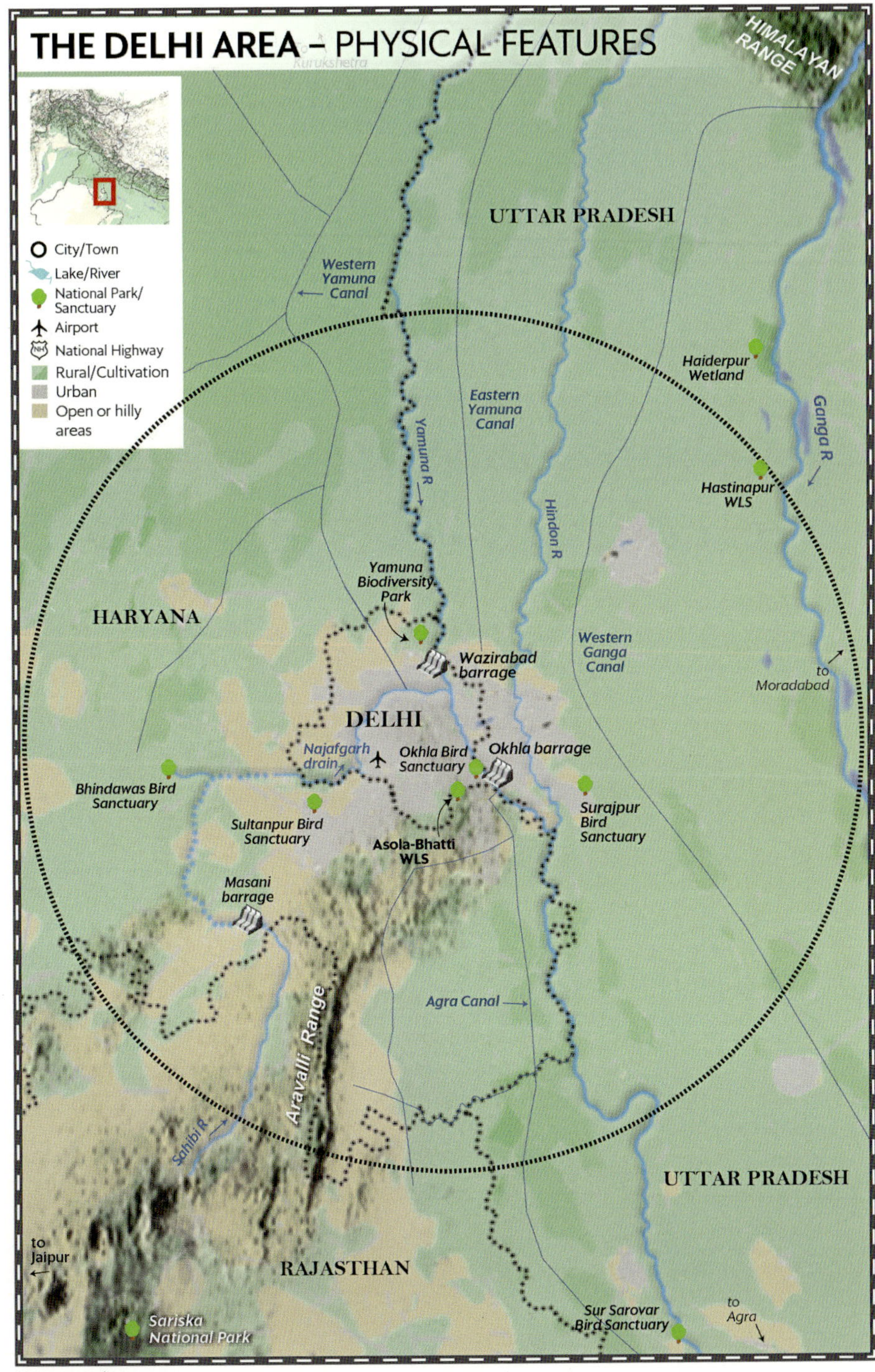

Map not to scale

The early planned colonies, catering to urgent institutional and housing needs after Independence, were designed for denser habitation than Edwin Lutyens' wooded New Delhi. Yet, they provided for recreational public spaces, city forests and other green areas, opening up opportunities for colonization by bird species adapted to wooded parkland, replacing those of the earlier, open landscapes.

The last seventy years, however, have seen the population of the Delhi area explode with regrettably poorly planned urbanization and consequently serious environmental damage. Delhi alone has a population of over 19 million people, according to 2021 estimates, of a total of over 32 million people in the NCT of 1,483 sq km, the world's second-most populous urban area after Tokyo. It now envelops within its urban sprawl several villages, agricultural land, historical sites and a large part of the low rocky landscapes of the Southern Ridge.

Even so, the area still holds a diversity of habitats that support a remarkably rich avifauna, a tribute to the ability of wildlife to survive as long as they have some space and to the abiding reverence Indians have for life and nature, which has contributed so much to this outcome.

Habitats of the Delhi Area

Two terrestrial ecoregions[4] of the Indian subcontinent overlap in the Delhi area, the Aravalli west thorn scrub ecoregion and the Upper Gangetic Plain ecoregion. From an ornithological viewpoint, however, the area's habitats fall into three natural divisions: (a) the hilly tracts, comprising the eroded northernmost outliers of the Aravalli Range (the 'Delhi Ridge') and isolated outcrops to the west, rocky with xerophytic thorny vegetation; (b) the fertile Upper Gangetic Plain in Haryana and UP, now urbanized or under cultivation; and (c) the low-lying saline and sandy plains southwest of the hilly tracts, which begin to show desert influences.[5] Delhi's urban habitats also offer a refuge for various bird species and should perhaps be considered as a fourth habitat type for the purposes of this study.

(a) The Aravalli Hills

The northwestern (or Aravalli west) thorn scrub ecoregion stretches from Rajasthan east to the Aravalli Ranges and covers their northern outliers in southern Haryana and the Ridge in Delhi. These rocky, arid tracts with thin soil naturally support an open, dry thorn flora dominated by species of *babool, kikar, kareel, peelu*, etc.

Delhi's Ridge landscapes (known as *kohi*) have been fragmented as the city expanded over the hilly tract into four parts: the Old Delhi Ridge near Delhi University, which is now just 87 ha (1.13% of the total Ridge area); the New Delhi or Central Ridge is 864 ha (11.1%); and further south, the Mehrauli or South-Central Ridge is 626 ha (8.05%). The fourth segment, the Tughlaqabad or Southern Ridge, is the last contiguous stretch of the hilly landform and is of special significance for the area's wildlife. It forms a broad belt of 6,200 ha (area within Delhi's state boundaries, 79.72%) of scrubby, wooded ravines that extend for 30 km from Tughlaqabad, Surajkund and the Asola-Bhatti

A White-browed Fantail dances in the Mangar Forest.

Wildlife Sanctuary, south through the Gurugram and Faridabad Aravallis to Damdama Lake near Sohna in Haryana. Though sandwiched between the rapidly expanding satellite towns of Faridabad and Gurugram to the east and west, crossed by roads and with much of its land being village owned or encroached upon, the Southern Ridge is still of high ecological value.

The Aravalli landforms extend further to the west and south in Haryana over six districts: Gurugram, Faridabad, Palwal, Nuh, Rewari and Bhiwani. Some of the isolated hill outcrops are clothed in dry, deciduous forest, as near Manethi, Rewari, the habitat for bird endemics such as the White-naped Tit *Machlolophus nuchalis* and Spotted Creeper *Salpornis spilonota*.

A few valleys, the best surviving example being the Mangar valley in Faridabad district, have deeper soil and support a more diverse climax forest community, dominated by *dhau*. Mangar Bani,[6] a sacred grove measuring 2.66 sq km, offers a glimpse of what some of these sheltered ravines may have been like in their natural state. It is a magical place, its ecological value demonstrated by the presence of apex carnivores like leopards *Panthera pardus* and hyenas *Hyaena hyaena*; as part of the Southern Ridge, it serves as a biogeographical corridor for the movement of wildlife between the Damdama Lake in the south of Gurugram (and even further south from the Sariska National Park in Rajasthan) to the Asola-Bhatti Wildlife Sanctuary in the north, enabling diverse mammal, bird and reptilian species – many with peninsular affinities – to establish themselves in the Delhi area. The expanse of forest is also critical for preserving the groundwater aquifers on which Gurugram and Faridabad are dependent.

As conceived by its planners and architects in the early twentieth century, the Ridge formed the backdrop to New Delhi and a programme of afforestation was initiated from 1914. In place of the stunted native vegetation, however, the Ridge was planted with exotic *vilayati kikar* and a few other species not native to the area. *Vilayati kikar* has since spread invasively around Delhi, naturally and as a result of ill-considered Forest Department programmes towards 'greening' hilly areas, tending to dominate the native flora and damage its ecological character. These hills have also been subjected to rampant stone quarrying, which is now prohibited by legal decree. Abandoned quarries have left the land scarred with pits that form seasonal ponds and a few deeper perennial lakes on the Southern Ridge.

THE SOUTHERN RIDGE

Map not to scale

It is only recently that public awareness, judicial directives and policy action have come together to check some of these damaging impacts. The Ridge is increasingly recognized as critical for the ecological security of the city and was notified as a reserved forest in 1994. Ambiguities, however, remain in several areas, and ground-truthing, validation and demarcation is still work in progress especially on the Southern Ridge.

Delhi's Biodiversity Parks

The Delhi Development Authority's (DDA) ongoing, and justly acclaimed, programme of developing biodiversity parks in and around urban Delhi has, as its 'prime goal ... the conservation and preservation of ecosystems of the two major life supporting systems of Delhi, the Yamuna River and the Aravalli hills. They seek to preserve the biodiversity of any habitat that is likely to be converted into urban infrastructure, to conserve keystone species and other threatened plant and animal species, establish field gene banks for threatened land races and wild genetic resources, promote education on environmental awareness and nature conservation, establish native communities of the Aravalli hills and the Yamuna River basin particularly of the Delhi region, develop mosaic of treatment and catchment wetlands that not only improve the water quality of untreated sewage but also sustain the rich aquatic flora and fauna of the Yamuna and monitor short term and long term changes in the ecology of the Delhi region.'
(Extract from the Delhi Biodiversity Foundation DDA website.[7])

Biodiversity parks are an innovative approach for the rehabilitation and conservation of damaged or lost natural heritage, and recreate the original ecological integrity of a landscape, simulating a sanctuary or nature reserve. Delhi's biodiversity parks are fully functional nature reserves and are useful models for replication elsewhere.

The contiguous Asola (1880 ha) and Bhatti (860 ha) wildlife sanctuaries (also referred to in the singular as the Asola-Bhatti sanctuary as they are contiguous) on the Southern Ridge were notified by the Delhi State Government in 1986 and 1991 to protect their characteristic semi-arid forest habitat and dependent wildlife. The ecological rehabilitation of these sanctuaries, carried out under the aegis of an eco task force since the year 2000, has shown very positive results. Asola also hosts the Bombay Natural History Society's (BNHS) Conservation Education Centre on its premises.

The Aravalli Biodiversity Park, Gurugram. Rewilding through ecological restoration.

Five of the seven biodiversity parks[8] currently being developed by the Delhi government are situated on the Ridge. These are the Kamala Nehru Biodiversity Park (122 ha), which includes the Old Delhi Ridge with its adjoining public areas and protected historical monuments; the Aravalli Biodiversity Park (283 ha) and Neela Hauz (part of the Sanjay Van Reserve Forest, 317 ha) which together protect over 600 ha on the South-Central Ridge; and the Tilpath Valley (70 ha) and Tughlaqabad Biodiversity Parks (81 ha) on the Southern Ridge.

The Asola-Bhatti Wildlife Sanctuary and these five biodiversity parks are further complemented by relict patches of natural ridge-type habitat that have survived in various places in Delhi. The Jawaharlal Nehru University (JNU) campus adjacent to Sanjay Van, for example, parts of the Hauz Khas and Jahanpanah city forests, the Mehrauli Archaeological Park and several 'wasteland' areas on the South-Central and Southern Ridges collectively create a mosaic of natural habitats, albeit fragmented, that offer a refuge for Delhi's birdlife.

In Haryana, non-governmental efforts with corporate and municipal support have also successfully restored 153.7 ha of hilly, rocky areas that were damaged and scarred by past mining and misuse as the Aravalli Biodiversity Park, paralleling Delhi's biodiversity parks initiative to re-create diverse natural habitats and regenerate aquifers. Elsewhere in the state, however, ecological degradation in the Aravalli landscapes remains a concern. While Asola–Bhatti is protected as a wildlife sanctuary in Delhi, the adjacent Faridabad and Gurugram Aravallis in Haryana are not. The administrative status of the Southern Ridge forest corridor is a patchwork; it does not enjoy full legal protection and remains threatened by real estate encroachment, fragmentation and land use change.

In 2016, the core area of Mangar Bani, measuring 270 ha. with a 500 m buffer zone comprising another 486 ha, was notified as a 'no construction zone', but it is yet to be formally granted reserved forest status. Nevertheless, this step, along with a 40.5 ha nature park set up at Bhondsi, hopefully signals a more far-sighted approach, but will require stronger legal enactments to protect the forest corridor so that these areas do not become ecologically unviable 'islands' isolated in an urban agglomeration.

Mangar Bani, salai tree overlooking the Bani

(b) The Upper Gangetic Plain

Beyond these hilly tracts, the landscapes of the Delhi area are part of the alluvial Upper Gangetic Plain ecoregion of the Ganga–Yamuna river system that covers most of the plains of UP and adjacent portions of Haryana. They include *bangar* lands, cultivated plains of old alluvium that define the majority of the ecoregion, alternating with *khadar*, the flood-prone lands of new alluvium enriched by fresh deposits of silt every year along the braided channels of the major rivers.

These plains are amongst the most intensely farmed areas in the world, rich in groundwater sources and irrigated through a network of canals. Extremely limited patches of the original flora remain, and it is difficult even to assign a particular vegetation type to it with any certainty. These relict patches support small areas of grassland and scrub in uncultivated or uncultivable areas, and the remaining deciduous woodland is dominated by *babool*/*ronjh*, *kikar* and *pipal*, with greater tree diversity in areas of richer soil, and *kareel*, *bistendu*, *neem* and *ber* in drier areas. The *semal*/silk cotton, *palash*/*dhak* and *amaltas* trees add a wonderful range of colours to the landscape as the weather warms up after the winter months. However, much of the vegetation now includes a number of non-native species, planted as ornamentals, avenue or shade trees along canals, in orchards and for commercial forestry. Uncultivated, seasonally flooded areas near wetland depressions and in the *khadar* have their own characteristic flora of various grasses

Asian Openbills at Dighal. Village lakes offer habitat for waterbirds.

like *kans*, *sarkanda*, *ulu*, *khus*, etc. with reedmace, tamarisks in sandy tracts and remnant riverine forest patches.

The richest and most biologically diverse habitats of these plains are their wetlands – the Yamuna and Ganga rivers (and their tributaries, notably the Hindon), the *khadar* floodplains and impoundments elsewhere, both natural and man-made. Many habitats have been lost to drainage, failing aquifers, encroachment or pollution, but the Delhi area still retains valuable remnants that can hopefully be conserved. As the demands for land and water intensify in the region, each wetland, barrage or weir, with its constituent habitats, will assume greater importance as a refuge for breeding, resting and feeding waterfowl. A sustainable conservation

The *Bangar* Plains: Farmlands, Fallows and Scrub

Cultivation on the *bangar* plains around Delhi is on relatively small, diversified landholdings, not extensive monocultures that spread over the countryside. These are traditional agricultural landscapes irrigated by old, well-established canal systems, alongside a long-standing practice of multifunctional agroforestry. They, thus, form a mosaic of varied, human-modified habitats with bits of remnant grasslands, fallows and overgrazed patches of ground interspersed amongst the farms (often with sparse *babool*, *ber*, *pipal* and other trees), village and temple groves and orchards, tree-lined roads and canal embankments, etc. Such localized habitat patches are quite widespread across the Delhi area, albeit scattered and usually limited in extent, but they offer a structural and ecological diversity that can support a range of wildlife. Unfortunately, they tend to be categorized as unproductive and unprotected 'wastelands', always at risk of encroachment and land use change and as dumping grounds. They nevertheless play their own role in the rural economy and need to be conserved as panchayat lands or community preserves.

Most birds of the farmlands tend to be common and generalized plains species, but the fallow patches, grazing grounds and grass flats attract specialized birds of open habitats, such as harriers *Circus*, quails *Coturnix*, buttonquails *Turnix*, plovers *Vanellus*, *Pluvialis* and *Charadrius*, coursers *Cursorius*, wheatears *Oenanthe*, several lark and pipit species etc. The 'flats' near the Sultanpur National Park, a popular birdwatching area close to Delhi, are an example of such a habitat and have consistently produced valuable bird records.

Relict patches of scrub, old roadside avenue trees and woodland patches in the middle of farmland, were protected by local sentiment or decree, may attract interesting migrants and raptor species that find shelter as well as hunting grounds in the surrounding habitat. Indian Spotted Eagles *Aquila hastata* breed in such woodland patches in Haryana, Brown Fish Owls *Ketupa zeylonensis* in eucalyptus plantations along canals, Woolly-necked Storks *Ciconia episcopus* frequent farmlands and breed on trees used for agroforestry and so on.

strategy should be formulated to support this functional diversity.

From where it enters the Delhi area, the Yamuna follows the Haryana–UP border from north to south for about 70 km, where it enters the NCT near Palla village in north Delhi. It then meanders over another 26 km before it reaches the barrage at Wazirabad in north Delhi. A second barrage at Okhla lies 22 km further downstream from Wazirabad, and 4 km south of this point is where the river exits the NCT at Jaitpur. It continues further south for another 75 km to leave our area somewhere east of Kosi Kalan, UP. Its embanked floodplains along this length vary in width from 800 m to 3 km and cover a total area of 9,700 ha within the NCT.

Sarus in the countryside

The *Bangar* Plains: Lakes and Wetlands

The alluvial floodplains in the Delhi area have facilitated the construction of an extensive irrigation network of canals and distributaries in the area, with most stretches being lined with embankments and tree plantations. Many natural waterbodies and marshes have, however, been regrettably drained, encroached upon, polluted or are subject to heavy disturbance. On the other hand, new wetland habitats have been created by seepage and flooded depressions along the major canals and for storage, offering productive habitats for birds.

East of the Yamuna in GautamBuddha Nagar district of UP, the Dhanauri wetland near Thasrana and the Surajpur lake in Greater Noida, on the margins of the urban sprawl, host a diversity of raptors and concentrations of waterfowl in the winter. The Dhanauri wetlands are well known as a breeding and roosting site for impressive numbers of Sarus cranes *A. antigone* and deserve to be protected as a community reserve. Surajpur, which receives a measure of protection, has an array of breeding marshland species and a large heronry.

To the west, in Haryana, the Western Yamuna Canal network irrigates large areas in the Sonipat, Rohtak and Jhajjar districts, and a number of natural wetlands, seasonally flooded depressions and reservoir storages, such as those in Dighal and Mandothi in Rohtak district, consistently produce many remarkable bird records.

The Sahibi river, a rain-fed tributary of the Yamuna, enters Haryana from Rajasthan and having traversed the Rewari, Jhajjar and Pataudi districts, it passes through west Delhi, where it has been channelized (as the Najafgarh drain) and is in a dismal condition due to the inflow of untreated sewage and pollutants from the surrounding populated areas. Nevertheless, the stream beds and dried lakes of the Sahibi form an ecological corridor of sorts, a seasonal wetland along its route, such as the Masani barrage wetland in Rewari district and the Khaparwas (83 ha) and Bhindawas (410 ha) wildlife sanctuaries in Jhajjar district. The man-made Bhindawas lake that stores irrigation canal waters in a former lake bed of the Sahibi, was declared a bird sanctuary in 1986 and recognized as an important bird and biodiversity area (IBA), attracting a range of waterfowl and raptor species.

The Yamuna in Delhi was a relatively healthy riverine ecosystem until the 1970s but is now in a deplorably degraded state. The Wazirabad barrage, constructed in 1957 in north Delhi to divert water to meet the city's growing needs, had drastically reduced flows during the non-monsoon months, leaving the stretch downstream contaminated by the increasing pollution ingress as the city expanded. Subsequently, a new barrage with increased water diversion capacities, constructed 224 km upstream in Haryana at Hathnikund in the 1990s, effectively terminated the flowing Yamuna at Wazirabad, leaving it with almost no fresh water downstream except in the monsoon season. In 1987, a new barrage was commissioned at Kalindi Kunj, downstream from the ageing Okhla weir (built in 1874), to store the waters of the Yamuna for the Agra Canal, and this has further altered the hydrology of the Yamuna. The barrages have stilled the river; with no fresh water inflow for much of the year, the Yamuna in Delhi is stagnant with very low dissolved oxygen levels, its water opaque with sewage and industrial effluents released into its channel as it passes through the city.

The situation has been the subject of much study and concern. Over the last two decades, various action plans have focused on effluent treatment infrastructure but with limited impact. The fundamental issue of supporting an ecological flow regime for the river to maintain its ecosystems, by which upstream water abstraction is managed to mimic the seasonal patterns and benefits of natural flows while balancing competing water uses, remains sidelined. More positive effects can be expected if the ecological character of the floodplain can be enhanced by reviving former channels and oxbow lakes and permitting a degree of controlled flooding to replenish aquifers. Encroachment in the *khadar*, illegal or misguided, has drawn public and judicial attention, but remedial measures remain inadequate.

Brown Fish Owls frequent canal-side plantations in Haryana

However, some efforts are being made. The National Institute of Hydrology in Roorkee, in a recent study commissioned by the National Mission for Clean Ganga (*Namami Gange*),[11] recommended an enhanced environmental flow regime for the Yamuna to sustain downstream riverine ecosystems; the recommendation was for more than twice the flows being presently released from the Hathnikund barrage during the lean season (between January and June) under a 1994 MoU for the sharing of Yamuna waters between the upper riparian states. These recommendations have, however, been rejected by the Haryana government as an 'environmental disaster' for the state if implemented.

The Delhi government's ambitious Yamuna Riverfront Development project named 'Asita' (another name for the Yamuna) has the objective of reviving and rejuvenating 1,476 ha of the floodplain/*khadar* tract, a 22 km stretch alongside the river between Wazirabad and Okhla, with an environmentally conscious and people-friendly approach involving the restoration of ponds and wetlands and the plantation of native species of plants, trees and grasses and the creation of public amenities. The geographical area has

Yamuna khadar *above Wazirabad*

The Yamuna *Khadar*: The Okhla Bird Sanctuary

Some of the city's well-known wetlands, though protected, have been altered by human agency with consequent impacts on their birdlife. The Okhla Bird Sanctuary on the Yamuna, listed as an IBA,[12] illustrates how human intervention can change the character of a specialized habitat dramatically, but the impact does not have to be entirely negative. Some losses can be replaced by other gains, provided the intervention is managed with care and an understanding of how the environment will cope with the change.

The Yamuna in Delhi has long been known for the numbers of waterfowl that have used it as a flyway on their way northwards in spring. Until the 1980s, when it still largely retained its natural flow patterns, a spectacular number of waders would gather in the evenings on the sandbanks exposed in Okhla as the summer advanced, calling excitedly as parties took off one by one for the next staging hop of their migration. Formations of cranes followed the river alignment northwards, their bugling heard clearly from up high. Unfortunately, we have lost these scenes, given the state of the river and the urban congestion around. The migrants probably still move up the river, too high to be recorded, however, and large wader flocks, with many glowing in their breeding finery, still gather at suitable sites elsewhere in the neighbourhood, especially in seepage marshes and inundated farmlands, such as the Chandu–Najafgarh complex of wetlands near the Sultanpur National Park to the southwest of Delhi.

But the nesting colonies of terns *Sterna* spp., skimmers *Rynchops albicollis*, pratincoles *Glareola* spp. and other birds that enlivened the sandbanks and channels of the Yamuna in past years are gone. The altered hydrology and sand mining downstream have destroyed the habitat, while the awful water quality cannot sustain the aquatic fauna on which these birds depend.

On the other hand, the Okhla barrage impoundment has created areas of marshes and reeds that are excellent habitat for breeding bitterns *Ixobrychus* spp. and amongst the best sites in India to study these elusive herons and other swamp species. In 1990, an area of 400 ha was designated as a bird sanctuary by the UP government, but it suffers because of high levels of pollution, urban encroachment, high-tension power lines that run along its eastern boundary and an ecologically sensitive zone restricted to just 100 m around the impoundage rather than considering the floodplains northwards and southwards of the sanctuary as one ecological unit.

The Ganga *Khadar*: The Hastinapur Wildlife Sanctuary

The Delhi area, as defined in this book, includes a part of the *khadar* floodplains of the Ganga just east of Hastinapur, Meerut district. The Hastinapur Wildlife Sanctuary was established in 1986 to protect 2,073 sq km of this unique riparian habitat on the western bank of the Ganga in the Meerut, Muzaffarnagar and Bijnor districts of UP. Recognized as an IBA, it comprises a variety of habitat types, including areas of grassland, woodland patches on elevated alluvial depositions, and swamps, marshes and lakes along the former channels of the meandering river.

Just upstream of the area, but within the boundaries of the sanctuary, is the Haiderpur barrage, constructed across the Ganga in Muzaffarnagar district in 1985, with a wetland impoundment of 1,214 ha, which has been identified under the *Namami Gange* Programme as a model wetland in the Ganga basin and declared a Ramsar site under the 1971 Convention on Wetlands adopted in Ramsar. Although subject to considerable disturbance, the sanctuary still hosts a rich faunal assemblage including endangered and vulnerable species, such as the swamp deer *Rucervus duvaucelii*, hog deer *Hyelaphus porcinus*, fishing cat *Prionailurus viverrinus*, smooth-coated otter *Lutrogale perspicillata* and Ganges river dolphin *Platanista gangetica*, and birds, including Sarus *Antigone antigone* and Indian Skimmer *Rynchops albicollis*. Finn's Baya *Ploceus megarhynchus* was reported breeding at this site in June 1979 but not since then. A programme to reintroduce gharial *Gavialis gangeticus* and freshwater turtle ssp. under the *Namami Gange* Programme has been under way since 2014.

Ganga khadar *landscape, Haiderpur*

been divided into ten parts and as of 2022, work on four has been initiated, with the clearance of encroachments, creation of waterbodies and plantation of natural riverine flora. Issues related to the management and removal of encroachments and reclaiming cultivated land in several places in the project area remain a challenge to implementation.

Two of the biodiversity parks that aim to conserve representative habitats as hubs for nature conservation and education are also situated in the river floodplain. The Yamuna Biodiversity Park, above Wazirabad in north Delhi, has been outstandingly successful in restoring 185 ha of active *khadar* floodplains, wetlands, marshes sand flats and native woodland, and has attracted a varied mammalian – apex carnivores such as leopards *Panthera pardus* – and a rich bird fauna, including a large heronry. Another ongoing initiative is the Kalindi Biodiversity Park being developed over 115 ha of floodplain in south Delhi adjacent to the Okhla Bird Sanctuary, forming part of the Delhi government's floodplain restoration project. The park will treat the raw sewage and industrial effluents discharged into the river by creating treatment wetlands for biological remediation, storage of flood water and restoring the ecosystem of the river *khadar*.

North of Wazirabad, where the river is relatively cleaner, in 2019, Delhi's Palla project was initiated as a three-year pilot scheme with the creation of a reservoir in the floodplains to harvest and store monsoon flows to replenish groundwater levels and the river itself in the lean months. As of June 2022, a 26 acre pond at Palla has already shown positive results and more reservoirs are being planned between Palla and Wazirabad.

(c) *Dabar* Areas West of the Aravallis

The sandy plains, known as *bagar*, covering parts of Rajasthan and the adjoining areas of the Sirsa, Hisar and Bhiwani districts of Haryana as well, begin to show desert influences and

Sultanpur Jheel in 1983–84. It had open, less wooded and wider horizons then.

The Sultanpur National Park landscape today, which is more wooded, greener and hemmed in.

Sultanpur National Park

The 143 ha Sultanpur lake southwest of Gurugram, protected since 1971 and declared a national park in 1989, is another well-known IBA. It used to form part of a remarkable complex of wetland ecosystems of the *dabar* in southwest Delhi that was centred on the Najafgarh Jheel, a seasonal lake created as the rain-fed Sahibi river spilled its overflow in the Najafgarh Lake basin as it entered Delhi. This extensive low-lying area was dredged in 1938, splitting it into three remnant lakes that continued to produce valuable ornithological records till the 1960s, when they were reclaimed by the widened Najafgarh drain.

A centre of salt production until 1923, this *dabar* area had a number of seasonal inundations (the 'Sultanpur lakes') till as late as the 1970s. Most have been drained and reclaimed since,[13] while the ecological character of the main lake has been altered in the 1980s by the Haryana Forest Department tree plantation programme to replicate a Bharatpur-style environment. With canal water brought in to compensate for reduced natural flows, an open, brackish inundation has been transformed into a wooded, freshwater wetland that no longer attracts the flocks of flamingos and pelicans it was famed for. In their place, the lake has gained perhaps the largest heronry of the Delhi area on the acacia-covered islands created by the intervention. Painted Storks *Mycteria leucocephala* first bred here in 1993,[14] and now do so annually with Asian Openbills and an array of other cormorant, heron and ibis species.

The bird-rich seepage marshes and inundations of the Najafgarh drain in the Chandu–Budhera villages and at Basai, an IBA, close to Sultanpur itself, still form an integrated ecosystem. The birds that breed at the Sultanpur lake depend on areas outside the national park for foraging and post-breeding dispersal, which underscores the importance of conserving this functional diversity to enhance their collective ecological value – losing one would adversely impact the others as well.

Heronries in the Delhi Area

Heronries are indicators of the health of wetland ecosystems and offer several valuable ecological services. The survival of a heronry is dependent on several factors: the availability of suitable nesting trees, food availability or a 'feeding habitat complex' and some protection from disturbance. The Delhi Zoo offers these three essential needs – nesting sites, food and safety – and has a long-established heronry that is a great asset to the Delhi Zoo.

There appear to have been no systematic surveys of the heronries in the Delhi area or a census of breeding pairs of selected species, a lacuna that should not be difficult to fill. It should be the responsibility of the wildlife departments of Haryana and Delhi to be well informed about these spectacular assets and any conservation issues faced, even as local birdwatchers can play a valuable supportive role.

Dozens of small heronries are scattered across the countryside near village or temple ponds or canals, usually comprising a few of the more common species nesting together. Some are sporadic and shift if the water levels or other factors change. The birds themselves are not usually harmed, but many of these village-owned ponds have multiple uses, including fish farming, where the nets can be a deadly hazard for the birds. If feeding sites in the area can be identified and set aside for the nesting birds with the cooperation and ownership of panchayats, these heronries can grow into major attractions and economic assets for the villages.

Heronry at the Sultanpur National Park

Some of the better-known heronries in the Delhi area include:

1. Surajpur (Asian Openbills, Black-necked Storks, Black-headed Ibis, Grey, Purple, Night and Pond Herons, Oriental Darters, cormorants and egret spp.)
2. Okhla Bird Sanctuary (Purple, Night and Pond Herons, Cattle and Little Egrets, Oriental Darters, Little and Indian Cormorants)
3. Yamuna Biodiversity Park (Oriental Darters, Grey, Purple, Night and Pond Herons, egret spp., cormorant spp.)
4. Delhi Zoo (Painted Storks, Black-headed Ibis, Little and Indian Cormorants, Grey and Night Herons, all three egret spp.)
5. Damdama Lake (Asian Openbills, a single species colony of 10–15 nests)
6. Sultanpur National Park (Painted, Asian Openbill and Black-necked Storks, Grey, Purple, Pond and Night Herons, Black-headed Ibis, darters, Little and Great Cormorants, all egret spp.)
7. Najafgarh Drain (Egret and heron spp., Oriental Darters, Little Cormorants)
8. Bhindawas Sanctuary (Grey, Purple and Pond Herons, Black-headed Ibis, Openbills, Oriental Darters, egret and cormorant spp.)
9. Tilyar Lake, Rohtak (Black-headed Ibis, heron and egret spp., Little Cormorants, darters)
10. Sonipat/Panipat canals (Grey, Purple, Night and Pond Herons, Black-headed Ibis, egret and cormorant spp., Oriental Darters)

are characterized by sand dunes alternating with scrubby vegetation and grass in rain-fed depressions. While these districts lie outside the Delhi area as defined in this book, the low-lying, poorly drained terrain to the southwest of the Aravallis, referred to as *dabar*, is essentially an extension of the same biotope – sandy and saline but waterlogged after rain. Trees characteristic of saline *dabar* lands include species of *peelu*, *babool*, *jhau* and date palm/*khajur*. *Dabar* areas, however, have been substantially altered since the 1960s; considered unproductive in the past, they have been drained – waterlogging is no longer an issue – and irrigated with sweet water and are under cultivation, or they have been subsumed by Delhi's urban and industrial expansion. The Najafgarh and Dwarka areas are prime examples.

(d) Urban Habitats

Within Delhi's urban limits are a number of public parks and facilities that still retain remnant, semi-wild patches of original habitat and well-wooded nooks and corners. Apart from the biodiversity parks, the others worthy of mention are the Delhi Zoological Gardens, with a large and active heronry and several waterbodies, and the adjacent Sundar Nursery; the Lodhi Gardens and Delhi Golf Club; the Buddha Jayanti and Mahavir Jayanti Vanasthali parks on the New Delhi Ridge; the Hauz Khas and Jahanpanah city forests, the Mehrauli Heritage Park adjoining Sanjay Van; and the Mughal Gardens and Shanti Van in north Delhi. These green 'islands' in Delhi's urban cityscape are a remarkable resource, a sanctuary for people and for wild creatures as well, and they will assume greater significance for biodiversity conservation in coming years. Many of the trees in the old established gardens have grown to maturity with great spreading crowns and thick trunks, providing nesting and roosting sites, abundant nectar (the silk cottons and other flowering species) and food (especially the native figs) for birds. They also offer refuge to a host of migrants which, disoriented by city lights as they overfly the sprawling metropolis, seek the dark, wooded security of these 'islands', which may teem with warblers, flycatchers and other species, playing host to spectacular murmurations of starlings and mynas in season.

Studies in Delhi demonstrate that maintaining green spaces of an adequate size, with high structural and native plant diversity and density, supports and enhances this avian biodiversity.[15]

Conservation

The scale and intensity of the competing demands placed on the landscapes of India's capital city are enormous, and the challenges to their ecological integrity will only grow. Sustainable development requires a bird's-eye view approach at the landscape level; understanding how the environment copes with change is key to keeping an ecosystem healthy and resilient. Wetlands of high ecological value cannot be conserved in isolation as biodiversity hotspots. They are functionally integrated in the landscape, and a nuanced grasp of the ecological operations that sustain them should provide the base for conservation strategies.

The Aravalli Biodiversity Park, Gurugram. Flooded mining pits now help recharge groundwater.

2 Birding around Delhi

The Delhi Bird List now stands at 471 species, depending on the taxonomic approach adopted (excluding another 22 species that have not been re-recorded since 1975), which has given India's capital the reputation of being one of the most bird-rich in the world. Delhi has also had the advantage of having hosted an almost continuous stream of ornithologists, mostly amateur, recording their observations of its birdlife over the last century or more, especially since the creation of New Delhi. Their published notes comment, to one extent or the other, on the status, populations and habitat of the various species and are of interest in their illustration of how Delhi's birdlife has evolved along with the city.

Some History

An extract from the *Delhi Gazetteer* of 1883–1884[16] gives us a glimpse of the wealth of wildlife in the Delhi area from a century and four decades ago.

> Pig abound all along the banks of the Jamná ... Black buck are found almost everywhere. *Chikára* in the range of hills which runs north-east of Delhi ... Wolves are not plentiful, but they are to be usually found in the neighbourhood of the old cantonment ... The *nilgai* is to be constantly found near the villages of Borari and Khadipur ... The mongoose is very common, and so is the hedgehog ... Snakes of every kind are plentiful, the cobra especially so. The old Fort called the Kotla is infested with them ... Leopards are found in the outlying villages ... *Pára* are abundant, especially in the neighbourhood of Borari on the bank of the Jamná ... *Mahsír*, *rohú*, and *batchwa* are found in the river Jamná and at Okhlah in the Agra Canal, and the entire river is infested with *muggurs*, the *gurryál* predominating; but the snub-nosed man-eater is also plentiful ... Between the old Fort and Okhlah ...
>
> Ducks of various kinds are found in the ponds in the cold weather; snipe in several places in marshes; quail are not uncommon in the fields; partridges, both black and grey, are abundant; and *kúlan* are fond of the fields of gram when the grain has not yet hardened.

According to a compilation by the Government of Delhi,[17] the Ridge still reported sightings of wolf, hyena, leopard, blackbuck and chinkara at the beginning of the twentieth century, but by 1908, blackbucks had become rare, while wolves and chinkaras have not been seen in the wild after 1940. A number of nineteenth-century bird records from around Delhi are scattered across ornithological or

White-headed Duck, artwork by J.G. Keulemans, from The Game-Birds of India, Burma and Ceylon: Indian Ducks and Their Allies, *E.C. Stuart Baker, 1921. A vagrant to the Delhi area, last recorded in 1882–83.*

sporting writings of the day, and they paint a remarkable picture. Consider this note[18] in the *Journal of the Bombay Natural History Society (JBNHS)*:

> In 'Stray Sport' by J. Moray Brown (1893) a chapter is devoted to small game shooting within a few miles of Delhi, from which attention might be drawn to statements which add to or differ from the recent lists of Delhi birds published in the *Journal* ...

Painted Sandgrouse by E. Neale: Antique print from Game Birds of India, Burmah and Ceylon, *Vol I, Hume & Marshall, 1879*

> *Botaurus stellaris*–The Bittern. Four were bagged in the course of a snipe shoot at Mongolpoor [*sic* – presumably Mangolpuri in west Delhi] where 3 guns got over a hundred couple of snipe on 26th November.

> Published in 1893, Brown refers to these notes having been made some years before, but it is interesting to remember that there was a time when a single gun could make a mixed bag of 60-odd head including antelope, hare, common and painted sandgrouse, black partridge, quail, snipe, mallard, teal, pochard and whistling teal, all on and within 3 miles of the Ridge!

Hume[19] has notes going back a century and more on the birds around Delhi, most notably those by Lt Col C.T. Bingham, an Irish military officer who served in India in the 1870s who, amongst his other accomplishments, authored two volumes on Hymenoptera (bees, wasps, ants and similar insects with four transparent wings) (1897, 1903) and Lepidoptera (butterflies and moths) (1905, 1907) in the *Fauna of British India* series.

The first attempt at drawing up a checklist of the birds about Delhi was by S. Basil-Edwards,[20] who studied the birds in the area over five months, November–March, of the winter of 1924–25, mainly in the *babool* jungles around Raisina when New Delhi was being built. In his words, 'Five months is too short a time ... The following notes can only be regarded as a nucleus for a more comprehensive paper ...' He lists some 202 species, and his writing makes for fascinating reading compared with Delhi's birdlife today.

Sir Norman F. Frome, a former postmaster-general of India and a keen ornithologist, compiled a second checklist of 300 species from Delhi and its neighbourhood within about 50 km, based on sight records maintained independently by him and six other observers over the period 1931–45.[21] By 1949, 10 additional species recorded by Maj Gen H.P.W. Hutson in 1943–45, 2 by Sir E.C. Benthall, businessman and public servant, who reported his observations over 1942–46 while stationed in Delhi as a member of the Viceroy's Executive Council, and 13 more by H.G. Alexander had raised the Delhi bird list to 325.

Horace G. Alexander, Quaker, teacher and ornithologist, spent much of his life in India and was a close acquaintance of Mahatma Gandhi and Indira Gandhi as well. His autobiographical memoir, *Seventy Years of Birdwatching*, devotes a number of pages to his time in Delhi. In 1950, he founded the Delhi Bird-watching Society (DBWS) along with Lt Gen Harold Williams, Engineer-in-Chief of the Indian Army and bird enthusiast, 'to encourage the study of birds in and around the capital city'.

The announcement in the *JBNHS* in December 1950[22] lists its founding members, a remarkable group of people in diverse positions and callings but united in their passion for birds and nature – Mr. H.G. Alexander (chairman), Capt H.C. Ranald (honorary treasurer and the Indian Navy's first chief of naval aviation in 1948), Mrs W.F. Rivers (honorary secretary), Mrs Indira Gandhi (destined to lead the country as prime minister in later years), Mr F.C. Badhwar (nature enthusiast, first Indian chairman of the Railway Board till 1954 and first Indian president of the Himalayan Club, 1964–67), Rev. J. Bishop, Mr C.J.L. Stokoe (of the mercantile firm Messrs. Bird & Co.) and Lt Gen H. Williams (as committee members).

It also counted amongst its members Dr Sálim Ali, Mrs Usha Ganguli (whose husband was then vice chancellor of Delhi University), Mr Lavkumar Khachar (of Jasdan State, Gujarat) and others, including members of the diplomatic community stationed in the new capital.

Indian Ducks and Their Allies *by E.C. Stuart Baker, 1921 (Title page)*

In 1950, the DBWS brought out a Delhi checklist totalling 356 species (351 according to current taxonomic classification), based on information available till then, as a working document for its members. It also published Maj Gen Hutson's bird notes from when he was stationed in Delhi (1943–45) as deputy engineer-in-chief of the Indian Army, supplemented by H.G. Alexander's records, in book form as *The Birds About Delhi*, edited by Lt Gen Williams, then president of the DBWS. With this publication, the Delhi list stood at 370 (actually 365, as per current taxonomy).

The impressive fieldwork over the 1950s, 1960s and into the 1970s, even as the city of Delhi was changing at a rapid pace, by a galaxy of competent ornithologists – H.E. Malcolm Macdonald (British High Commissioner from 1955–60), Julian P. Donahue (professional lepidopterist with an interest in Asian birds since his time in India, 1958–62), Victor C. Martin (then with the British High Commission in New Delhi), Peter Jackson (Reuters correspondent in Delhi and honorary secretary of the DBWS in the 1960s, later of the International Union for Conservation of Nature [IUCN]), Capt N.S. Tyabji of the Indian Navy; Mrs Usha Ganguli and others – produced a flowering of new information about Delhi's birds.

Some wrote about their passion with a flair and competence that stands out to this day, such as Macdonald's superb *Birds in My Indian Garden*. Much of this material, along with the new species recorded till 1970, was collated in Usha Ganguli's *A Guide to the Birds of the Delhi Area*, published posthumously after her untimely demise in 1970, which lists 408 species (including historical records). This was a major addition to the literature on Delhi's birdlife and remains a vital reference for any serious student of Delhi's birdlife today.

Antony J. Gaston conducted intensive research between July 1971 and June 1974 on the New Delhi

Ridge for his doctoral thesis on *Turdoides* (now *Argya*) babblers, when he recorded, and mist-netted, several species for Delhi, some of which have not been reported since. Abdulali and Panday's *Checklist of the Birds of Agra, Delhi and Bharatpur*, published by the Bombay Natural History Society (BNHS), was another useful resource that appeared around this time.

After a very productive innings over two decades, the DBWS seemed to lose steam sometime in the 1970s, at least as a formal forum. Ali & Ripley's seminal ten-volume *Handbook of the Birds of India and Pakistan* appeared during these years, and individual birdwatchers, both residents of the Delhi area and visitors from outside, continued to work in the field and made substantial new discoveries. Many recorded their findings in articles in the *JBNHS*, its monthly outreach magazine *Hornbill* and the more informal but lively and popular *Newsletter for Birdwatchers* that began publication in 1960 under Zafar Futehally's keen editorship and remained a useful platform for birdwatchers across the country for the next four decades and more. Two useful, and more comprehensive, efforts to catalogue the avifauna of Delhi were published – an updated checklist by the NGO Kalpavriksh in 1991 and a compilation of past literature and field observations during 1994–95 by Tak & Sati, published in 1997.

Growing Interest

In the 1990s, attempts to bring the growing number of people interested in birds in the Delhi area together under the DBWS, revived as the Delhi Bird Club, were rewarded with outstanding success in the year 2000 with the club's reformation as an internet-based forum, Delhibird, for reporting, discussing and creating site checklists, and for organizing regular field outings. Along with the rapidly increasing penetration of electronic media and internet connectivity, by the early 2000s, there was a parallel boom in the ease and affordability of high-quality optical and digital photography equipment and illustrated field guides of an international standard. Grimmett and the Inskipps' *Birds of the Indian Subcontinent*, Kazmierczak's *A Field Guide to the Birds of India* and others fortuitously came together around the turn of the millennium and led to an explosion of interest and infectious enthusiasm, especially among Delhi's educated youth. Rasmussen and Anderton's *Birds of South Asia: The Ripley Guide*, first published in 2005 with an updated edition in 2012, is the most recent guide now available for the birds of the Indian subcontinent, its path-setting taxonomic approach and detailed morphological and plumage data being of exceptional value.

To these resources must be added journals and biannual bulletins including *Forktail* and *BirdingASIA*, of the Cambridge-based Oriental Bird Club, along with its very comprehensive collection of splendid images from the field, orientalbirdimages.org, which are now permanently archived at the Macaulay Library of the Cornell Lab of Ornithology, Cornell University, USA. The South Asia–focused, well-researched, peer-reviewed and widely cited bimonthly *Indian BIRDS*, published since 2004 by the New Ornis Foundation, is a worthy successor to the former *Newsletter for Birdwatchers*.

Horizons were also broadened with the launch of several online public forums and social media networks offering valuable platforms for Delhi's birders to document their records, comments and conservation issues, provided that the wealth of data being generated from the field could be retrieved in a timely manner, validated and permanently recorded for appropriate use. Few things capture the imagination like the outstanding colour photographs of birds spread across these forums; with many more eyes out in the field, better identification skills and exposure to the joys of birding, an increase in travel and the exploration of new, little-known birdwatching sites, Delhi's birders have produced several remarkable finds. Delhibird's *Official Checklist of the Birds of the Delhi Region* listed 415 species with an additional 47 historical records of birds seen only before 1970. *The Atlas of the Birds of Delhi and Haryana*, published in 2006, was an attempt to bring together this wealth of new data from Delhi's growing birding community and update the available information, and Nikhil Devasar and Rajneesh Suvarna's *Birds about Delhi* is the latest effort at producing a comprehensive field guide to the birds of the area.

Citizen Science

Almost 80 years ago, James Fisher commented on the nature of birding in his delightful little book, *Watching Birds*, that 'the observation of birds may be a superstition, a tradition, an art, a science, a pleasure, a hobby, or a bore; this depends entirely on the nature of the observer.'

With the growing numbers of people interested in birds in and around Delhi, birdwatching is no longer dismissed as a puerile, meaningless obsession. It's a fascinating hobby, bringing to the watcher's consciousness a wondrous world of colour, song, grace and vivacity. It can lead us to some of the most wonderful places, providing freedom from restriction and routine and an excuse to be out in the open. To many, it opens new vistas of investigation and study – professional, scientific, behavioural, biogeographical – and others make a career of it in art, writing and photography, or by organizing specialist birding and nature trips.

Some of the finest contributions to Indian ornithology have been made by amateurs through their observations and writings from the field. These are as good as the reliability of the data on which they are based, and the validation of field data is thus crucial. Birders today are fortunate in having easy access to top-class optical equipment, field guides, sound recordings and so on, but bird identification is a skill best learnt from field experience, and the habit of keeping field notes is the best training for this.

The availability of good digital equipment has led, unfortunately, to a reliance on photographic evidence alone to document records, and the decline in note-taking has resulted in a noticeable decline in field identification skills. While a still photograph of a bird in all its detail is a great learning tool, it can be difficult to identify, or validate, a record from photographs alone without the benefit of field notes on size, shape, behaviour, voice and jizz.

Most of Delhi's birds are, fortunately, not difficult to identify. Some groups require more scrutiny, and a few others are genuinely challenging, such as raptors, for which individual, sexual, or age-related variation, moult, abrasion, or fading of feathers are all relevant in clinching an identification. But above all, an assessment of jizz, a combination of features and behaviour that is unique to a species, is an excellent pointer, and there is really no substitute for this.

Birders are usually keen conservationists: some devote their time and effort to conservation projects, others to advocacy, spreading awareness and alerting the authorities, media and relevant institutions about developments in the natural environment around us. Socially conscious with wider horizons, the growing tribe of birders tends to be educated and aware and in a position to make a difference. For these birds that give us so much pleasure, the least we can do is help them by protecting them and their habitats, and – as we watch, photograph and enjoy them – ensure that their interests remain foremost.

3 The Delhi Checklist: An Overview

Taxonomy

The taxonomy used in this book follows *A Checklist of the Birds of India v7.0*, dated 28 February 2023.[23] This *Checklist*, first published in 2016 and annually revised, meets a long-awaited need for an independently reviewed, definitive list of bird species reliably recorded from within the political boundaries of India till date. The *Checklist* itself draws upon the increasing convergence between two primary global resources on the taxonomy and biogeography of the birds of the world, eBird/Clements[24] and the International Ornithological Congress (IOC),[25] which are themselves regularly updated in line with global taxonomic research. The Clements checklist was developed to support eBird, the world's largest biodiversity-related citizen science project.

There are various ways in which birds, or any other group of organisms, can be classified, but all taxonomic classifications have to address the definition of a species. Some global avian checklists follow the biological species concept, which is centred on reproductive isolation and based on a range of phenotypic characters – morphological, behavioural, vocal and ecological – and rely on quantitative criteria for species delimitation (Tobias approach[26]).

The phylogenetic species concept, on the other hand, attempts to reconstruct the evolutionary ancestry of a species, using genetic or genomic (molecular) data, and reflect them in its taxonomic approach as a hierarchy of nested clades or monophyletic groups that reflect the evolutionary process. In many cases, the two concepts yield similar conclusions, though molecular data can highlight genetic divergences and convergences that lead to some interesting and, at times startling, results in defining monophyletic taxa that upturn conventional taxonomic arrangements.

Phylogenies based on genetic studies alone provide useful data, but the results vary depending on what genetic material is used and which analytical methods are employed. In addition, they could have significant limitations because many species are less reproductively isolated than previously thought. Avian taxonomy is now increasingly focused on the concept of 'integrative taxonomy', where multiple lines of evidence, including genotypic, phenotypic, behavioural and acoustic information, are used to define species limits.[27] Molecular data should thus be considered along with other morphological and behavioural information in working out the evolutionary ancestry of a taxon and should not be considered the final word. Even so, all our conclusions remain hypotheses, which is only to be expected as we try to fit a continuous evolutionary process into an artificial taxonomic straitjacket. The laws of nature are premised on change; taxonomies cannot be static if they are to reflect this living, changing world.

The Delhi Checklist

Currently, 23 orders and 79 families of birds are represented in the Delhi area (out of a total of 26 orders and 114 families in India), and 471 species (of 1,353) have been recorded over the past half-century or so since 1975, which has been selected as the cut-off year; any records prior to that are considered historical. Four species included in the list probably pertain to escapes or released birds; the illegal trapping of several species remains rampant for the pet and wildlife trade, or release by the devout as auspicious acts. Twenty-two species have not been re-reported from the Delhi area since 1975: some were accidentals that may or may not reappear; for some, no suitable habitat now remains; for others, the records are so sketchy that their claimed presence must remain hypothetical.

Of this total figure of 471 species, the inclusion of six species requires reconfirmation mainly because of issues with identification or taxonomy. Knowledge of the bird helps; photographs are very useful, but field notes and documentation of jizz, vocalizations and behaviour can clinch an identification with greater certainty, and especially if it is made independently by more than one observer. Their inclusion here casts no aspersion on the observers; checklists are inevitably added to or pruned as better, and more accurate, information becomes available. Twenty-one species have been excluded after examination of their claimed occurrence.

The Annotated Checklist relies on several sources, either published or otherwise in the author's knowledge, for records of species and associated data. Relevant literature was retrieved from the online 'Bibliography of South Asian Ornithology',[28] which proved immensely useful. References have been quoted for any published data used in preparing the Checklist.

Many ornithological observations, however, by the growing tribe of birdwatchers and bird photographers in the Delhi area appear in eBird field checklists, often accompanied by photographs and/or comments. These are permanent records, and such citizen science data has been trawled for relevant records for the purpose of the checklist. Social media forums are also popular amongst Delhi's birder and bird photographer communities for posting pictures and discussion and outreach, and valuable records may be gleaned from their archives. But social media is difficult to reference as an academic source; it is also ephemeral and permanent access cannot be guaranteed. It is always advisable to publish a note on any interesting observations in a peer-reviewed journal for record. A few social media posts have, nevertheless, been cited in this text to validate interesting records.

For data gleaned from eBird or social media groups, references and observers' names have not been specified for each record as much of this information can be accessed online. Exceptions have only been made for records of exceptional interest. However, the author reiterates that each individual record has been examined and assessed for correctness and authenticity before inclusion in the Annotated Checklist. If a specific source/ reference for any record is required, the author may be contacted for details.

In the Checklist, the presence of each bird species about Delhi is first contextualized within its overall **global distribution and movement patterns**, marked with the icon 🌐 to place the avifauna in perspective, followed by a brief note on its **status in the area** marked with the icon 📍. The terms 'South Asia', 'Indian subcontinent' or 'subcontinent' are used interchangeably here to denote the area

Definitions of Abundance

Common: Found in moderate to large numbers and easily found in appropriate habitat at the right time of year

Fairly Common: Found in small to moderate numbers and usually easy to find in appropriate habitat at the right time of year

Uncommon: Found in small numbers and usually – but not always – found with some effort in appropriate habitat at the right time of year

Scarce: Occurs, not even annually, in very small numbers. Not to be expected, usually located only after diligent search in the appropriate habitat and season

Rare: Occurs sporadically and erratically, in most years absent

covered by the political boundaries of India and its neighbours, Bangladesh, Bhutan, Maldives, Nepal, Pakistan and Sri Lanka, while 'southern Asia' is a wider geographical construct encompassing the area extending from the Persian Gulf across the Indian subcontinent to Southeast Asia.

The status of each species in the Delhi area is indicated by an impressionistic assessment of the bird's **primary seasonal occurrence** indicated by such as Breeding Resident, Summer Visitor, Passage Migrant, Winter Visitor, Casual Visitor, i.e., not limited to a season and Vagrant/Accidental; and its **relative abundance**, indicated by .
A question mark (?) following an assessment of a bird's status indicates some uncertainty about the status assessment; Data Deficient would indicate that insufficient information is available to attempt even a tentative assessment for the species.
It should be recognized that a species tagged, for example, Summer Visitor or Passage Migrant, may not be entirely absent from the area in other seasons. Further, abundance tags are species and habitat dependent, thus 'common' will have different connotations for a predator, such as a shikra, which naturally occurs at lower densities, and a passerine, for example, a myna, which would be numerically more abundant. Second, birds are not spread evenly over the landscape, and the tags reflect a bird's status in the appropriate habitat at the right time of the year.

The **IUCN Red List of Threatened Species** is the most comprehensive, regularly updated database available on the global conservation status of all species, classifying species at risk of extinction in eight categories: Not Evaluated (NE), Data Deficient (DD), Least Concern (LC), Near Threatened (NT), Vulnerable (VU), Endangered (EN), Critical (CR) and Extinct (EX). When applying these categories to the 471 bird species in the Delhi area checklist, 5 are Critical, 8 Endangered, 15 Vulnerable, 25 Near Threatened, and the rest, 418, are all Least Concern.

IUCN Red List Categories of the Species at Risk on the Delhi List (2022)

CR (5 sp.): Baer's Pochard *Aythya baeri*, Sociable Lapwing *Vanellus gregarius*, Red-headed Vulture *Sarcogyps calvus*, White-rumped Vulture *Gyps bengalensis*, Indian Vulture *Gyps indicus*
EN (8 sp.): Lesser Florican *Sypheotides indicus*, Black-bellied Tern *Sterna acuticauda*, Indian Skimmer *Rynchops albicollis*, Greater Adjutant *Leptoptilos dubius*, Egyptian Vulture *Neophron percnopterus*, Steppe Eagle *Aquila nipalensis*, Pallas' Fish Eagle *Haliaeetus leucoryphus*, Finn's Weaver *Ploceus megarhynchus*
VU (15 sp.): Lesser White-fronted Goose *Anser erythropus*, Marbled Duck *Marmaronetta angustirostris*, Common Pochard *Aythya ferina*, Horned Grebe *Podiceps auritus*, Sarus Crane *Antigone antigone*, Sharp-tailed Sandpiper *Calidris acuminata*, River Tern *Sterna aurantia*, Indian Spotted Eagle *Clanga hastata*, Greater Spotted Eagle *Clanga clanga*, Tawny Eagle *Aquila rapax*, Eastern Imperial Eagle *Aquila heliaca*, White-naped Tit *Machlolophus nuchalis*, Bristled Grassbird *Schoenicola striatus*, Stoliczka's Bushchat *Saxicola macrorhynchus*, Green Munia *Estrilda formosa*
NT (25 sp.): Falcated Duck *Mareca falcata*, Ferruginous Duck *Aythya nyroca*, Lesser Flamingo *Phoeniconaias minor*, Great Thick-knee *Esacus recurvirostris*, Northern Lapwing *Vanellus vanellus*, River Lapwing *Vanellus duvaucelii*, Eurasian Curlew *Numenius arquata*, Bar-tailed Godwit *Limosa lapponica*, Black-tailed Godwit *Limosa limosa*, Curlew Sandpiper *Calidris ferruginea*, Asian Dowitcher *Limnodromus semipalmatus*, Woolly-necked Stork *Ciconia episcopus*, Black-necked Stork *Ephippiorhynchus asiaticus*, Painted Stork *Mycteria leucocephala*, Oriental Darter *Anhinga melanogaster*, Dalmatian Pelican *Pelecanus crispus*, Black-headed Ibis *Threskiornis melanocephalus*, Himalayan Vulture *Gyps himalayensis*, Pallid Harrier *Circus macrourus*, White-tailed Sea Eagle *Haliaeetus albicilla*, Lesser Fish Eagle *Haliaeetus humilis*, Grey-headed Fish Eagle *Haliaeetus ichthyaetus*, Red-necked Falcon *Falco chiquera*, Laggar Falcon *Falco jugger*, Alexandrine Parakeet *Psittacula eupatria*

The local status – both in terms of occurrence and abundance – of a bird in the Delhi area may well differ from that for India as a whole or from its IUCN-determined global conservation category. It may appear odd to Delhi's birdwatchers, for instance, that some species present in reasonable numbers around Delhi appear in the Red List, such as the Steppe Eagle (EN) or Alexandrine Parakeet (NT). However, populations of these birds in other parts of their range, outside India, may have sharply declined – for example, Steppe Eagles in Eastern Europe or Alexandrine Parakeets in Southeast Asia – and hence they have been included.

These lists also serve to highlight, however, some of the factors that are putting birds at risk. The preponderance of raptors in the lists may point to the increasing use of agricultural pesticides, loss of habitat and possibly persecution; declines in skimmer and inland tern species illustrate the risks posed by the pollution of freshwater habitats, changes in river hydrology and incessant encroachment and disturbance of their sandbank nesting areas, and so on.

The abundance of a particular bird species depends on various factors and falling numbers may indicate that it is either struggling globally or the decline could be localized, perhaps due to loss of habitat. Unusual concentrations may suggest improved habitat quality or better protection, or something more sinister – that we have lost other suitable habitats in the neighbourhood, forcing the birds to concentrate in the few that remain.

Patterns in Delhi's Avifauna

The core composition of the avifauna of the Delhi region is typical of the plains of north India, mainly widespread (though not necessarily common) birds of agricultural biotopes including some spectacular raptors, herons, storks and cranes. But Delhi's geography also brings in other factors that impart more varied flavours to its birdlife.

(a) Breeding Species and Residents

The list of breeding species has 197 species that are believed to breed, or have bred, in the area: their nesting either confirmed by direct observation or circumstantial evidence (198, if the breeding colony of feral Great White Pelicans at the Delhi Zoo is included).

It includes the following 12 species that probably breed in the Delhi area, for which circumstantial evidence is available, and it is expected that breeding will be confirmed for most of them sooner or later.

1. Ruddy-breasted Crake *Zapornia fusca* (Present and calling in suitable habitats in the breeding season)
2. Brown Crake *Zapornia akool* (Present and calling in suitable habitats in the breeding season)
3. Baillon's Crake *Zapornia pusilla* (Chicks photographed, needs reconfirmation)
4. Glossy Ibis *Plegadis falcinellus* (Circumstantial evidence)

A Ruddy-breasted Crake amongst water hyacinth leaves

5. Little Tern *Sternula albifrons* (Formerly bred; courtship feeding by pair in suitable habitat)
6. Whiskered Tern *Chlidonias hybrida* (Formerly bred; present in suitable habitat in breeding season)
7. Short-toed Snake Eagle *Circaetus gallicus* (Formerly bred; circumstantial evidence, territorial behaviour)
8. Brown-capped Pygmy Woodpecker *Dendrocopus nanus* (Resident in suitable habitat)
9. Rufous-tailed Lark *Ammomanes phoenicura* (Pairs in suitable habitats in breeding season)
10. Pale Martin *Riparia diluta* (Circumstantial evidence; mating, nest material collection)
11. Asian Brown Flycatcher *Muscicapa dauurica* (Present in post-breeding season, juveniles recorded)
12. Tickell's Blue Flycatcher *Cyornis tickelliae* (Present males singing in suitable habitats in breeding season)

Yellow-footed Green Pigeon, a common Delhi resident

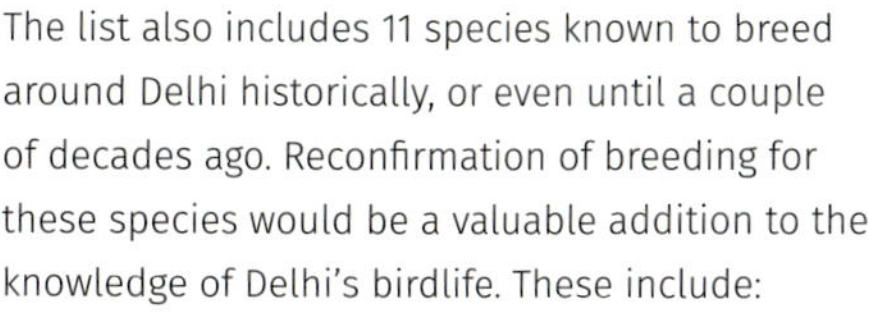

The list also includes 11 species known to breed around Delhi historically, or even until a couple of decades ago. Reconfirmation of breeding for these species would be a valuable addition to the knowledge of Delhi's birdlife. These include:

1. Indian Nightjar *Caprimulgus asiaticus*
2. Great Thick-knee *Esacus recurvirostris*
3. Indian Skimmer *Rynchops albicollis*
4. River Tern *Sterna aurantia*
5. Black-bellied Tern *Sterna acuticauda*
6. Red-headed Vulture *Sarcogyps calvus*
7. White-rumped Vulture *Gyps bengalensis*
8. Tawny Eagle *Aquila rapax*
9. Pallas' Fish Eagle *Haliaeetus leucoryphus*
10. Laggar Falcon *Falco jugger*
11. Indian Nuthatch *Sitta castanea*

There are four species for which data currently available is insufficient. They have been categorized as Scarce visitors in this Checklist but may possibly breed here and present a challenge to birdwatchers to confirm their status.

1. Rock Bush Quail *Perdicula argoondah*
2. Yellow-legged Buttonquail *Turnix tanki*
3. Brown Hawk Owl *Ninox scutulata*
4. White-bellied Minivet *Pericrocotus erythropygius*

Of these 197 breeding species, about 35 are mainly local migrants, summer/monsoon breeding visitors from their winter quarters further south in the peninsula, such as the Blue-tailed Bee-eater *Merops philippinus* for instance, or even from as far as East Africa, such as the Pied Cuckoo *Clamator jacobinus*. Some on this list may not vacate the Delhi area entirely in the winter, perhaps increasingly so, with global warming; the Common Hawk Cuckoo *Hierococcyx varius* are an example.

Certain summer movements are yet to be understood fully. Bristled Grassbirds *Schoenicola striatus* breed in Delhi through the monsoon, but where do they go the rest of the year?
Small Buttonquail *Turnix sylvaticus* have been

recorded here in the summer; they are known to be nomadic depending on the creation of suitable habitat, but the patterns of their movements remain opaque.

The outliers of the Aravallis, which form the Ridge landscape of Delhi, act as a biogeographical corridor for some peninsular taxa to access and establish themselves as breeding species in patches of suitable habitat here, introducing a distinctive element into Delhi's birdlife. Several species, including Indian Pitta *Pitta brachyura*, Marshall's Iora *Aegithina nigrolutea*, White-bellied Drongo *Dicrurus caerulescens* and Black-headed Cuckooshrike *Lalage melanoptera*, have been only recently discovered nesting in these previously poorly studied Southern Ridge habitats.

The birdlife of any given area is in a constant state of flux, evolving with and adapting to the changing environment, exploiting new opportunities as they appear, and in this context, the apparent range expansions reflected in Delhi's birdlife are of interest. The Yamuna and Ganga rivers, and their irrigation canal systems, have probably facilitated the spread of some sedentary species of wet Indo-Gangetic habitats, such as the White-tailed Stonechat *Saxicola leucura*, Yellow-bellied Prinia *Prinia flaviventris* and Bengal Bushlark *Mirafra assamica*, into the Delhi area in the 1970s; as also Sind Sparrows *Passer pyrrhonotus* of the Indus floodplains from the west. From the other direction, the Bronze-winged Jacana *Metopidius indicus* is an 'eastern' species that has spread in the Delhi area as a breeding resident in the last 50 years.

A final comment about Delhi's nesting species: they display the entire gamut of reproductive strategies known in the bird world – lifelong monogamy, seasonal monogamy, simultaneous and successive polygyny, simultaneous and successive polyandry, successive bigamy, casual mating, parasitism, cooperative breeding, etc.

(b) Non-breeding Visitors

There are 190 species categorized as non-breeding visitors to the Delhi area, most being winter migrants from the Himalayas and beyond, but these also include some 40-odd species that are 'mainly' passage migrants and a few, such as the Greater Flamingo *Phoenicopterus roseus*, whose movements tend to be irregular over the seasons and are categorized as Casual Visitors.

Migrants adopt various strategies to complete their migrations. Some species move gradually through the landscape, halting to rest and feed as they go, while others travel rapidly, either on a broad front or along well-defined, regular paths. Some are long-distance travellers, migrating non-stop to and from their winter quarters, or between one or more staging posts en route to replenish their fuel reserves. Some may follow different strategies and routes during their southward and northward passages. Migration may require physiological and even behavioural changes in some species, such as accumulating reserves of fat prior to starting their migratory journey or flying in flocks to reduce the energy used in migration or the risk of predation.

The definition of flyways has been useful in developing a broader perspective of migration patterns and routes, especially for long-distance migrants and for conservation management over their entire migration space. Initially conceived for waders and waterfowl, the concept proves useful for migratory species from other families as well, both passerines and non-passerines. But flyways remain generalizations, and each individual species tends to follow its own migration strategy and route, though usually within the broad pattern. Birdlife International recognizes eight overlapping flyways,[30] three of which cover Asia between them and are relevant to the Delhi area.

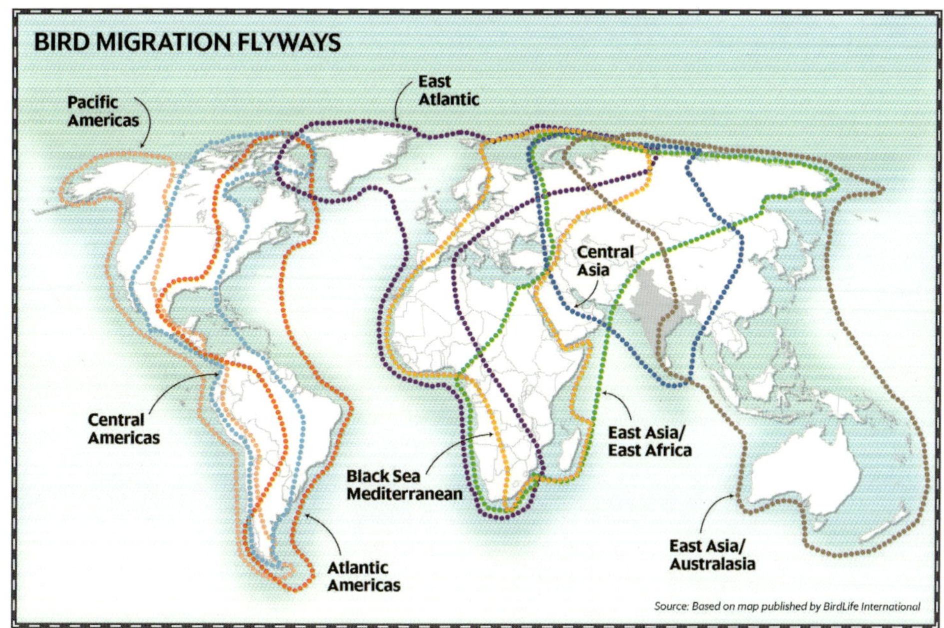

Map not to scale

The Central Asian Flyway[31] spans continental Eurasia between the Arctic Ocean and Indian Ocean, with the Indian subcontinent as its core wintering area. The Eurasian–African Flyway overlaps with it on the west and the East Asian–Australasian Flyway on the east; all three flyways have geographically separate wintering areas but share their breeding catchment areas in northern and Central Asia.

Delhi is the wintering ground for many bird species that breed in Siberia and Central Eurasia and follow the Central Asian Flyway to winter in the Indian subcontinent. They face a major obstacle in the high ranges of the Karakoram and the Himalayas, and many – especially raptors – are forced to funnel their migration routes through the deep river gorges that cut through these ranges, including of the Indus and Satluj rivers in the west, the Karnali, Kali Gandaki and Arun rivers in Nepal, the Kuri Chhu in Bhutan and the Brahmaputra with its major tributaries in the east, or even directly over the range, over the high passes and shoulders, as Bar-headed Geese *Anser indicus* and some other wader and waterfowl species are known to do. A flood of migrants sweeps over the Delhi area in season, mostly over a broad front, but the north–south flowing Yamuna also tends to funnel migrants along its alignment as a flyway for many waterbirds and a host of land birds as well.

The term 'passage migrants' requires elucidation. Most passage migrants spend the winter south of Delhi, in the peninsula and even beyond, and pass through the area on their way to and from their winter quarters. Every species that winters south of the Delhi area (and summers north of it) will naturally be more common on passage than in winter/summer, but odd individuals may remain through these seasons as well. Some species are regular and common on passage, for example, Blyth's Reed Warblers *Acrocephalus dumetorum*. Others are irregular and scarce: Rusty-tailed Flycatchers *Ficedula ruficauda* overfly Delhi on passage but are seen here only as straggling migrants.

Wild ducks flood Delhi's lakes and marshes in winter.

Migratory paths play a role. Black-headed Buntings *Emberiza melanocephala* that breed in West Asia and winter in the Deccan, migrate through Baluchistan, Sindh, Gujarat and Rajasthan. Their migratory stream passes largely south of Delhi, and they are far rarer here than their sibling species, the Red-headed Buntings *E. bruniceps* that breed in Central Asia and take a more northern path. Tickell's Warblers *Phylloscopus affinis* that breed in the Himalayas take a more easterly migratory path as they move northwards from their winter quarters in southern India and then spread west along the Himalayas (and vice versa in autumn). They are unusual about Delhi on passage, while the closely related Sulphur-bellied Warblers *P. griseolus* take a direct route and are regularly seen.

The Sulphur-bellied Warbler, a ground haunting species

Delhi also lies on the eastern fringe of the migration paths of a number of species that breed in Central Asia and the northwest of the subcontinent and migrate through the Indus plains and parts of western India in the autumn to winter in Arabia and East Africa. Eurasian Rollers *Coracias garrulus* follow this pattern; they are regular on passage through the Delhi area in the autumn, though not on their return journey in the spring when they take a route further to the west. Some occurrences are merely accidental, such as the odd Rufous-tailed Scrub Robins *Cercotrichas galactotes* that have strayed here during their autumn passage south.

(c) Vagrants

Birds are living creatures with extraordinary aerial mobility and are prone to vagrancy. But avian vagrancy remains an enigma, and various hypotheses have been put forward to explain it.[32] Different species have different migration strategies, believed to be largely genetically determined. Possible causes for birds deviating from the norm include sickness or damage to the bird's navigation system, perhaps some genetic variation leading to 'compass' errors, or reverse migration, a bird flying south in spring, for example, when it should be flying north to its breeding grounds. Young birds of the year tend to be frequent vagrants during their first southward movement in autumn, either because of inexperience or perhaps their genetic programming fails to work properly.

A bird may deviate from its usual migration pattern simply to survive a natural obstacle or event. Migrants can get blown off course or be compelled to descend and take shelter from bad weather, or suffer health conditions, which can result in unexpected sightings or even 'falls' of numbers of migrants at likely sites. They may overshoot their usual migration path or wander further south from their normal wintering ranges during a harsh winter. The link between unusually heavy snowfall during cold weather snaps in the Himalayan ranges and the occurrence of vagrants further south in the north Indian plains is well known. A related phenomenon involves birds from one migration flyway joining birds that share common breeding or wintering grounds from another flyway and then migrating back along with the other species. This seems especially frequent in waterfowl and is one explanation for the occasional hybridization in this family. Some species are naturally prone to wander, certain waders for example, or to erupt at irregular or regular intervals, perhaps linked to the cyclical abundance of a food item. Others are sedentary, such as woodpeckers or babblers, and records far from their known range would be truly extraordinary.

Thus, 78 species have been classified as vagrants, or accidental visitors, about Delhi. These include four species that had probably escaped captivity, or were released, but there is an unlikely chance they may have been genuine vagrants. The six species classified as 'Vagrant/Reconfirmation desirable' are also essentially vagrants, and if they were confirmed and included, the list of vagrants would reach 84. The term 'Data Deficient' is self-explanatory; at least one species, Yellow-legged Buttonquail *T. tanki*, thus tagged, may yet be proven to breed, but the information available is too sparse and sketchy to draw conclusions.

A number of species are genuinely exceptional in their occurrence in the Delhi area: many undertake long migratory movements in single hops, such as the Terek Sandpiper *Xenus cinereus* for instance, which has appeared in Delhi on passage, forced down by bad weather or a chance mix-up with migrating parties of other species. Almost any migrant that breeds in the western Himalayan region or further north and usually overflies Delhi to winter in the southern peninsula is possible here as an accidental waif.

Several species that breed in Central Asia and use the Eurasian–African Flyway[33] are forced to fly west and then south en route to East Africa to skirt the formidable obstacles presented by the Tibetan Plateau and the Himalayas, which effectively bar direct migration south. They may, however, appear in mainland India exceptionally and have been recorded as vagrants in the Delhi area. Caspian Plover *Charadrius asiaticus*, Rufous-tailed Rock Thrush *Monticola saxatilis* and Song Thrush *Turdus philomelos* are examples.

Waterfowl enliven the scene at Okhla.

Eastern influences in the Delhi list include a smattering of East Asian breeders that winter in eastern or southern India or further east, in Southeast Asia and Australasia[34], and may overshoot as far west as Delhi on their migrations. Vagrancy records of Brown-breasted Flycatcher *Muscicapa muttui*, Baikal Teal *Sibirionetta formosa* and Sharp-tailed Sandpiper *Calidris acuminata* show this pattern.

In addition, Delhi's proximity to the Himalayan foothills can bring in unusual visitors during heavy snowfall in the mountains. Altitudinal migrants that regularly descend to the foothills, White-capped Redstarts *Phoenicurus leucocephalus* for example, may drift considerably further south during such cold snaps. Some are far more unlikely, but even a high-altitude Horned Lark *Eremophila alpestris* has nevertheless been reported once from Delhi.

Vagrants can appear in the most unexpected places, and Delhi is no exception. But they are, in a sense, peripheral to the local avifauna, accidental strays that provide a chance to 'tick off' a species and add to one's life list, though they do not really contribute to the intrinsic biodiversity or ecological richness of the area. The lure of the list, the thrill of finding a rarity, is ingrained in most birdwatchers, but how do we interpret these records? Is the rarity a seasonal but erratic visitor to the Delhi area, such as a Grey-winged Blackbird *Turdus boulboul*, or a genuine vagrant such as a Mistle Thrush *T. viscivorus*? For the purposes of this Checklist, any species that could theoretically occur in the Delhi area, given its presently known range and movements, and has been recorded here, no matter how rarely or irregularly, is considered a visitor, but one that has strayed from its usual range or migratory pattern is referred to as a vagrant.

Yet vagrants are, in a manner of speaking, like pioneers. Having lost their way, many perish, but a few may survive, and this can have significant consequences. The impact of changes in land use or climate on bird populations are also likely to first show up in their unexpected movements. These can be drivers for the evolution of migration patterns and the distribution of species. Is the vagrant a harbinger of range expansion or new colonization? The development of a new migration route or wintering grounds?

Looking back at the last two decades in the Delhi area, attention should be drawn to the recent flurry of records of a few species that would not be considered anything other than vagrants in the normal course. Black Eagles *Ictinaetus malaiensis* that had never been conclusively reported from Delhi earlier first appeared in 2012, followed by scattered but regular records since 2014, all between October and March, from the wooded hilly areas about Delhi. Oriental Pied Hornbills *Anthracoceros albirostris*, a few individuals of which had last been reported in 1973, reappeared, first in 2013 and then in quick succession on several occasions since 2016; Whistler's Warblers *Phylloscopus whistleri* were recorded for the first time in the winter of 2009–10, but there have been almost annual winter records of the species since then. Neither the great increase in the number of observers in the Delhi area nor the possibility of escapes can fully explain these sudden bursts of records.

While many, if not the most recent, reports of Oriental Pied Hornbills may pertain to the same female bird and evidently an escape (given its apparent familiarity in urban landscapes), the status of Black Eagles and Whistler's Warblers may well have changed in recent years, and both have been included in the Checklist as rare, irregular winter visitors to the Delhi area, a few birds drifting south from their usual wintering range in the Himalayan foothills. Grey-hooded Warblers *P. xanthoschistos*, never recorded from Delhi previously but being reported over the last five or six years in winter, even a party of five or six on one occasion, are, however, still categorized as vagrants, though this may require to be reviewed if they continue to be regularly reported in coming years.

Bird populations have a plasticity which enables them to adapt to a changing environment. What does a White-naped Tit *Machlolophus nuchalis*, or a Tickell's Blue Flycatcher *C. tickelliae*, unrecorded from Delhi earlier, but now seen in the breeding season in its appropriate habitat, indicate? Is this is a scarce breeder in the Delhi area that has probably been overlooked all these years, or is the species expanding its range? Are the growing but yet enigmatic records of Stoliczka's Bushchat *Saxicola macrorhynchus* from the Delhi area an indication of vagrancy, passage movement, occasional or regular wintering, or an attempt to establish a breeding population? Only continued observation in the coming years may find answers to some of these questions.

Tickell's Blue Flycatcher, a male photographed at Bhondsi, Gurugram

4 The Birds of the Delhi Area: An Annotated Checklist

ANSERIFORMES: WATERFOWL

ANATIDAE: SWANS, GEESE AND DUCKS

Most of the 28 species on the Delhi list are winter visitors from beyond the Himalayas with a breeding catchment that spans Europe and Asia; they pour into our wetlands by October, leaving again by April, and their many colours, their whistling and honking and their spectacular flights add drama and a special thrill to an early morning at one of our lakes in winter. The waterfowl are a diversified group, and there are a number of evolutionary clades within the family, of which three are represented about Delhi: the whistling ducks in their own subfamily Dendrocygninae; the swans and geese in Anserinae; and the remaining ducks on our list in the omnibus Anatinae.

Most geese and 'dabbling ducks' prefer shallower waters and are mainly herbivorous, feeding by grazing or from the surface of the water. 'Diving ducks' and mergansers are mainly carnivorous and feed by diving in deeper water. Most nest on the ground, but some, including three of our resident species, may nest in holes and hollows of trees, or by appropriating other birds' stick nests. Both the clutches and the eggs themselves are often large, and in some species more than one female may lay eggs in one nest if sites are scarce. Usually only the females incubate and care for the chicks, which are highly precocial. Naturally occurring hybrids are not uncommon among Anatidae and have appeared in the Delhi area as well. Only four species breed here, and of these, three show tropical affinities: the Whistling- and Knob-billed Ducks being pan-tropical, Pygmy Geese restricted to the Old World tropics, while the Spot-billed Duck is a typical representative of the dabbling ducks.

India: 43 species, Delhi: 28

▲ *Ganga khadar landscape with Greylag Geese at Haiderpur*

Lesser Whistling Duck *Dendrocygna javanica*

Mainly Summer Breeding Visitor · Fairly Common

A Lesser Whistling Duck in flight

A widely distributed throughout the Oriental region and resident subject to local movements, from the Indian subcontinent to China and south through Southeast Asia. Monotypic.

This duck is mainly a summer/monsoon breeding visitor to the Delhi area from about April till November, although a few may remain through the winter. They appear to have been scarcer in the past; there are historical records of sporadic nesting, yet they were considered 'very uncommon' till the 1960s.[35] They have increased visibly since, affecting *jheels* and well-vegetated wetlands, and nesting between May and October, often in reed beds of the river *khadar* and other marshes of the area and occasionally in abandoned nests of raptors and crows in trees, or of cormorants and egrets, as at the heronry in the Surajpur wetlands, Greater Noida. The typical clutch is of 7–12 eggs,[36] and the birds may nest in loose colonies in the larger, more extensive reed beds. Pairs or small parties keep separate from the mixed waterfowl flocks during the day and fly out to feed mainly in the late evenings or at night.

Bar-headed Goose *Anser indicus*

Winter Visitor · Common

A distinctively Asian goose, breeding at high altitudes in Central Asia, Ladakh, Tibet and Mongolia and wintering widely in the subcontinent. Bar-headed Geese migrate directly over the Himalayas between their summer and winter quarters, and flocks have been reported at very high elevations of up to 9,000–10,000 m during passages. Monotypic.

These geese arrive in the Delhi area in November, and most birds leave in March, with stragglers remaining till May. Characterized as only 'fairly common' till the 1960s, they are now regular and commonly recorded at their favoured wintering sites, with up to 150–300 birds at the larger wetlands, though spectacular flocks of up to 1,000 have also been reported. Both Bar-headed and Greylag Geese may frequent the same wetlands but do not mix. Bar-heads tend to prefer more open habitats as daytime refuges, flying out in gaggles at dusk to favoured grazing grounds where they may feed in close proximity of villages and habitation.

Bar-headed Geese grazing in a grassy patch

A pair of Greylag Geese

Greylag Goose *Anser anser*

Winter Visitor Common

Widespread across the Palaearctic, breeding at temperate latitudes from Europe to northeast China and wintering in the Mediterranean region, West Asia and the Indian subcontinent to southern China. There are two subspecies, nominate *anser* in Europe and *rubrirostris* in Asia, the latter population being paler with distinctive pink bills and legs.

The Asian race *rubrirostris* is a common winter visitor to the Delhi area between October and April–May, while unusually early arrivals have been recorded even in August. Greylags have maintained their numbers in the area, and flocks from the low 100s to 1,000 or more are regularly reported at the larger waterbodies. The impoundments at Okhla, the Dhanauri wetlands in the GautamBuddha Nagar district, the Sultanpur National Park neighbourhood and several other well-vegetated wetlands may hold large concentrations. Flocks spend the day loafing on the *jheels* and fly out in the evenings to feed in neighbouring grazing grounds and cultivation.

Greater White-fronted Goose *Anser albifrons*

Winter Visitor Scarce

Circumpolar as a breeding species, in open tundra at Arctic latitudes; five subspecies are recognized, of which three are limited to North America and two to the Palaearctic. Nominate race *albifrons* nests in tundra and taiga habitats in Russia across eastern Siberia, wintering in Europe, West Asia, Japan and China. Its main winter quarters in Asia thus lie to the west and east of the subcontinent, only small numbers enter northwestern India.

Small numbers of these elegant geese have occurred irregularly about Delhi, mainly between November and April, with recent records from the river *khadar* and wetlands in Gurugram, the Rohtak and Jhajjar districts in Haryana and the GautamBuddha Nagar and Meerut districts in UP. Single birds, or rare parties of up to five or ten, are the norm in the area, with one report of 23 birds at the Sultanpur National Park in 1993.[37] In many years, they are absent.

Lesser White-fronted Goose *Anser erythropus*

Vagrant

A globally vulnerable species breeding across Arctic Eurasia, wintering mainly on the coastal plains of the Black Sea and Caspian Sea and south to Iraq and in eastern Asia; a rare straggler to the northern subcontinent. Monotypic.

Lesser White-fronts have been recorded sporadically in winter from the area at long intervals: there are three nineteenth-century specimens from Sultanpur, Haryana, collected by N.W. Chill in March 1879,[38] and four later records; from the Bhindawas Lake, Jhajjar district, in the early 1980s; one with five Greater White-fronts at the Sultanpur lake in March 1981,[39] and single birds in the Dighal wetlands, Rohtak district, in November–December 2016 and in February 2019. Two were also reported from the Haiderpur Wetland, and the Hastinapur Wildlife Sanctuary, UP, but just outside the limits defined for this Checklist, in early February 2021.

Knob-billed Duck *Sarkidiornis melanotos*

Breeding Resident · Fairly Common

Resident in sub-Saharan Africa and tropical Asia from the Indian subcontinent to Indochina, subject to seasonal movements. Affects riverbanks, lakes and marshes, ideally with some trees in the vicinity, but not dense growth. Monotypic.

Often referred to as *Nukta* locally, this large, handsome species is a fairly common resident in the area, with larger numbers seemingly being recorded in recent years than in the past. Small parties of up to 5–10 birds may be found in the larger wetlands all year, but their numbers build up in the pre-monsoon hot season, the males in iridescent finery with impressive, enlarged combs on their bills. As the rains break, females may be seen flying around lakes and over the surrounding countryside, looking for nest sites in large trees: abandoned kites' nests or a hollow large enough to hold their eggs, which are then padded with grass and feathers. A typical clutch is of 7–15 eggs, and more than one female may lay in one nest, perhaps the 'harem' of a single male, or the result of the limited availability of nest sites for such cavity-nesting ducks. Post-breeding, loose gatherings of 30 or more may collect at favoured locations. These ducks feed principally on vegetable matter by grazing or wading in shallow waters, but they also take insects and other invertebrate food items.

Three male Comb Ducks in breeding condition

A pair of Ruddy Shelducks

Ruddy Shelduck *Tadorna ferruginea*

Winter Visitor · Common

Conspicuous and striking birds, nesting in holes and crevices in cliffs, even in man-made structures, often near inland lakes but sometimes far from water and usually in mountainous areas, ranging from northwest Africa and southeastern Europe to Central Asia, Ladakh and Tibet, and migrating southwards, including to the Indian subcontinent and Southeast Asia, in winter. Monotypic.

Ruddy Shelducks, known locally as *Chakwa Chakwi*, are well-known visitors to the open margins of rivers, reservoirs and lakes throughout the Delhi area between October and April, straggling into May. They have largely maintained their numbers; the Okhla sandbanks were a favourite haunt in the past, though the Yamuna's dismal state has taken its toll. However, the river's cleaner stretches above Wazirabad, the Ganga at Hastinapur, UP, and most large lakes still attract numbers of this species. They stay in pairs or parties that can build up to 100 or more birds prior to emigration in March. They are a curious mix of wariness and familiarity, feeding and resting in human proximity but the first to sound the alarm at any sign of danger.

Common Shelduck *Tadorna tadorna*

Winter Visitor · Scarce

Largely resident from Europe and Central Asia and south to Iran, reaching North Africa, West Asia and the northern subcontinent in winter. Affects coasts, estuaries and brackish lakes in parts of its breeding range, but mainly fresh waters in its inland winter quarters in India. Monotypic.

These handsome shelducks are scarce winter visitors in small numbers, typically not exceeding 10–15 birds, to Delhi's wetlands, including the Yamuna *khadar* in its cleaner stretches above Wazirabad in Delhi, the Okhla reservoir, the Sultanpur National Park and wetlands in its neighbourhood, the Bhindawas Wildlife Sanctuary in Jhajjar district and other open waterbodies and riverine habitats in the area.

Cotton Pygmy Goose

Nettapus coromandelianus

Mainly Summer Breeding Visitor

Uncommon

A pair of Cotton Pygmy Geese in flight

Distributed largely across the Oriental Region, where the nominate subspecies *coromandelianus* is widespread, with another race extending its range to eastern Australia.

Also called the Cotton Teal, these little ducks are rather uncommon summer breeding visitors about Delhi and may have declined in recent decades as their numbers rarely exceed the low tens even where they are regular. They typically appear in the area in April and nest during the monsoon months in holes in trees or derelict buildings near water, laying 6–14 eggs in a cavity lined with feathers and down by the birds themselves. A few may remain in the area all winter. They prefer the smaller ponds with floating and submerged vegetation or weedy channels on the margins of larger waterbodies, where they can be very inconspicuous in non-breeding plumage. In the breeding season, this character changes, and pairs make a pretty picture as they chase each other with their typical cackling calls over the water lilies, handsome in the striking black-and-white contrasts of their summer plumage. They are mainly vegetarian in diet, dabbling and grazing on floating vegetation.

Baikal Teal *Sibirionetta formosa*

Vagrant

This dabbling duck with its striking green and yellow face pattern is an East Palaearctic speciality, nesting in eastern Siberia and wintering mainly in Japan, Korea and eastern China; very small numbers straggle irregularly as far as Southeast Asia and northeastern India. Monotypic.

Only a vagrant as far west as Delhi. Usha Ganguli, in her *Guide to the Birds of the Delhi Area* (henceforth Ganguli), mentions two historical records from November 1879 and May 1947. Recent records include single males at the Okhla reservoir in February 2006 and February 2011; at the Dighal wetlands, the Rohtak district, in March 2013; at Sultanpur National Park in April and again in December 2013; and an apparent hybrid Baikal X Common Teal at Dighal in January 2018. These records are of the easily identifiable adult males, and it is possible that vagrant females have been overlooked.

Garganey *Spatula querquedula*

Passage Migrant Common

Pretty, but soberly dressed, little ducks that breed right across the Palaearctic at northern and temperate latitudes, migrating in winter to Africa, India and Southeast Asia. Monotypic.

Garganeys are common passage migrants about Delhi, wintering usually further south in peninsular India. They are amongst the earliest migrant ducks to arrive from their breeding areas in Central Eurasia, with the vanguards coming in as early as July and the movement continuing till November. Spring passage is from end February through April/May, with late birds well into June. Their numbers are typically in the low hundreds, well distributed in the post-monsoon season but in spring, when the countryside is drier, they tend to concentrate on the larger lakes and rivers of the area.

Northern Shoveler *Spatula clypeata*

Winter Visitor Common

A dabbling duck with a morphologically specialized bill, spatulate and broadened distally with comb-like lamellae along its lateral edges for sifting out small food items, small invertebrates and plant seeds, etc., from the surface of the water and mud while swimming or wading. Northern Shovelers nest in mid- to high latitudes across the Holarctic, wintering to central America, parts of sub-Saharan Africa and southern Asia. Monotypic.

Shovelers are arguably the most abundant wintering duck in the Delhi area, arriving in September and leaving by April, though stragglers may linger till June. Flocks of a few thousands may gather at favoured lakes and wetlands, especially in spring when passage migrants augment their numbers. They prefer shallow waters with floating algae and vegetation, which suits their feeding method. A few Shovelers sometimes join Spot-billed Ducks in some of the waterbodies within Delhi urban limits, such as at Hauz Khas or the Delhi Zoo.

Northern Shoveler

Gadwall

Gadwall *Mareca strepera*

Winter Visitor Common

Widespread as a breeding species across the Holarctic, in temperate latitudes in both North America and Eurasia and wintering further south, including in North Africa, the Middle East and southern Asia. Monotypic.

Gadwalls are amongst Delhi's commonest dabbling ducks in winter, arriving by October and leaving by March or April. They seem to be more abundant now about Delhi than past reports would suggest. Widespread in all the larger wetlands, they prefer well-vegetated waters and are commonly in company of other dabbling ducks. Flock numbers in the low hundreds are typical, but up to 5,000 have been recorded at a site, both in mid-winter and on migration.

Falcated Duck *Mareca falcata*

Winter Visitor Rare

An East Palaearctic speciality like the Baikal Teal, breeding in eastern Siberia, wintering in Japan, eastern and southern China, up to Myanmar; also winters in small numbers in northeastern India – albeit rather more often than that species – straggling westwards. Monotypic.

These very handsome ducks are rare, erratic visitors to the area, with less than ten recorded occurrences over the years to date. There are five historical, pre-1975, records: from Sultanpur, Haryana, in February, probably in the 1870s;[40] from north Delhi in February;[41] from Badshahpur, near Gurugram, in February 1958, one shot out of six at Sonipat in 1969; and one in November 1974 at Sultanpur. Since 1975, there have been a few more records about Delhi – at least one male at Okhla on 2 February 1979 with a party of Gadwall and another at the Sultanpur National Park on 5 March 2013;[42] at Surajpur Lake in third week of February 2023; and hybrids, probably with congener Gadwall, have been recorded twice, at the Dighal wetlands, Haryana, on 11 December 2016 and at Sultanpur National Park in late February–early March 2020. It is likely that females, very similar to female Gadwalls, may have been overlooked.

Eurasian Wigeon

Eurasian Wigeon *Mareca penelope*

Winter Visitor · Common

Nest at high latitudes across Eurasia, wintering to southern Europe, North Africa and southern Asia. Monotypic.

These strikingly coloured ducks are common winter visitors to the Delhi area between early October and April, frequenting most of the larger wetlands, including rivers, *jheels* and irrigation reservoirs, where the males' characteristic whistle '*whee-OOO*' is an unmistakable indicator of their presence in a large mixed mass of ducks on the water or in the air. Essentially vegetarian, they feed on floating water vegetation by dabbling on the surface of the water, even upending at times, and grazing along the banks. Counts may reach a few hundred at a given location, these numbers remaining seemingly unchanged over several decades.

Indian Spot-billed Duck *Anas poecilorhyncha*

Breeding Resident · Common

A resident dabbling duck of South Asia and Myanmar to Southeast Asia. Two subspecies: nominate *poecilorhyncha* in the Indian subcontinent and *haringtoni* in eastern Assam and further eastwards in Myanmar and Indochina.

These handsome ducks are common breeding residents of well-vegetated lakes, ponds and wetlands of the area, occurring even within the larger urban parks as long as there is some waterside cover where they can nest. As the breeding season approaches, pairs wander over the countryside looking for suitable sites. They may start nesting as early as March (when their wintering cousins are still around) through the summer and early monsoon, building a pad of grass and reeds lined with feathers in pondside vegetation, in which between 8–12 eggs may be laid, and a pair may raise two broods every year. Essentially vegetarian, they feed mainly by dabbling in shallow waters, or grazing in paddy cultivation and harvested fields. The flocks are normally small, but larger numbers – hundreds – may collect at favoured locations such as Okhla in the summer before breaking up into pairs for breeding, though they do not behave as a cohesive flock.

Spot-billed Duck

Mallard *Anas platyrhynchos*

Winter Visitor Scarce

Widespread Holarctic ducks, breeding across temperate Eurasia, as close as Kashmir. Largely resident in Europe but migratory in continental Asia, wintering in the Mediterranean region, West Asia, northern India and southern China. The quintessential 'wild ducks' from which most domestic breeds are derived. Two subspecies, the nominate over most of its range.

Mallards are rather scarce visitors to the lakes and *jheels* about Delhi, from late November to March, their numbers usually in single digits or the low tens at a site. They appear to have been scarce about Delhi even in the past, so there may not have been any marked change in status.

Northern Pintail *Anas acuta*

Winter Visitor Common

Widespread, with a circumpolar distribution as a breeding species at mid- to high latitudes across North America and Eurasia, wintering as far south as central America, Africa and southern Asia. Monotypic.

These graceful ducks are common winter visitors to the area from September/October till April, rarely even May, but may have become rather less abundant in recent decades. They were considered the commonest of the migrant ducks in the past, with peak counts of 5,000 birds on the Yamuna and at the Sultanpur National Park in the 1980s. They remain well distributed, but their numbers appear to fluctuate year on year from the low hundreds to a couple of thousands at the major duck refuges. They prefer waters with plenty of submerged vegetation where they feed on aquatic plants or insects by dabbling, grazing along the edge, or upending in the shallows, the males' long tails pointing comically upwards as they do so.

Northern Pintail

Common Teal in flight

Common Teal *Anas crecca*

Winter Visitor · Common

Another very widespread species, breeding in wooded wetlands throughout northern Eurasia and wintering to southern Europe, North Africa, the Middle East and southern Asia. Monotypic, although the closely related Green-winged Teal *A. carolinensis* of North America is recognized as a subspecies by some taxonomies.

True to their name, these small ducks are amongst the commonest and most widespread wintering waterfowl in the Delhi area, usually arriving in September, sometimes even as early as July/August, and leaving by April/May. They prefer a variety of productive, shallow waters and seepage marshes, both large and small, with at least some floating and emergent vegetation. Counts in the low hundreds are typical, but peaks of over a thousand have been recorded in mid-winter.

Marbled Duck *Marmaronetta angustirostris*

Winter Visitor · Rare

A Mediterranean and West Asian duck with a fragmented distribution in shallow, reed-fringed wetlands in arid habitats, dispersive and partially migratory, that nests and winters as close as Sindh, Pakistan, but is a much less regular, occasional visitor further eastwards into northwestern India. Monotypic.

These ducks are rare, sporadic winter visitors to the area. Historically, W.M. Chill obtained a specimen in April 1881 from 'near Gurgaon', and Ganguli details a few records in January, April and November between 1947 and 1967.[43] Post-1975 records include single birds at the Sultanpur lake in January 1990; at the Basai marshes, Gurugram district, in late April 2003; at the Dighal wetlands, Rohtak district, in December 2012 and March 2013 (possibly the same individual) and two in Dighal in late February 2014, with one staying on till at least the first week of April; and in Jhajjar district, singles at the Mandhoti wetlands, in the first half of February 2021 and the Bhindawas Wildlife Sanctuary in mid-December 2022.

Red-crested Pochard, male, at the Budhera Wetlands, Gurugram

Red-crested Pochard *Netta rufina*

Winter Visitor · Fairly Common

Breeds very locally in Europe but widely across Central Asia at temperate latitudes, wintering from the Mediterranean basin eastwards to the Indian subcontinent. Monotypic.

These striking pochards, with their glowing orange heads and red bills, are present about Delhi between November and March, though individuals have lingered on till as late as 11 May. They favour open, relatively deeper waters where they can feed by diving, including the cleaner stretches of the Yamuna above Wazirabad, the Haiderpur Wetland upstream of the Hastinapur Wildlife Sanctuary, UP, and the larger, more open lakes of the area. Their numbers now are usually in the tens, compared to counts of 100 or more from the Delhi area in the 1980s and earlier; nevertheless, larger gatherings have also been reported at times from favoured habitats, such as the Haiderpur impoundment on the Ganga, where 1,200 were counted in February 2022.

Common Pochard *Aythya ferina*

Winter Visitor · Common

Widespread breeder in temperate Eurasia, from Europe to Central Asia and northern China, wintering to southern Europe, Africa and southern Asia. Though Common Pochards are very widely distributed with large populations globally, they have declined rapidly over parts of their range and are considered globally vulnerable.

Common Pochards arrive in the Delhi area in October and leave again by April/May. Like other diving ducks, they prefer open, deeper waters with some shoreline vegetation and are well distributed in the area in the larger lakes and irrigation barrage impoundments. Flock sizes may reach a hundred or more, and they are a usual component of the rafts of ducks and coots that while away the day on our wetlands.

Ferruginous Duck at a village pond in Dighal, Haryana

Ferruginous Duck
Aythya nyroca
Winter Visitor · Uncommon

A handsome pochard with a wide but fragmented breeding range across temperate Eurasia from Central Europe east to Mongolia, wintering to Africa and southern Asia. Small numbers nest within Indian limits as well as in the lakes of the Kashmir Valley. Monotypic.

These are the least numerous of the regularly wintering pochards in the area and are present in the area from November to April, preferring sheltered, vegetated ponds and lakes deep enough for diving to feed. Their numbers rarely exceed 20–25 at a site, significantly less than recorded earlier, perhaps reflecting declining populations across their Eurasian breeding range.

Baer's Pochard *Aythya baeri*
Vagrant

An East Asian pochard that nests in the Amur basin in northeastern China and winters in the coastal plains of eastern China and Southeast Asia, with small numbers reaching eastern India as well, but has undergone an extremely rapid population decline, resulting in its being progressively declared Endangered in 2008 and Critically Endangered in 2012. Monotypic.

This pochard is a vagrant to the Delhi area with just two records – from Okhla in January 2001 and Bhindawas in February 2002[44] – before their population collapsed.

Tufted Duck *Aythya fuligula*
Winter Visitor · Common

Widespread pochards that breed at mid- to high latitudes across the Palaearctic from Europe to eastern Siberia, wintering in Europe, Africa and southern Asia. Monotypic.

Tufted Ducks are common about Delhi in winter, arriving in October and leaving in March/April. They prefer deeper, open waters that are free of surface vegetation and mix freely with other pochards that frequent similar habitats. They are well distributed, including on the barrage impoundments on the Yamuna above Wazirabad and Okhla, and the Hastinapur Wildlife Sanctuary on the Ganga, the larger, more open lakes, sewage treatment plants, etc. There has been no substantive change in wintering populations over the last half-century at least.

Tufted Duck

Greater Scaup *Aythya marila*

Winter Visitor Rare

Diving ducks of the Holarctic far north, essentially maritime, with two subspecies: nominate *marila* in Eurasia and *nearctica* in North America. Eurasian populations winter along the coasts and sheltered bays of Europe and in the Black Sea and Caspian Sea. Very scarce further south, with only sporadic records inland from the northern Indian subcontinent.

Scaups have been reported on a mere handful of occasions about Delhi, with two historical specimen records from 'near Gurgaon' in March 1881[45] and verified observations from the Yamuna at Okhla in February 1998, November–December 2002, October–November 2004 and January 2011, and from the Bhindawas Wildlife Sanctuary in February–March 2004.

Smew *Mergellus albellus*

Vagrant

Small mergansers, the males unmistakable in the black-and-white contrasts of their breeding plumage, nesting in tree holes in the Palaearctic taiga belt, and wintering in lagoons and inland lakes in Europe, Central and East Asia. Only small numbers winter in the northern subcontinent, scarce in the west in the Indus and Punjab plains and even rarer eastwards. Monotypic.

Ganguli mentions nineteenth-century records from Najafgarh and one from February 1922. The only two later reports from about Delhi are from Jharli, in Jhajjar district, and the Bhindawas Lake, Haryana, of single birds in female plumage on 16 November 2014 and 8 March 2015. There are recent reports in January and November 2021 and December 2022 – single males observed over three consecutive winters – from Haiderpur Wetland as well, which form part of the Hastinapur Wildlife Sanctuary on the Ganga but are just beyond the limits of the area covered by this Checklist.

Red-breasted Merganser *Mergus serrator*

Vagrant

Widespread as a breeding species along the forested rivers and lakes of the northern Holarctic but wintering mainly in saltwater habitats along the coasts of North America and Europe, in the Black Sea and Caspian Sea and eastern Asia. Small numbers reach the Persian Gulf and the Makran coast but are only vagrants in India with rare records from Bengal and near Mumbai. Monotypic.

A vagrant female/immature was recorded on 6 December 2020 at the Mandothi Jheel, Jhajjar district, Haryana, that remained till the third week of that month and was photographed and documented by several observers.

GALLIFORMES: LANDFOWL

PHASIANIDAE: PHEASANTS AND ALLIES

Landfowl, or 'game birds' as they are usually called, are a large, diversified family, exemplified in popular conception by domestic fowl. They are usually divided into three subfamilies based on molecular criteria, all of which are represented in India, but only one, Pavoninae, in the Delhi area. They are primarily terrestrial, with strong legs adapted for a life on the ground, and plumages that range from resplendent, often embellished with crests, combs and ornamental feathers, in the males to cryptic in the wonderfully camouflaged females. They exhibit varied breeding economies, many species being polygynous, the males performing spectacular displays or crowing loudly from conspicuous posts in the breeding season to advertise territory and attract mates. Females, as in most polygynous species, take on all duties of incubation and care for the precocial chicks. Most – but not all – nest on the ground, usually in a concealed depression under a bush or in a grass clump, and lay clutches of four to nine, or even more, uniform brownish-buff or olive-coloured eggs.

India is rich in landfowl species and is home to some of the world's most brilliant and colourful members of the family. However, most of these are forest or mountain dwellers, and Delhi's open habitats host just seven species. These include our national bird, the Peafowl; two quails, one widespread migratory species and one nomadic regional representative of the genus; two bush quails representing a small group of four Indian endemics; and two francolins that represent a group spread widely across Asia and Africa.

India: 42 species, Delhi: 7

▲ *Black Francolin*

Indian Peafowl *Pavo cristatus*

Breeding Resident · Common

Universally known, a South Asian endemic. Monotypic.

Peafowl are familiar, semi-feral breeding residents in Delhi and the surrounding countryside, where they are protected by law and religious sentiment. Single males, a cock with a 'harem' of hens, or larger droves of females and immatures frequent fallow ground about villages and temple groves, wooded areas of the Ridge, the larger urban gardens and city forests, where their loud calls can be heard in the mornings and evenings over the din of traffic. They roost in tall trees, the males often flying up in the gathering darkness to perch on large branches with their spectacular trains drooping behind them. Peacocks are polygynous, mating successively with a number of hens. They breed through the monsoon, when females build bulky nests in the undergrowth, in creeper-covered or thorny bushes or hedges, or even on low roofs of village houses, in which they lay 4–6 eggs as a norm.

Indian Peacock

Common Quail *Coturnix coturnix*

Passage Migrant · Scarce

Widespread across the Palaearctic region from Europe to Central Asia, also very locally in India and Africa. A strongly migratory species (an unusual trait in the family, most phasinids being short and broad-winged, to enable quick getaways when flushed from cover but do not support long migration flights), with Common Quails breeding in Europe and Asia wintering in the Mediterranean region, Africa and South Asia. Monotypic.

Common Quails that winter in central and southern India presumably migrate through the Delhi area. A century ago, they were netted for food about Delhi, but are now rather scarce passage migrants with sporadic reports of birds calling in wheat *Triticum* fields in spring (February–April) or flushed from grassy patches in autumn (October–November). They are rarely reported in winter. Its unmistakable trisyllabic whistle is usually the first indication of its presence, unless flushed. The few reports are from the Yamuna floodplain and grassy flats such as around the Sultanpur National Park.

A male Rain Quail announcing its territory

Rain Quail *Coturnix coromandelica*

Summer Breeding Visitor · Fairly Common

Also called Black-breasted Quail; variously resident, local migrant or nomadic in grasslands, dry crops and mixed grass-and-scrub country across much of the subcontinent and eastwards to Thailand. Monotypic.

Rain Quails are summer/monsoon breeding visitors to the Delhi area between May and September, preferring open grassy areas interspersed with cultivation, nesting usually in a shallow depression scraped by the females themselves and lined with dry grasses, in which 6–8 eggs are laid. Their abundance seems to vary, rather common in some years and less so in others, possibly dependent on the monsoon. The males' loud double-noted whistles are then an unmistakable indication of their presence in a locality, and the fact that these are being heard less often about Delhi in recent decades suggests that the species' numbers – and suitable habitat – may be declining in the area.

Jungle Bush Quail *Perdicula asiatica*

Breeding Resident · Scarce

Endemic to India and Sri Lanka, with four subspecies. Of these, *P. a. punjaubi* is a widespread resident in the grass and bush and dry deciduous woodland habitats in northern India.

There was just one record of this bush quail from the Delhi area earlier, of two birds shot in Sohna in November in the 1940s.[46] Harvey, however, describes them in the *Atlas of the Birds of Delhi and Haryana* (henceforth, *Atlas*), published in 2006, as resident in broken scrub country in Haryana and on the Ridge in Delhi, mentioning Asola and the Badkhal Lake on the Southern Ridge as specific sites, and there is an older report from Tughlaqabad as well. Despite extensive coverage of likely terrain, the presence of this bush quail could not be reconfirmed until recently, in April 2017.[47] However, several records since then from a number of sites on the South-Central and Southern Ridge in Haryana, over most months of the year, are evidence of their resident status in these 'Aravalli' habitats of the area. There are also reports from the Ganga *khadar* near the Hastinapur Wildlife Sanctuary. There is no reason why they should not occur more widely in these habitats as the race *punjaubi* is resident and common along the Aravalli Ranges in Rajasthan and elsewhere in Haryana and UP as well.

A male Jungle Bush Quail on the Southern Ridge

Rock Bush Quail *Perdicula argoondah*

Resident? Data deficient

Endemic to India, widely distributed in the western half of the country in dry, stony and scrubby habitats, from Haryana southwards into the peninsula. Three subspecies, race *meinertzhageni* in the northern part, the nominate to its south and *salimalii* on laterite soils in Karnataka.

These bush quails have been recorded from Narnaul, Mahendragarh district, just beyond our limits, but the first, and only, record to date from within the Delhi area itself was of a male photographed on 2 March 2023 on stony ground at Bandhwari, Gurugram district; they are presumably scarce residents of Southern Ridge habitats, where they co-exist with their sibling species, Jungle Bush Quail.

Black Francolin *Francolinus francolinus*

Breeding Resident Common

A resident from West Asia across Iran and the northern Indian subcontinent to Assam, but also introduced and naturalized elsewhere. Six subspecies, of which two occur within Indian limits, *asiae* across the Indo-Gangetic Plain and *melanonotus* in the Northeast.

These handsome, well-known birds are common residents about Delhi (race *asiae*) in their preferred habitats: a mosaic of damp reedy ground, cultivation and marsh, rough grass-and-bush country, or the taller crops such as wheat *Triticum*, mustard *Brassica*, or sugarcane *Saccharum*. Though they may be found in the same general area, Black Francolins prefer lusher habitats and avoid the dry scrublands beloved of the congeneric Grey; unlike that species, they are usually seen singly or in pairs, not in coveys. They do not usually roost in trees as the latter often do, though males commonly call from elevated perches. They breed between April and September, when the cock's spirited crow is an arresting feature of its habitat, though this has shrunk in recent decades and led to a certain diminution in their numbers.

Grey Francolin *Ortygornis pondicerianus*

Breeding Resident Common

Ranges from the Persian Gulf and southern Iran to the Indian subcontinent, also naturalized on some Indian Ocean islands. Three subspecies, of which two in India: the pale race *interpositus* in the northern plain and nominate *pondicerianus* in the peninsula and Sri Lanka.

Grey Francolins are numerous breeding residents (race *interpositus*) in the area in open scrubland, village outskirts, even in larger rambling city gardens and vacant plots within urban limits. They are usually in pairs, coveys or often family parties with 6–8 half-grown chicks in the breeding season and are fast runners, only flying with a whirr of wings if necessary. They breed freely throughout the area, raising two broods from April to October. They have retained their abundance wherever their preferred habitat remains, their ringing calls a familiar backdrop to a morning excursion in the countryside.

Grey Francolin

PHOENICOPTERIFORMES: FLAMINGOS

PHOENICOPTERIDAE: FLAMINGOS

A small family of six species of unique, specialized waterbirds that live their lives in brackish lakes and lagoons and are found on all the continents except Australia. They are remarkable in many ways, with their proportionately small bodies; extremely long and slender sinuous necks and long legs with webbed toes, enabling them to even swim; and extraordinary angular bills that are bent downward – flamingos feed with their bills immersed upside down in shallow water – and furnished with comb-like lamellae used to filter food items from water or mud. Flamingos are also well known for their pink plumage colours, derived from carotenoid pigments in the food they eat, which includes plankton, small aquatic invertebrates and algae. Flamingos nest in very large and dense colonies of distinctive, cone-shaped mud nests built by the birds themselves in shallow, saline lakes. This iconic family is represented by two species in the Old World, both of which occur in India and are recorded in the Delhi area as well.

India: 2 species, Delhi: 2

▲ *A small party of Greater Flamingos at Najafgarh*

Greater Flamingo

Phoenicopterus roseus

Non-breeding, Year-round Visitor

Common

Greater Flamingos

Widespread across Africa, the Mediterranean region and India, with their nesting colonies highly localized over their range. The nearest large colonies to us are in Kachchh, Gujarat, whence flocks disperse widely over western and southern India, though they have been suspected of breeding opportunistically elsewhere, even closer to Delhi, as well. Monotypic.

Small to very large flocks may be seen all year at the larger inundations in the Delhi area, often shifting from one site to another depending on local hydrographic conditions. They were historically considered uncommon winter visitors, but by the 1950s, flocks of 500–4,000 or even more began to be recorded at the Najafgarh lakes and other large wetlands in the area, with the largest numbers congregating in summer.[48] The Sultanpur National Park remained a reliable site, hosting up to 2,000 birds till the 1980s when the character of the lake changed from an open brackish *jheel* to a more wooded, freshwater wetland, which made it unattractive for flamingos. They also frequent various other shallow, often seasonal, lakes in the area, notably in the Rohtak and Jhajjar districts, Haryana, and the Yamuna River as well, where the impoundments at Okhla and Wazirabad attract up to 200–400 birds when water levels are low. In recent years, the largest numbers, 1,000–1,200, have been reported from inundations along the Najafgarh drainage where they are present through the year.

Lesser Flamingo *Phoeniconaias minor*

Vagrant

Large concentrations occur and breed locally, in the Rift Valley salt lakes of East Africa, with three other, more or less discrete, populations in South Africa, West Africa and northwestern India. Monotypic.

Lesser Flamingos are mere stragglers to the Delhi area, mainly on the Haryana side and possibly from its nearest regular site at the Sambhar Salt Lake in Rajasthan. There are a few historical records, and a few more recently, of single birds in the Greater Flamingo flocks at the Sultanpur National Park in March 1988; from the same general area at Basai in September 2003; a group of at least four juveniles and an adult at the Najafgarh wetlands in February 2020; and one bird there in March 2022. In addition, there are plausible but unverified reports from Sultanpur in June 1991; the Dighal wetlands, Rohtak district, in June 2012; and the Bhindawas Wildlife Sanctuary in December 2012.

PODICIPEDIFORMES: GREBES
PODICIPEDIDAE: GREBES

These birds are a cosmopolitan family of highly specialized divers, who have structurally adapted to a life in water. Almost tail-less, their legs with lobed toes are set so far back on their bodies that they can barely walk, but are very effective for rapid propulsion in water, both in swimming and diving. They are inextricably bound to freshwater habitats, feeding on fish, aquatic insects and other invertebrate life, even anchoring their floating nests to fringing or emergent vegetation. Grebes take off from the water clumsily, pattering along till airborne, but can then fly long distances as shown by three of our grebes, of which two are migrants and one vagrant, from temperate Eurasia.

India: 5 species, Delhi: 4

▲ *Little Grebe, in breeding plumage*

Little Grebe *Tachybaptus ruficollis*

Breeding Resident · Common

Widespread in temperate and tropical latitudes throughout the Old World. Nine subspecies recognized, with race *capensis* in South Asia and Africa.

Little Grebes are common breeding residents in wetlands large and small, rivers, lakes and even village ponds throughout the Delhi area. Single birds or small parties may be found in scattered association, their only requirement being open waters that are deep enough for their hunting technique or to dive and vanish underwater when they apprehend danger. They avoid choked waters but tolerate open patches amid floating vegetation, such as the water chestnuts cultivated in village ponds. They breed during the monsoons, building a floating nest of weeds attached to emergent pond-side vegetation to hold a clutch of four to five eggs.

Horned Grebe *Podiceps auritus*

Vagrant

Horned, or Slavonian, Grebes breed in inland lakes in North America and northern Eurasia, but winter mainly in maritime habitats. They have two subspecies; the Palaearctic birds belong to the nominate race that winters mainly along European and East Asian coasts. The closest regular wintering area is the Caspian Sea, probably the source of vagrants to the subcontinent.[49]

Six birds were reported from the Dighal wetlands, Haryana, on 14 December 2017 with at least two remaining there till late January 2018[50] and the others (presumably) dispersing to other waterbodies in the immediate neighbourhood, from which they were reported over the same period. At that time, this was only the third record from India.

Great Crested Grebe *Podiceps cristatus*

Winter Visitor Uncommon

Breeds widely across Eurasia, mainly in temperate latitudes but even as close to Delhi as in Gujarat and Ladakh. Three subspecies; nominate *cristatus* in Europe and Asia is resident in most of Europe but migratory in colder continental Asia, with these latter populations wintering to the south of their breeding range. Two other races extend its range to Africa and Australasia.

A Great Crested Grebe in breeding dress

These elegant grebes are present in the area in small numbers from November to March, typically in ones and twos often associated with diving ducks which have similar requirements of relatively deeper, clearer waters for feeding. The cleaner stretches of the Yamuna above Wazirabad, Okhla, the man-made water treatment impoundments in Gurugram district and elsewhere, the Ganga *khadar* in UP and the larger lakes in the area are preferred localities. They seem to have been more frequent in the 1950s and 1960s when they were 'generally found in small groups of less than a dozen' with peak counts of 30 at Najafgarh and 100 near Faridabad.[51] No dates are mentioned for such large gatherings, which may possibly have been migratory flocks.

Black-necked Grebe *Podiceps nigricollis*

Winter Visitor Scarce

Nests in western North America, much of north temperate Eurasia and in East and South Africa as well; Palaearctic populations winter in southern Europe, the Middle East and southern and eastern Asia. Three subspecies, with the Eurasian birds being of the nominate race.

These grebes are irregular, usually scarce, winter visitors to the Delhi area with just one historical record from Sultanpur, in late November 1969, but several scattered reports over recent decades since 2001, from the Yamuna at Okhla and the deeper lakes and ponds, both natural and artificial. They are not reported in most years but may be overlooked or confused with Little Grebe in non-breeding plumage on a casual look.

COLUMBIFORMES: PIGEONS AND DOVES
COLUMBIDAE: PIGEONS AND DOVES

A large and cosmopolitan family, most diversified in the Australasian Region, whose members show remarkable variations in adaptability to human-modified habitats. The Rock Pigeon has gone feral and is totally at home in cities around the world. On the other hand, some of the most dramatic and unfortunate bird extinctions have taken place in this family, the Dodo *Raphus cucullatus* and the Passenger Pigeon *Ectopistes migratorius*, being two iconic examples.

Three broad clades or subfamilies are recognized within the family, two of which, Columbinae (which includes typical pigeons and doves) and Raphinae (green pigeons, etc.) have representatives in India and in Delhi as well. Pigeons and doves are primarily granivorous or frugivorous, often swallowing large-seeded fruit like *ber* or *jamun* whole. They are also one of the few birds that can suck up water with their bills, rather than having to raise their heads to gulp each beakful. Monogamous, they typically build a flimsy platform of twigs (two crossed sticks) that serves as a nest, the rock pigeons on a ledge and others on a fork of a branch, that just manages to hold the two eggs of a clutch. Chicks are fed 'crop milk' regurgitated by the parents, that consists mainly of water, protein and fat.

India: 34 species, Delhi: 8

▲ *Spotted Dove*

Rock Pigeon *Columba livia*

Breeding Resident · Common

Ancestor of the domestic pigeon but restricted as a wild species to parts of Europe and North Africa to Central Asia and India. Thirteen subspecies recognized, of which two occur within Indian limits: *neglecta* in the northwest Himalayas and *intermedia* in the rest of the country. Even so, the degree to which seemingly natural populations have been affected by mixing with feral birds is often unclear.

Birds about Delhi have the characters of race *intermedia*, the dark rumped race of South Asia, though some show evidence of mixed domesticated blood. They are prolific breeders through the year and have found the perfect substitute for their native rock faces and cliffs in city buildings and other structures. Over and above this capacity to adapt to human-modified habitats, pigeons are fed daily at temples, parks, public squares, etc., resulting in the pigeon population growing to pest proportions. Large flocks flying out of the city in the mornings to feed in harvested fields in the countryside used to be a common sight in the past but are less visible now with birds finding adequate food in the city itself.

Oriental Turtle Dove *Streptopelia orientalis*

Passage Migrant · Uncommon

Oriental Turtle Dove

Widespread over much of Asia, a summer breeding visitor to northern parts, while southern populations are mainly resident. Six subspecies are recognized, of which three occur in India: *meena* breeding in Central Asia and the western Himalayas and wintering mainly in South Asia; *agricola* in the eastern Himalayas; and *erythrocephala* in the peninsula.

Oriental Turtle Doves are regularly recorded on passage through the Delhi area, usually in small numbers in open woodland and gardens, mainly in March–April and September–November. There are scattered mid-winter reports from the area as well. The pigeon-like, shorter-tailed jizz of these birds is distinct from other Indian doves even in a distant view.

Eurasian Collared Dove *Streptopelia decaocto*

Breeding Resident · Common

Collared Doves show strong dispersive tendencies, having naturally expanded their former native range in West and South Asia to colonize much of Europe and North Africa in the course of the twentieth century. There are two subspecies: nominate *decaocto* in India (and westwards to the Middle East and beyond) and *xanthocycla* in Myanmar.

Common residents about Delhi, in urban parks and everywhere in surrounding rural areas, breeding mainly in the summer months. Their trisyllabic 'coo' becomes a familiar sound of the city as the weather warms up in March. In the non-breeding season, they collect in flocks that roam the countryside and feed on spilled grain in freshly harvested fields.

Red Collared Dove *Streptopelia tranquebarica*

Breeding Resident · Fairly Common

A male Red Collared Dove showing its pastel colours

Ranges from the Indian subcontinent to China and Southeast Asia; two subspecies: nominate *tranquebarica* in most of India, the deeper coloured *humilis* in the northeastern parts of the country.

These attractive doves, though present about Delhi all year, do show some local, seasonal movements in the area. They prefer well-wooded groves near open country, scrub and dry woodland rather than the proximity of built-up areas, unlike some of their congeners. Yet, till about 30 years ago, birds used to enter New Delhi to breed in parks and large roadside trees from March to September, dispersing thereafter over the surrounding countryside to feed with Eurasian Collared Doves in harvested fields. There are a few recent reports from the city, but they still commonly nest in groves and orchards in rural areas, though they tend to avoid areas overrun by *Prosopis juliflora*.

Spotted Dove *Spilopelia chinensis*

Casual Visitor Scarce

Resident from India eastwards across mainland Southeast Asia, but introduced and now established in many other parts of the world, including islands of the Indian Ocean and Pacific Ocean. Three subspecies: nominate *suratensis* on the subcontinent, *tigrina* in Southeast Asia and nominate *chinensis* in China.

There are sporadic records from about Delhi scattered over the year, of singles or pairs in parks and woodland. Delhi's natural habitats are rather dry for its liking, but it appears that a few birds occasionally wander from the Himalayan foothills where it is a common breeding species. More frequent further to the east, as in the Ganga *khadar*, the Hastinapur Wildlife Sanctuary, where they are probably resident.

Laughing Dove *Spilopelia senegalensis*

Breeding Resident Common

Widespread and common over much of Africa, and West Asia eastwards to Pakistan and India; five subspecies recognized over this large range, of which race *cambayensis* is resident in India.

These are typically doves of drier country and scrubland, and Delhi's habitats suit them admirably. Laughing Doves are widespread and familiar, breeding freely in city and village alike over much of the year except the coldest months, November to February. They do not usually flock with Eurasian Collared Doves and other congenerics to feed in the winter months in rural areas.

Ashy-headed Green Pigeon *Treron phayrei*

Vagrant

Resident in the lowland forests along the Himalayan foothills from Nepal eastwards across Southeast Asia. Monotypic.

A male individual, photographed in the Delhi cantonment on 14 and 23 December 2015, was probably an accidental waif from this pigeon's usual range in the Nepal *terai*. Pigeons are known wanderers, and this species has appeared as a vagrant as far west as the Uttarakhand foothills in winter[52] and Dehradun in late June 2021, and as far to the south as Agra in August 2019.

Yellow-footed Green Pigeon *Treron phoenicopterus*

Breeding Resident Common

Widespread across the Indian subcontinent and Southeast Asia; five subspecies, two of which are resident in India: nominate in northern and eastern India and *chlorigaster* in the peninsula.

These are common frugivorous pigeons of gardens and parks, the Ridge and generally any well-wooded patch, even isolated groves of large fig trees in open countryside. They are usually seen in small flocks, but larger numbers may gather at fruiting fig trees, especially banyan. Birds in the area do not show the clearly demarcated grey underparts of the nominate race of the Himalayan foothills; an occasional bird may show grey tones on the flanks and belly, but in the main, the Delhi population appears decidedly closer to *chlorigaster* of peninsular India. They breed from March to June, when they are also most visible and vocal.

PTEROCLIFORMES: SANDGROUSE
PTEROCLIDAE: SANDGROUSE

A family of strong-winged, terrestrial, granivorous birds of deserts and dry scrub habitats. They keep in pairs and small flocks and are well known for their regular drinking habits, commuting considerable distances daily to favoured pools at regular times in the morning and evening, converging from all directions to quench their thirst. Some desert-haunting species are also known, in the breeding season, to carry water to their nests in their soaked belly feathers, presumably to cool their eggs, usually three in a clutch, and prevent them from overheating, or for their chicks to drink. (Such belly-soaking behaviour for thermoregulation, both for incubating adult birds themselves and their eggs, is also known in some families of the Charadriiformes.)

Two species are resident in the Delhi area. The Black-bellied Sandgrouse *Pterocles orientalis* was formerly an uncommon visitor to the area but appears to have changed its main wintering areas in Rajasthan over the years, possibly because of persecution and habitat loss, and seems unlikely to reappear here.

India: 7 species, Delhi: 2

▲ *A pair of Chestnut-bellied Sandgrouse*

Chestnut-bellied Sandgrouse, male

Chestnut-bellied Sandgrouse *Pterocles exustus*

Breeding Resident Uncommon

A widespread sandgrouse of semi-desert and arid habitats in the African Sahel, Arabia and eastwards to Pakistan and India; six subspecies, of which *P. e. hindustan* ranges from southeastern Iran to the subcontinent.

The Indian race, *hindustan*, is a bird of open dry country and short grass flats, much of which has been cultivated, reclaimed or built upon in the Delhi neighbourhood. They are now restricted to fragments of suitable habitat, areas of low scrub and fallow ground of the Southern Ridge landscapes and the Sultanpur National Park and Najafgarh neighbourhoods, where 100 or more birds have been counted coming to water on a morning. The Sultanpur area is their only presently known breeding site, with juveniles recorded in June, but suitable habitat is still available, especially on the Haryana side of the Delhi area; their apparently restricted distribution perhaps only reflects better coverage of the few sites where they are known to occur.

Painted Sandgrouse *Pterocles indicus*

Breeding Resident Uncommon

An Indian near-endemic, affecting stony, broken ground with sparse, thorny scrub growth and Deccan Plateau habitats. Monotypic.

A handsome resident of the dry, stony, broken country, typical of the Aravalli habitats in the Delhi area, Painted Sandgrouse are fairly widely distributed, albeit in small numbers, over the more open portions of the Southern Ridge from Gurugram, the Asola–Bhatti Wildlife Sanctuary and Badkhal in Haryana and southwards in suitable habitats of the range. Areas further north, such as the South-Central Ridge habitats of the JNU campus, Vasant Kunj and Mehrauli, from where they were recorded till the 1980s, are either built upon or too densely overrun with *Prosopis juliflora* to support this species. They are most easily seen as they fly, usually in ones and twos and not in flocks, to their favoured watering places to drink at dusk.

A pair of Painted Sandgrouse

OTIDIFORMES: BUSTARDS

OTIDIDAE: BUSTARDS

Impressive, and often large, birds of open country, that have adapted to a life on the ground with long, strong legs, but are capable of flying long distances when required. They are most diversified in the deserts and grasslands of the African region. Bustards have elaborate territorial and courtship displays, on the ground in the case of open country bustards, or aerial – for better visibility – in those that live in grasslands, such as the floricans. Most bustards are polygynous, the females choosing their mates from amongst the displaying males, but thereafter assuming all parental duties. There are historical records of the Great Indian Bustard and Macqueen's Bustard from the Delhi area, and the Lesser Florican has been recorded in recent years as a vagrant.

India: 5 species, Delhi 1

▲ *Lesser Florican, male in display (near Ajmer, Rajasthan)*

Lesser Florican *Sypheotides indicus*

Vagrant

Breeds in western India (mainly Rajasthan and Gujarat) and disperses into peninsular India in the non-breeding season; subject to nomadic movements that are not entirely understood. Monotypic.

Lesser Floricans arrive to breed in western India as the rains rejuvenate the grasslands, and they have reached the Delhi area as rare vagrants in this season. There are anecdotal accounts of floricans being seen displaying near Pataudi, Haryana, in the past, but the only recent records are of a female at the Sultanpur National Park on 4 June 2006;[53] a male at Dadri, UP, on 17 July 2014; and a male and at least one female near Sultanpur in the second week of July 2018.

CUCULIFORMES: CUCKOOS
CUCULIDAE: CUCKOOS

Cuckoos are well known for their distinctive calls that are associated with seasons in many cultures as they can be very vociferous when breeding. The family includes several distinct clades, often designated as subfamilies, of which three are represented in India: Coucals Centropodinae, Malkoha Phaenicophaeinae and the brood-parasitic cuckoos Cuculinae. Some cuckoo species build their own nests, incubate their eggs and care for their young, as is the normal practice in most bird families, but many others are parasitic, laying their eggs in the nests of other species and abdicating these responsibilities to the hosts. Some cuckoos parasitize a variety of species and lay eggs that mimic the eggs of their host. The chicks themselves may eject the eggs of the host from the nest and appropriate their foster parents' attentions for themselves.

Of the nine cuckoos on the Delhi list, two are non-parasitic, a coucal and a malkoha, both sedentary, terrestrial birds with short, rounded wings and laboured flight; six are brood parasites, of which five breed in the area. These are longer-winged, strong fliers that show local, or long-range – for the Pied Cuckoo – migratory movements.

India: 24 species, Delhi: 9

▲ *Asian Koel, a female in a mulberry tree*

Greater Coucal *Centropus sinensis*

Breeding Resident · Fairly Common

Greater Coucal

Resident across the Indian subcontinent, southern China and Southeast Asia; six subspecies recognized, of which three occur within Indian limits: nominate *sinensis* in the northern plains, *intermedius* in northeastern and *parroti* in southern India.

Greater Coucals are fairly common as breeding residents about Delhi in a variety of brushy habitats including gardens, tall crops and woodland, and they are often found in reed beds. These are monogamous, non-parasitic cuckoos, and pairs work together to build a large, globular nest ball of grass with an opening on the side, concealed in a thicket or reed clumps, in which 3–4 eggs are laid. Two races may occur; some in city gardens show the brownish forecrowns of the peninsular race *parroti*, while many others belong to the northern *sinensis* with all-black heads and distinctively barred juveniles; the latter are most often reported from wetter areas and reed beds, such as at Okhla, the Hastinapur Wildlife Sanctuary, or Dhanauri wetlands, UP. It is unclear whether the two races show different habitat preferences.

Sirkeer Malkoha *Taccocua leschenaultii*

Breeding Resident · Uncommon

A Sirkeer Malkoha feeding a chick in thorn scrub habitat

An endemic malkoha of the subcontinent. Three subspecies are recognized: *sirkee* in the northwest, *infuscata* in the Himalayan foothills and in eastern India and nominate *leschenaultii* in southern India and Sri Lanka.

The paler race *sirkee* represents the species in the Delhi area as a breeding resident. Ganguli had considered these 'not-too-common' residents in the 1960s, and their status does not seem to have changed materially wherever their preferred dry thorn scrub habitat is still intact – as in Aravalli habitats of the Ridge in both Delhi and Haryana as well as isolated dry woodland patches and mixed scrub along canal banks elsewhere. They breed from April to July in the area when their peculiar courtship calls and displays may attract the observer's attention in their habitat. Pairs build a shallow nest of twigs, usually low in a thorn bush, that holds their clutch of usually two eggs.

Pied Cuckoo

Pied Cuckoo *Clamator jacobinus*

Summer Breeding Visitor Common

Widespread across sub-Saharan Africa and the Indian subcontinent, with three subspecies over this range, *pica* is a transcontinental migrant race that breeds during the monsoons in the subcontinent and crosses the Arabian Sea to spend the remaining months in Africa; nominate *jacobinus* is mainly resident in southern India and Sri Lanka and *serratus* in South Africa.

The Pied Cuckoo is famous in most of India as the *Chatak*, harbinger of the monsoon, and these conspicuous, crested birds are common and noisy monsoon visitors to the Delhi area between late May and early November. They are found in a range of habitats including city gardens, woodland and the floodplain, wherever its preferred babbler hosts occur. A.J. Gaston, in his studies on babblers on the New Delhi Ridge in the early 1970s, found Pied Cuckoos parasitizing Jungle Babblers *Argya striata*, Common Babblers *A. caudata* and Large Grey Babblers *A. malcolmi*, in that order of preference, with up to 71% of the first, 42% of the second and 29% of the third species' nests cuckolded in his study area. He found that the cuckoo chicks hatched first, after 11–12 days incubation compared to 14–16 days for the babblers, to dominate the nest in competition with the babbler chicks. Most adults seem to leave the area by September, and birds in October–November appear to be mainly juveniles of the year.

Asian Koel *Eudynamys scolopaceus*

Mainly Summer Breeding Visitor Common

Largely resident across South Asia, eastwards to China and Southeast Asia. Some populations show local movements, vacating the northern parts of their range after the breeding season. Five subspecies have been recognized over their range, with the nominate in the subcontinent.

The beloved Koel of Indian poetry and folklore is widespread in parks, groves of mango, figs and other fruit trees, wooded patches about villages, etc., but it is far more in evidence and vocal as the weather warms up in March and then all through the monsoon. Most birds appear to move out from the Delhi area during the coldest part of the year. The Koel is parasitic on crows (Corvidae) in India, though elsewhere it is known to parasitize shrikes (Laniidae) and mynas (Sturnidae) as well. Koel eggs hatch after 13 days of incubation as compared to 16–17 days for crows' eggs, and Koel chicks outcompete the hosts' chicks for food. This is one of the few cuckoos that is largely frugivorous,[54] though its chicks would surely be fed on all manner of food by their corvine foster parents.

Grey-bellied Cuckoo

Cacomantis passerinus

Summer Breeding Visitor · Uncommon

Grey-bellied Cuckoo, male

Endemic to the subcontinent, a seasonal visitor in the northern parts and generally resident in the peninsula. Monotypic.

This small cuckoo appears to have expanded its range in the area over the past three decades and established itself as a regular but localized monsoon breeding visitor about Delhi between May and October. Ganguli did not include it in her Delhi list, and the first record in the area dates from August 1978; by the 1990s, these cuckoos had spread much more widely about Delhi in woodland patches and groves in rural areas.[55] Their characteristic calls are an unmistakable indicator of their presence from early June, after their arrival. Believed to parasitize mainly Ashy Prinias *Prinia socialis* and Tailorbirds *Orthotomus sutorius*, cuckoo eggs resemble their relevant hosts'. It appears that cuckoo females may be genetically programmed to lay eggs resembling those of the species that fostered them.

Drongo Cuckoo *Surniculus dicruroides*

Vagrant

The Drongo Cuckoo of South Asia have been split into two species based on vocal and morphological differences:[56] the Fork-tailed *S. dicruroides* endemic to the Indian subcontinent and the Square-tailed *S. lugubris* of the Himalayas and further east.

The only record of a drongo cuckoo in Delhi is one observed on 10 May 1986 on the New Delhi Ridge, a day after a violent storm, possibly a migrant blown off course.[57] The species cannot now be determined as the split had not been formalized at that time; it is likely to have been *S. dicruroides* as the species shows migratory movements between the Himalayan foothills and peninsular India, whereas *S. lugubris* is believed to migrate eastwards to Southeast Asia.

Common Hawk Cuckoo *Hierococcyx varius*

Mainly Summer Breeding Visitor · Common

Endemic to the subcontinent and familiar as the *papiha* of northern Indian poetry and folklore; two subspecies; nominate *varius* in India.

These hawk cuckoos are summer breeding visitors to the area from March onwards till October; they are common and very vociferous in city parks and gardens, wooded patches and generally wherever their preferred Jungle Babbler *Argya striatus* hosts occur. Some remain throughout the year, with a handful of records from mid-winter too. Like many other cuckoos, they relish the hairy caterpillars that abound in the monsoon, which most birds avoid.

Indian Cuckoo *Cuculus micropterus*

Vagrant / Reconfirmation desirable

Resident from the Indian subcontinent to China, northwards to the Amur region and across Southeast Asia. A summer breeding visitor to the northern parts of its range, including in the Himalayan foothills and central Indian ranges; two subspecies; the nominate in India and mainland East Asia.

Indian Cuckoos have been reported from the area as stragglers; a bird seen in July in the 1940s,[58] and the *Atlas* mentions one in mid-August 2004, with the comment 'occasional records of birds on passage'. There are plausible records of sightings, but these are unverifiable in the absence of any descriptions or of calling birds, and risk of confusion in identification is high. They have, however, been reported from just to the northeast of the limits of this work, from Haiderpur Wetland in the Hastinapur Wildlife Sanctuary on the Ganga in April and June–July.

Common Cuckoo *Cuculus canorus*

Summer Breeding Visitor and Passage Migrant Uncommon

A widespread Old World species and known to breed in India in the Himalayas and the central Indian ranges, including the Aravallis. Winters mainly in Africa and in smaller numbers in southern Asia as well. Four subspecies; *bakeri* breeding in the Himalayas and locally in central India; Central Asian birds *subtelephonus* are also believed to winter.

In the past, both Ganguli and Bill Harvey[59] had considered Common (or Eurasian) Cuckoos as stragglers during their passages in March–April and August–October, with an old report of a bird calling in July. From 2015, however, and annually since, birds have been recorded calling freely between April and August, with pair chases and other breeding behaviour in the wooded habitats of the Southern Ridge and outliers of the Aravalli Ranges in Haryana. These hilly areas have only lately begun to receive intensified ornithological coverage, and this cuckoo's status as a breeding summer visitor seems to have been overlooked earlier. It is not known which species are their potential hosts in the Delhi area, but these cuckoos are known to parasitize various insectivorous passerines elsewhere. They are also reported more widely on passage, both in spring and autumn. This is one of the several Aravalli species that were known to breed as close to Delhi as the Sariska National Park, and they have now been found to extend into the Delhi area along the broadly contiguous landscapes that form the Southern Ridge.

Common Cuckoo, a summer visitor to the area

CAPRIMULGIFORMES: NIGHTJARS, SWIFTS AND ALLIES
CAPRIMULGIDAE: NIGHTJARS

Nightjars have characteristic loud calls, but the high sound levels on city roads makes it difficult to locate them by ear unless one has the opportunity to live, or travel at night, in rural areas. They live in a variety of both open and wooded habitats and are aerial hunters, catching night-flying insects in their wide gapes in flight. Their legs are too short even to make a scrape on the ground, and they lay their two eggs on a bare patch in the grass, or directly on the leaf litter, without any attempt to build a nest. Their cryptic plumage and nocturnal habits make them difficult birds to study, and our knowledge of their status and movements about Delhi is quite incomplete.

The Caprimulgidae comprise three subfamilies: eared-nightjars Eurostopodinae of tropical Asia, the widespread typical nightjars Caprimulginae and the nighthawks Chordeilinae restricted to the Americas. Only the typical nightjars are found in the Delhi area.

India: 10 species, Delhi: 5

▲ *A Savanna Nightjar, dozing on a rock with eyes half-closed, Gurugram*

Jungle Nightjar *Caprimulgus indicus*

Passage Migrant · Uncommon

Endemic to India and Sri Lanka with two subspecies: nominate *indicus* is mainly resident in woodland and plantations from the Aravallis east to Odisha and south through peninsular India, with some local movements. In the Aravalli Range, this race breeds north to at least the Sariska National Park, less than 150 km from Delhi. *C. i. kelaarti* is limited to Sri Lanka.

These nightjars' movements and status in the Delhi area remain unclear, but they are, in all likelihood, mainly passage migrants here. The first record of this species was a specimen in non-breeding condition collected in New Delhi in late July 1962.[60] Gaston[61] reported birds calling 'regularly in the evenings' on the New Delhi Ridge between 20 August and 17 September 1973. There were just two records thereafter, in July 1988 and early October 2011 from the North Delhi Ridge, until 2014, when the Aravalli habitats of the Southern Ridge in Delhi and Haryana began to be explored intensively by local birdwatchers; these nightjars have been regularly recorded every year since then, mostly single birds roosting on horizontal branches of trees, from wooded habitats along the Ridge and elsewhere, between mid-May and early October. Jungle Nightjars breed from March to May, and there is a recent report of birds calling in early March near Rewari, Haryana; thus, while there is an evident post-breeding dispersal of these nightjars into the Delhi area, it is possible that a few breed locally in 'Aravalli' habitats as well.

A Jungle Nightjar, roosting on a branch on the Southern Ridge

Sykes's Nightjar *Caprimulgus mahrattensis*

Winter Visitor · Rare

Diffuse into western India in winter from their breeding areas in the desert parts of Afghanistan and Pakistan. Monotypic.

These stragglers have reached Delhi, typically single birds flushed from dry, relatively bare ground in the area: Ganguli refers to two historical records from the scrub country near the Yamuna in October; one was caught below the Okhla weir in December 1978; at Dighal on 3 February 2016; near the Sultanpur National Park on 4 August 2018 and again on 23 November 2018.[62]

Large-tailed Nightjar

Caprimulgus macrurus

Passage Migrant · Uncommon

A Large-tailed Nightjar, roosting on a branch

Widespread from India to Australia; six subspecies recognized over this range, of which *albonotatus* represents the species in moist wooded landscapes of northern and northeastern India. They nest in the Himalayan foothills between March and May and are partially migratory, spreading southwards into the northern plains thereafter.

The large nightjars appear to be mainly passage migrants in the Delhi area, where records cluster in late February–March and late June–August. A non-breeding female was collected in Delhi in end August 1961, the first record from Delhi;[63] spring records include birds calling in March 1973 on the New Delhi Ridge[64] and single birds or pairs observed, or heard calling, in late February and March at the JNU campus and other locations on the South-Central Ridge. There have been several records of these nightjars roosting on branches or amidst leaf litter in the woodland on the Southern Ridge in July-August in recent years. One trapped by A.J. Gaston on 19 December 1972 on the New Delhi Ridge is the sole winter record from the Delhi neighbourhood, and it is possible that a few birds winter here as they do at the Keoladeo Ghana National Park in Bharatpur, 160 km south of Delhi.

Indian Nightjar *Caprimulgus asiaticus*

Former? Summer Visitor · No recent records

Small nightjars that range from India to Southeast Asia and are fairly catholic in their choice of habitat, typically in light scrub jungle, fallow wasteland, stony nullahs and overgrown gardens. Though claimed to be resident throughout the country, from Jammu in the Himalayan foothills through the peninsula, they appear to be largely sporadic or absent across the Gangetic Plain. There is evidence of local movements, at least in the northern parts of their range. Three subspecies recognized: *asiaticus* in India and most of Southeast Asia, and two others in Sri Lanka and Thailand.

Indian Nightjars had been recorded in past years in stony scrub habitats of the Ridge and even in the large, rambling gardens of New Delhi. Frome, in the 1940s, considered them 'chiefly summer visitor(s) with some residents.'[65] H.P.W. Hutson, in his book *The Birds about Delhi* published in 1954 (henceforth, Hutson), had noted gatherings of up to a dozen from 20 September till mid-October in New Delhi, probably parties of this locally migratory nightjar in transit to warmer parts of central India. Gaston heard them calling regularly on the New Delhi Ridge in August and September in the early 1970s, while Ganguli[66] considered them 'possibly resident but locally migratory' uncommon summer visitors from February to October. Birds have also been heard calling in March–April in past years on the JNU campus on the South-Central Ridge, but records are conspicuously lacking from the Delhi area since the 1990s at least, with no indication of breeding. Unless they are being regularly overlooked, it can only be assumed that there has been a sharp decline in their numbers about Delhi.

Savanna Nightjar *Caprimulgus affnis*

Summer Breeding Visitor and Passage Migrant Uncommon

Widespread across southern Asia, a polytypic species with 10 recognized subspecies that fall into two or three groups which were formerly considered distinct species because of differences in size and colour. Subspecies *monticolus* of India and Southeast Asia belongs to the northern group of three races that were formerly considered to comprise a separate species, Franklin's Nightjar.[67]

The bare stony landscapes and outscoured ravines of the Aravallis in the Delhi area suit these nightjars' requirements perfectly. Birds have been recorded calling in the breeding season, March–June, in such habitats in both Delhi and Haryana, and breeding has been confirmed at Surajkund, a female with a chick seen on 10 August 2012,[68] and at the Aravalli Biodiversity Park, Gurugram (nest with eggs in June 2018). Savanna Nightjars are also regular passage migrants through the area in autumn with a noticeable influx in August, presumably of birds that have bred in the Himalayan foothills where they are common summer breeding visitors. Observers, both in the past[69] and in recent years, have noted a clustering of records between mid-July and September, loose gatherings of up to 20 or more birds roosting in rocky scrub country and in smaller numbers till mid-October. They may migrate in parties, and communal roosting in the post-breeding period, July to October, has been recorded elsewhere as well.[70] Other nightjars, e.g. the Grey, roost singly, and communal roosts are possibly characteristic of some species, including the Indian Nightjar.

A Savanna Nightjar, roosting on an exposed rock, as it often does

CAPRIMULGIFORMES: NIGHTJARS, SWIFTS AND ALLIES
APODIDAE: SWIFTS

A cosmopolitan family of small birds superbly adapted to an aerial lifestyle with their long wings and very short legs that only enable them to cling to vertical surfaces. They glue their nests, made of vegetable down and feathers mixed with saliva, under ledges and corners of monuments, bridges, or rock cliffs; in palm swifts, to the undersides of dangling fan-palm leaves. Powerful fliers, their capacity to range far from their usual haunts enables them to make sudden appearances almost anywhere depending on the weather and abundance of flying insect food, disappearing just as rapidly as conditions change. Although four swift species have been recorded, only one actually nests in the Delhi area.

India: 16 species, Delhi: 4

▲ *A pair of Little Swifts*

Alpine Swift *Tachymarptis melba*

Casual Visitor Rare

Widespread in mountainous regions from southern Europe and Africa to India. Ten subspecies recognized, of which *T. m. nubifugus* nests in the Himalayas and disperses southwards into the northern plains and central India, depending on local weather conditions, and *dorabtatai* is mainly resident in the hills of the western peninsula.

This spectacular swift has appeared only sporadically in the Delhi area. Ganguli mentions August records from the 1950s, and occasional parties have been reported recently as well with a mere handful of records from the Basai marshes, Haryana, in September 2003; over the Yamuna at Delhi during a very cold February 2005; near Jhajjar in March 2019 and at Manethi, Rewari, in late February 2021, both in Haryana.

Blyth's Swift *Apus leuconyx*

Vagrant

Endemic to the subcontinent, Blyth's Swifts breed in the Himalayas, a part of the population spreading in winter into peninsular India, particularly the Western Ghats. Monotypic.

One bird was reported from Sultanpur 'feeding with House Swifts brought low by a cold front over Sultanpur in February 2001'.[71] There have been no subsequent reports.

Little Swift *Apus affnis*

Summer Breeding Visitor Common

Widespread from Africa to India, with 'some northern populations showing migratory movements. Six subspecies are recognized, of which two occur in India: nominate *affnis* in the north and the darker *singalensis* in southern India and Sri Lanka. Himalayan breeders from Nepal eastwards to Japan, formerly lumped as a race of this swift, are now split as the Nepal House Swift *Apus nipalensis*.

Little Swifts, also called Indian House Swifts, are common summer visitors about Delhi, but become very scarce in the coldest months, November to February. They nest between March and August, typically in small colonies or even singly, under eaves and corners of occupied and derelict buildings, historical monuments, under bridges and canal structures, etc. Old, unoccupied nests of Dusky Crag Martins *Ptyonoprogne concolor* or Streak-throated Swallows *Petrochelidon fluvicola* are sometimes appropriated. Flocks 'ball' in the air in the evenings with high-pitched screams that, to quote Hugh Whistler from his *Popular Handbook of Indian Birds*, 'so aptly seem to express the fierce joy of an aerial creature in its element.'

Asian Palm Swift *Cypsiurus balasiensis*

Vagrant

Resident across much of India and Southeast Asia, Palm Swifts are inseparable from fan palms, particularly palmyra palm of eastern and southern India, under the leaves of which they glue their remarkable nests. Fan palms are not common in the Delhi area, though a few have been planted as ornamentals in gardens. Four subspecies are recognized, of which two occur in India: the nominate over most of their Indian range and the darker *infumatus* in Assam and further eastwards in Southeast Asia.

Delhi is towards the western fringe of their (and the palmyra's) usual range. They are vagrants here; Ganguli mentions sightings in 1952–53 by H. Alexander at Qudsia Gardens in Old Delhi, and the only records since then are of small parties of 4–6 again at the Qudsia Gardens and Raj Ghat in November 1977, single birds at Okhla and in a New Delhi garden in March and July 1979,[72] at Okhla again in September 2022 and three birds at Hiranki, Alipur, in north Delhi, on 27 November 2022. Palm Swifts, in fact, are now visibly scarce in parts of the Upper Gangetic Plain with few reports from areas where they used to be frequent three or four decades ago.

ORDER GRUIFORMES: RAILS, CRANES, ETC.
RALLIDAE: RAILS, CRAKES AND COOTS

These are a large family of small to medium-sized, large-footed, wetland birds adapted for life in the marshes, often elusive, overlooked and poorly known. Some are thin enough to move easily through dense reed beds, others are plump, short-winged and seemingly weak fliers, but many do migrate considerable distances. Coots prefer open waters and have lobed toes adapted for swimming. Most rails and crakes stay singly or in loose parties, or they keep in pairs when breeding, though here again, coots are an exception and commonly gather in flocks like ducks on lakes. They are mostly omnivorous, though some, like coots, are largely vegetarian. They typically nest amid thick vegetation in a pad constructed of grass and reeds, either on the ground or raised in low bushes; three to eight eggs in a clutch are the norm in the family, though moorhens and coots may lay up to 12; newly hatched chicks are precocial, capable of walking about and feeding by themselves in proximity to their parents.

India: 19 species, Delhi: 10

▲ *The Water Rail is a secretive winter visitor.*

Water Rail *Rallus aquaticus*

Winter Visitor Scarce

As a species, widespread from Europe to Central Asia; three subspecies are recognized, with the Central Asian race *korejewi* nesting as close to the Delhi area as Kashmir and wintering in the Middle East, the Indus plains in Pakistan and northwestern India.

Water Rails had not been recorded from the Delhi area, presumably overlooked due to their furtive, skulking habits, prior to the winter of 2002–03, when their characteristic squealing calls first drew attention to their presence at the Basai marshes in the Gurugram district. They have since been found to be regular visitors in winter, November to March, at a number of suitable wetland sites with soft, muddy ground fringed by aquatic vegetation, but they are easily missed.

Spotted Crake *Porzana porzana*

Winter Visitor Scarce

A widespread crake, breeding across Eurasia and wintering in Africa and northern India. Monotypic.

First recorded from the area only in March 1979 at the Sultanpur Lake, these crakes have been found, with more intensive ornithological coverage in recent years, to be regular but scarce visitors about Delhi between October and April. Possibly overlooked earlier, there are now almost annual records of Spotted Crakes from ponds and wetlands across the area, even at small seepage marshes and borrow-pits, with dense fringing vegetation and reeds with some adjacent muddy ground for them to feed.

Common Moorhen *Gallinula chloropus*

Breeding Resident Common

Widespread in Europe, Asia and Africa in a variety of freshwater habitats, but requires some open water with fringing vegetation and cover. Five subspecies recognized, of which nominate *chloropus* is resident across the Indian subcontinent.

A common gallinule, nesting throughout the area in most wetlands, both large and small, seepage ditches, even artificial waterbodies in parks and other urban locations, as long as there are patches of open water with some floating vegetation and surrounding cover in the form of reeds or bushes. The resident population is probably augmented by migrants from Central Asia in winter. Common Moorhens often swim out into the open and feed while swimming and walking about on open ground, but they need cover to retreat into when necessary. They build a basket nest of dry twigs in reed beds, bushes and similar waterside thickets, sometimes raised on a creeper-covered tree stump, breeding from May to September, but some may start much earlier as downy chicks have been seen as early as March.[73]

Common Moorhen

Eurasian Coot *Fulica atra*

Mainly Winter Visitor Common

A very widespread species, breeding over much of Europe and Asia to Australia and wintering south to Africa and southern Asia. Four subspecies are recognized, and the nominate race is both locally resident in the Indian subcontinent and a common winter visitor as well.

Primarily winter visitors to the Delhi area between October and March, affecting lakes, rivers and open marshes where they may form large rafts mixed with ducks. At some of the larger lakes, winter counts may exceed 1,000. Eurasian Coots prefer waters that are not too deep but still have room to dive and some floating or submerged vegetation on which they feed. A few birds remain through the summer, with evidence of at least occasional breeding, as at Surajpur in GautamBuddha Nagar district in August 2016 and at Sampla, Rohtak district, in late November 2021 (family party photographed with juveniles).

Grey-headed Swamphen *Porphyrio poliocephalus*

Breeding Resident · Common

Grey-headed Swamphen

Formerly known as Purple Swamphens, with six distinct groups of subspecies, which together range from southern Europe, Africa and Asia to Australia. Based on morphological and molecular evidence, specific rank has now been given to each of these six lineages, including the Grey-headed Swamphen, whose range then gets restricted from Turkey and Iran, across the subcontinent to continental Southeast Asia. Four subspecies are recognized, with the nominate race resident from India eastwards to southern China.

These are common gallinules in most wetlands of the Delhi area with sufficient cover, preferring a mix of reed beds, grassy patches and open water channels with some floating aquatic plants. Flocks of 40 or more may feed in the open in such habitats, when their blue coloration glows in the sun and makes them very conspicuous. They breed from May to September, building a pad of twigs, reed stems, etc. in tangled clumps of waterside vegetation, when the precocious chicks at various stages of growth may commonly be seen.

Watercock *Gallicrex cinerea*

Summer Breeding Visitor · Uncommon

Resident from the subcontinent to East and Southeast Asia; the northern populations of China and Korea are migratory, while Indian birds are subject to local movements during the monsoon. Monotypic.

The handsome Watercocks are summer/monsoon breeding visitors to the Delhi area between June and September, affecting the larger marshes with a mosaic of reed beds and open grassy patches or wet cultivation. They were first recorded only in 1963 in north Delhi, but they have spread rapidly since then, having colonized suitable habitat created at Okhla by the building of the barrage in the 1980s and are found today in several other wetland sites in the area. As they settle down to establish territories on arriving in their breeding grounds, the males become active, noisy and more visible, and their unique three-part booming territorial calls are commonly heard in these marshes.

White-breasted Waterhen *Amaurornis phoenicurus*

Breeding Resident · Common

Familiar residents of the subcontinent, eastwards to China and Southeast Asia. Four subspecies are recognized, of which three occur in India, nominate *phoenicurus* on the mainland.

These waterhens are just as common in the area as elsewhere in India in damp habitats of all sorts – reed beds, ditches, seepage marshes, village ponds and urban parks. They nest freely about Delhi with an extended breeding season that extends from March/April up to September and can be very vocal at this time. Their nests are pads of twigs and green vegetation concealed in the undergrowth or up in a bush, and the precocial chicks, covered with black down, may often be seen accompanying females during the monsoon months.

Ruddy-breasted Crake *Zapornia fusca*

Breeding Resident · Uncommon

Widely distributed residents, but subject to local movements, from the subcontinent eastwards to China, Japan and Southeast Asia. Four subspecies are recognized, with two in India: nominate *fusca* in northern India and *zeylonica* in the southwest and Sri Lanka.

Ruddy-breasted Crakes were first recorded in Delhi in June 1962 from the south of Okhla but then were lost sight of until relocated at the same location after over two decades in 1985–86.[74] Their status in the area remained unclear at that time, but they have since been found to be fairly widely distributed in reed beds and marshes about Delhi, recorded in most months and presumably breeding, as suggested by birds calling in July in the paddy fields south of Okhla.

Brown Crake *Zapornia akool*

Breeding Resident · Uncommon

Locally but widely resident in the subcontinent and southeastern China. Two subspecies are recognized, with the nominate in India.

Brown Crakes were recorded in Delhi for the first time only in June 1962, having probably been overlooked earlier. They are widespread but less frequently reported than Ruddy-breasted Crake *Z. fusca* in the area. They seem to move around, appearing sporadically even at isolated small ponds, reedy irrigation channels and isolated seepage marshes. They have been recorded off and on through the year and heard calling in the summer, in May and June, which suggests they probably breed in the area.

Baillon's Crake *Zapornia pusilla*

Mainly Passage Migrant · Common

Widespread from Europe across temperate Asia and parts of Africa and Australasia, nesting commonly in Kashmir and 'very probably also elsewhere within our limits', with 'insufficiently documented' reports of breeding at other locations in India, including UP.[75] Six subspecies are recognized; Indian birds are subspecies *pusilla*, the paler eastern race of Asia.

Baillon's Crake, a juvenile moulting into adult plumage

These diminutive crakes are common during their spring and autumn passages through the Delhi area, in March–April and September–October, affecting well-vegetated pools and swamps with surrounding reedy cover, but they are evidently scarcer and more localized in winter. A few probably breed here as well, perhaps opportunistically. Records from the Basai marshes in Gurugram district as early as 3 August and very young juveniles netted there with adults in early September suggest breeding,[76] and a downy chick accompanying a pair of adults at the Dhanauri wetlands, GautamBuddha Nagar district, on 29 September 2007[77] provides further evidence.

ORDER GRUIFORMES: RAILS, CRANES, ETC.

GRUIDAE: CRANES

The Delhi area hosts three species of these tall and graceful birds, the resident Sarus with its evocative bugling, and two migratory cranes whose spectacular migratory flight formations are now a regrettably rare sight here. Historically, one of the most elegant of the group, the critically endangered Siberian Crane has also occurred in the Delhi area. They live in a variety of open habitats, usually close to wetlands in which they roost and spend the day, flying out to feed in the mornings and evenings in the surrounding countryside, in rice fields, fallows and grasslands, and often in the proximity of villages and habitation.

India: 5 species, Delhi: 3

▲ *A Sarus family*

Sarus Crane *Antigone antigone*

Breeding Resident · Fairly Common

These spectacular cranes are distributed from northern India across Southeast Asia (where they are rare) to Australia. Three subspecies are recognized: nominate *antigone* in India, *sharpii* in Southeast Asia and *gillae* in northern Australia. Though protected by law and popular sentiment, they are declining over much of their Indian range due to loss of habitat, pesticides and disturbances while nesting.

The pairs of Sarus that grace our countryside impart a charm to the scenery that can scarcely be surpassed. They may be seen in cultivated countryside in the vicinity of *jheels*, marshy wetlands, village ponds and paddy fields, where they breed between July and November, raising usually two precocial chicks, collecting in family parties thereafter and even larger groups by late winter. These gatherings disperse again as the rains break in June/July. They were considered fairly common about Delhi a few decades ago, and though they are still well distributed in appropriate habitats, several former sites no longer hold Sarus. The largest numbers in the Delhi area have been reported from the Dhanauri wetlands, GautamBuddha Nagar district, which may hold up to 15 breeding pairs and over 100 birds annually as they gather at this favoured wetland in the dry season, with a maximum of over 140 recorded in April 2017.

Demoiselle Cranes halt on passage through the area.

Demoiselle Crane *Grus virgo*

Passage Migrant · Scarce

The main breeding grounds of this elegant crane are in central Eurasia, and this population winters in northern and West-Central India. Monotypic.

Demoiselle Cranes are passage migrants through the Delhi area mainly in February/March to early May and late September to November. There are pre-1975 historical records of 'flocks near the river in November and April'; passage in later decades has usually involved flight formations of cranes passing overhead, with autumn maxima of 550 at Sultanpur on 2 November 1991, 126 along the Yamuna in end September 1996 and over 100 on 26 September 2020, and 200 near the Sultanpur National Park on 3 October 2020. On spring passage, numbers are usually smaller – often in single digits – but larger parties have also been recorded on occasion with a maximum of 264 birds counted on 4 April 1986 at Sultanpur and 92 cranes moving north in early March 2015 over the Asola Wildlife Sanctuary. There are a few mid-winter reports as well, typically involving single birds or very small groups.

Common Crane *Grus grus*

Winter Visitor · Uncommon

Breeds across temperate Europe and Asia, wintering in the Mediterranean basin, the Middle East and southern Asia. Monotypic.

Local winter visitor to the area between September and April. The Yamuna seems to act as a migratory flyway with observers having recorded formations of cranes high over the river floodplain, moving southwards in autumn (late September and October) and northwards in spring (March and April). This is less evident now as the floodplain has become increasingly hemmed in by urban sprawl, and the cranes tend to fly higher as they cross the city. The larger lakes at the Sultanpur National Park and Bhindawas Wildlife Sanctuary, Haryana, are used as day refuges in the winter, the cranes dispersing to feed in the surrounding countryside in the mornings and evenings. Such flocks involved as many as 200–300 cranes at Sultanpur in the 1960s and through the 1980s,[78] but numbers have since reduced to typically less than 100 birds, although maxima of 164 were reported at the Bhindawas lake in 2001, and 358 on the Ganga sandbanks at Haiderpur Wetland, Hastinapur Wildlife Sanctuary, in March 2020.

CHARADRIIFORMES: WADERS, GULLS AND ALLIES
BURHINIDAE: THICK-KNEES OR STONE-CURLEWS

A cosmopolitan family of plover-like birds, large-billed, large-eyed and mainly nocturnal, whose strong legs and prominent knees (actually ankles!) give them their name, 'Thick-knees'. They inhabit rather dry, open and lightly wooded country, even bare sandy or rocky courses of the larger rivers, nesting in a scrape in the ground in the shade of a bush, or on a bare sandbank, in which typically two eggs are laid.

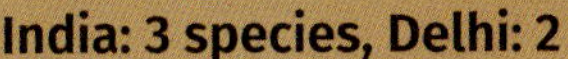

A pair of Indian Thick-knees

Indian Thick-knee *Burhinus indicus*

Breeding Resident · Common

Resident east of the Indus River throughout the subcontinent eastwards to Indochina. Monotypic.

These thick-knees (or stone-curlews as they are also referred to) are common residents about Delhi in sparsely wooded dry countryside, fallow grazing grounds, stony scrubland or open woodland. They also occur within urban limits in large gardens or vacant plots in suburban colonies. Pairs and parties (up to 30 together have been recorded after breeding) pass the day in the shade, or even in the open on baking sandbanks or ploughed fields. Their ringing calls resound at dusk and dawn in the darkness, all through the year, but they are especially vociferous in the breeding season between March and November. Birds display standing erect, tail spread and depressed, calling continuously and flashing open the black-and-white contrasts in their wings.

Great Thick-knee *Esacus recurvirostris*

Former? Breeding Resident · Rare

Resident from the subcontinent eastwards to Indochina. Monotypic.

This charismatic wader was, from past accounts, 'not uncommonly seen' on sandbanks and fields by the Yamuna, with Ganguli mentioning non-breeding flocks of '20 to more than 50' in the 1960s near Okhla and Wazirabad. Hutson, too, recorded flocks of 10–30 and found a nest on a Yamuna sandbank in April. Such numbers[79] now sound incredible, but birds continued to be recorded sporadically in ones and twos on Yamuna sandbanks in late winter and summer till the 1990s, with circumstantial evidence of breeding south of Okhla in May 1996.[80] They were seen again at that site in August 2004, possibly displaced by the flooding further south. Although crepuscular in habit, they are not secretive birds and may be out in the open in the daytime, hiding behind a sand heap with only their heads and eyes showing. Harvey, in the *Atlas*, restricted their distribution in 2006 to the Yamuna north of Wazirabad, but there have been no reports since even from there. Pollution, encroachment and rampant sand mining has sadly destroyed their riverine habitat. They continue, however, to be recorded in small numbers throughout the year on the sandbanks of the Ganga.

CHARADRIIFORMES: WADERS, GULLS AND ALLIES

RECURVIROSTRIDAE: STILTS AND AVOCETS

A cosmopolitan family of elegant, boldly patterned, wading birds, with long slender bills and very long legs that enable them to forage in deeper waters. Stilts feed on insects and other small invertebrates by picking them from the water or mud, while avocets locate their food items by scything their upcurved bills side to side as they wade through the water. As a family, they tend towards opportunistic breeding, often colonial, dependent on local conditions and the availability of sites. Their nests, raised pads of water weeds lined with dry grass, etc. are built on mudflats or in shallow water to hold their clutch of camouflaged eggs, usually four. Their precocial chicks may often be seen feeding with their parents. The Delhi area hosts two members of this family, both widely distributed species in Europe, Asia and Africa.

India: 2 species, Delhi: 2

▲ *Black-winged Stilt*

Black-winged Stilt *Himantopus himantopus*

Breeding Resident and Winter Visitor · Common

Resident, and migratory in part, in the Mediterranean basin, Africa, the Middle East and much of temperate and tropical Asia. Monotypic.

Stilts are local breeding residents in the Delhi area and common winter visitors as well. Wintering numbers begin to build up from August in the wetlands and marshes of all sorts, large or small, even artificial waterbodies in city parks. Their numbers reduce by April, but many remain to breed through the summer, between May and September–October. In the 1880s, large numbers of stilts bred at the salt works south of Sultanpur until they were closed down in 1923.[81] In later years, stilts have been found breeding in the same general area in small, scattered colonies at Najafgarh, Sultanpur and Basai, Haryana, the nests spaced 2–3 m apart in shallow, weedy inundations, each normally holding four eggs. Stilts prefer saline ground and drying mudflats for nesting but may breed opportunistically wherever they find conditions suitable.

Pied Avocet *Recurvirostra avosetta*

Winter Visitor Common

These elegant waders nest across Europe and Asia, locally in Africa and the Middle East, wintering further to the south. They are mainly winter visitors to India, though they have been recorded breeding in the Rann of Kachchh in the past. Monotypic.

Avocets were considered only uncommon passage migrants about Delhi in April–May, but Ganguli records that they had become much commoner and more regular by the 1960s. Presently, the largest numbers appear between September and June, though they are occasionally reported in other months as well. They are regular in winter by the rivers in the area, *jheels*, village ponds and seepage inundations along canals, in flocks of up to 100 birds, but peak autumn passage counts of 500 or more had been recorded on the Yamuna in September–October in the early 2000s, with a record count of 1,200 in October 2003.[82] Avocets are fairly catholic in choice of habitat but prefer shallow water in which they can sweep their upturned bills from side to side, like a hockey stick, to search for edible morsels; their toes are partially webbed, enabling them to swim in deeper water as well and even upend to feed.

A Pied Avocet and its reflection

CHARADRIIFORMES: WADERS, GULLS AND ALLIES

CHARADRIIDAE: PLOVERS

A number of our waders, some highly migratory species amongst the plovers Charadriidae and sandpipers Scolopacidae in particular, breed at high latitudes in Asia and winter along our coasts. They are strong fliers, and many of them overfly the Delhi area on their migration routes in direct flights from their staging to their wintering grounds. They are consequently rarely recorded here except as vagrants forced down by bad weather, poor health or chance mix-ups with migrating parties of other species, while the main movement passes over unnoticed. The months of May and June, as the northward migration is drawing to a close, is when these vagrants are most often recorded – these are probably weaker birds, straggling migrants which cannot undertake the demanding long hop and need to rest midway, or merely because they are less likely to be overlooked once the main flood of migrants has passed.

Plovers, as a family, show two lineages: the larger lapwings in the subfamily Vanellinae and the smaller plovers in the Charadriinae. These are birds of the grasslands, mudflats, river sandbanks, even lawns and open grounds within urban areas with a distinctive manner of feeding, running and stopping abruptly to dip and pick up some food, insect or other small invertebrate from the ground. They are excellent fliers, the smaller plovers with pointed wings for direct, fast flight, and one, the Pacific Golden Plover *Pluvialis fulva*, reputedly makes some of the longest annual migrations in the world, often with extensive non-stop flights over the ocean. Nests, for most Charadriiformes, are simple scrapes in the ground lined with scant grass, in which they typically lay a clutch of four eggs, and both eggs and the precocial chicks are cryptically patterned for camouflage. Many smaller plovers are also adept at diverting the attention of any potential threat, human or animal, from their nests by performing highly realistic 'broken wing' displays, while the larger lapwings will dive bomb a predator mercilessly, with loud screams that can be just as effective.

India: 19 species, Delhi: 15

▲ *Grey Plover*

Grey Plover *Pluvialis squatarola*

Passage Migrant and Winter Visitor Rare

From their circumpolar breeding range in the Arctic tundra, Grey Plovers winter widely along coastal habitats, beaches and tidal estuaries across the globe. Migration is primarily coastal; they are far scarcer and irregular inland, but odd individuals are reported even in mid-winter by freshwater lakeshores and sandbanks of large rivers. Monotypic.

In the Delhi area, Grey Plovers have been recorded mainly on river sandbanks during both migration periods and rarely in winter as well. Historically, Ganguli noted only four occurrences – November, December, January and June – between 1947 and 1969. A smattering of later reports follows the same broad pattern: single birds or pairs by the Yamuna with at least six records of birds during northward passage in May and three mid-winter records in December–January in 1998, 2002 and 2016; four at the Dighal wetlands, Rohtak district, in November 2018; and two on flooded farmland at Chandu, Gurugram district, in November 2021. One wintering individual was reported from the sandbanks of the Ganga at the Hastinapur Wildlife Sanctuary in early February 2019. The characteristic three-noted whistle of this bird locates its presence in a large mixed flock of waders.

Pacific Golden Plover

Pluvialis fulva

Passage Migrant Scarce

Breeds in arctic Siberia and Alaska and winters from South and Southeast Asia across Australasia and Pacific Oceania. In India, they are most common in winter in the eastern and southern parts of the country, especially in coastal areas and short-grass habitats, such as grazing grounds, playing fields, ploughed land, mudflats, etc. Monotypic.

Pacific Golden Plovers have been reported on passage through the area in autumn and late spring, with scattered records of single birds or small parties by the river and wetland margins, often with migrating parties of other waders. Historically, Ganguli considered them vagrants and mentions just two pre-1975 records, in May 1964 and 1970. However, recent observations are certainly more frequent and widespread, mainly late April through May, and in autumn between late September and November. Most spring migrants are in partial or full breeding plumage.

A Pacific Golden Plover in winter plumage

Northern Lapwing in winter plumage

Northern Lapwing *Vanellus vanellus*

Winter Visitor · Uncommon

Breeds across temperate Eurasia from Europe to China in open grassy habitats, cultivation and wetlands and winters from Europe and the Mediterranean region through the Middle East to northern India and southern China. Monotypic.

These lapwings visit the Delhi area in small numbers between October and March. While nowhere regular, single birds or parties of up to 50 birds together may appear at wetlands, lakes and the river *khadar*, preferring to feed in scattered flocks in damp grassy areas and ploughed fields in the neighbourhood.

River Lapwing *Vanellus duvaucelii*

Breeding Resident · Fairly Common

Inhabit sandy and pebbly banks of rivers, both Himalayan as well as in the plains, from northern India to mainland Southeast Asia. Monotypic.

These noisy, but often somewhat furtive, lapwings are fairly common residents along both the Ganga and Yamuna Rivers, nesting on islets and sandbanks between March and May–June when water levels are at their lowest. The nest is a shallow unlined scrape on the ground. The loss and disturbance of the sandbanks because of the expansion of melon cultivation, barrage operations and sand mining has led to a diminution in their numbers, especially along the Yamuna in Delhi. The *Atlas*, published in 2006, had estimated a population of 50 scattered breeding pairs along the Yamuna within the limits of the NCT and, as a general observation, it seems this figure may have dipped a bit since then.

River Lapwings can be seen near the Yamuna.

Yellow-wattled Lapwing *Vanellus malabaricus*

Breeding Resident · Fairly Common

A pair of Yellow-wattled Lapwings, denizens of grasslands

A subcontinental endemic lapwing of dry, open habitats. Monotypic.

Yellow-wattled Lapwings are fairly common residents of the semi-arid landscapes of the Delhi neighbourhood and would certainly be more common if it were not for the continuing diversion of their dry, short-grass habitat to urbanization and agricultural expansion. They may be seen in pairs or small parties in over grazed flats and fallows near villages and bare, dry ground in general, even around the IGI Airport. They nest in such open ground during the summer months, March to June, before the rains break, in a shallow scrape on the ground, laying a clutch of four well-camouflaged eggs. When Delhi was smaller, less congested and less wooded, this plover actually nested on New Delhi's Central Vista in 1942.[83] Many birds move locally to drier, short-grass areas as their habitat tends to get flooded during the monsoon.

Grey-headed Lapwing *Vanellus cinereus*

Winter Visitor · Scarce

An East Asian lapwing that nests in northeastern China and Japan and winters from India to northern Indochina; commonly in eastern India, straggling westwards in progressively smaller numbers through the Gangetic Plain as far as Delhi and Bharatpur. Monotypic.

These lapwings were believed to occur about Delhi sporadically and were first recorded in the area at Okhla only in early December 1972, but with intensified ornithological coverage, they are now recorded in small numbers almost every other year for the last couple of decades, from the wetlands and seepage marshes between September and April. Specific locations include the Yamuna *khadar*, the Sultanpur National Park neighbourhood and various other wetland habitats in both Haryana and UP, including several from the Ganga floodplains in the Hastinapur Wildlife Sanctuary, where they are being reported regularly in small parties.

Red-wattled Lapwing

Red-wattled Lapwing

Vanellus indicus

Breeding Resident Common

A common resident from Iraq to Southeast Asia. Four subspecies are recognized, of which two occur within Indian limits, the nominate in most of the country and *atronuchalis* in Assam and further eastwards.

These vociferous, striking lapwings are a familiar species in areas of open ground everywhere about Delhi, preferably, but not necessarily, near water of some sort – a pond, ditch or sewage trickle. It is of little importance whether such a habitat is out in the countryside, near villages or within urban conglomerations, where their loud calls and aerial displays never fail to attract attention. They breed between March and September, building a nest on the ground or even on flat rooftops, often quite out in the open. The nest is typically a shallow scrape lined with dry grass holding a clutch of four eggs. After breeding, they may gather in loose flocks of 15–20 birds; an unusually large one of 142 was seen at Basai, Haryana, in September 2003.[84]

Sociable Lapwing *Vanellus gregarius*

Winter Visitor Rare

Critically endangered lapwings, having undergone drastic population declines driven by hunting pressures and habitat loss; the subject of extensive ongoing conservation efforts. They nest in Central Asia, their breeding range now almost entirely restricted to the steppes of central and northern Kazakhstan but winter mainly in two disjunct areas: in northeastern Africa and the Middle East (on the west), and Pakistan and northwestern India (on the east). The birds following the latter route migrate over 2,800 km between their breeding and wintering quarters, with just one staging halt en route on the Turkmenistan–Afghanistan border.[85] Monotypic.

These lapwings were always considered scarce, erratic visitors to the Delhi area, recorded on dry mudflats and fallow fields in the vicinity of wetlands between December and March. Frome reported small flocks in January; Hutson saw it between December and February, and Ganguli mentioned two observations of small parties in January and March 1954. There are reports from the Sultanpur neighbourhood in Haryana in the winter of 1969–70 and in November 1974 and 1984 and, remarkably, a flock of 55 was seen in February 1993, probably on passage;[86] it seems to have been last reported from the Delhi area in early March 2012.[87]

White-tailed Lapwing

Vanellus leucurus

Winter Visitor · Fairly Common

Breeds in West and Central Asia, from Turkey to the subcontinent (in Baluchistan and at least, occasionally, as far east as Gujarat) and winters from northeastern Africa across to the northern Indian subcontinent in the Indo-Gangetic Plain. Monotypic.

These graceful lapwings, with their subdued colours and diagnostic white tails, are fairly common migrants about Delhi between September–October and April, well distributed in *jheels* and swamps, flooded grassland and wet mudflats. White-tailed Lapwings usually keep to themselves, singly or in scattered parties of up to 10–15 birds, and they do not usually mix with other waders.

White-tailed Lapwing

Lesser Sand Plover *Charadrius mongolus*

Passage Migrant · Uncommon

The five subspecies normally recognized for this species bunch into two groups: the distinctive '*atrifrons* group' comprising three races (*pamirensis*, *atrifrons* and *schaeferi*) that breed inland on the Pamirs, Ladakh, the Himalayas and Tibet, and winter mainly along the East African, and South and Southeast Asian coasts; and the disjunct '*mongolus* group', with two races (*mongolus* and *stegmanni*) breeding in eastern Siberia and wintering mainly in East Asia and Australia. Most birds wintering in India are believed to belong to race *atrifrons*, but the other four have been recorded as well based on measurements of birds in winter plumage and breeding plumage patterns in some cases.[88]

Lesser Sand Plovers, very common winter visitors to the Indian coasts, are reported regularly on their spring and autumn passages from the Delhi area – there are odd reports from March onwards, but the records are largely clustered towards the end of the spring passage, mid-May to mid-June, when small numbers, rarely parties of as many as 30, may be seen on the Yamuna sandbanks and wetlands such as the Sultanpur and Najafgarh lakes. At this time, many birds are in nearly complete breeding dress and appear to be of the '*atrifrons* group'. There are a few records for autumn as well, in September to November. Even on migration they seek out bare sandbanks or mudflats by desiccating lakes and inundations that substitute for their typical sandy coastal habitats.

Greater Sand Plover *Charadrius leschenaultii*

Vagrant

Breeds on inland lakes in Asia, from Turkey and Transcaspia to Mongolia, and winters along the coasts of East Africa, Arabia and Asia to Australasia. Three subspecies are recognized, of which *C. l. leschenaultii* winters around the Indian Ocean coastline.

Greater Sand Plovers are rarely reported inland in India during passages, but vagrants have appeared in the Delhi area historically. They have been observed between April and early June (though doubts have been expressed about these old records) and one in near-breeding plumage in early June 1965. One at the Sultanpur National Park in March 1988 and two at the Basai marshes in the same general area in November 2002[89] are the only later records.

Caspian Plover *Charadrius asiaticus*

Vagrant

Breeds in Central Asia and winters in East Africa; accidental in the Indian subcontinent on passage and once or twice in winter, with only a few scattered records. Monotypic.

One individual, in its unmistakable breeding plumage, was observed on the short-grass flats near the Sultanpur Lake on 1 March 1986.[90]

Kentish Plover *Charadrius alexandrinus*

Winter Visitor Fairly Common

Breeds over a wide range spanning Europe, North Africa and Asia, wintering in Africa and southern Asia. Four subspecies are recognized, of which the nominate race *alexandrinus* is a widespread winter visitor in India and also breeds locally in the Indus plains in Pakistan and in Gujarat. A second race, *seebohmi* is resident in southeastern India and Sri Lanka.

Although largely a coastal species elsewhere, Kentish Plovers occur fairly commonly inland in India and about Delhi as well. Wintering numbers at a site are typically in the tens, but as for many other waders, the river may be an important migration flyway for this species too. The numbers increase in March–April when migratory flocks often gather in the late evenings by the Yamuna. Pairs of *alexandrinus* in breeding dress have remained on an Okhla sandbank as late as mid-June, defending territory but with no other evidence of nesting.[91]

Kentish Plover, in winter plumage

Common Ringed Plover *Charadrius hiaticula*

Vagrant

Breeds across northern Eurasia and winters mainly in Africa and West Asia. Only small numbers visit the subcontinent, mainly on the Sindh and Gujarat coasts, with wintering records from southern India and Sri Lanka as well. Several scattered records inland from northern India as well, most probably of birds on passage. Two subspecies, of which the eastern *tundrae* that nests in northern Asia, occurs in southern Asia.

Ganguli mentions just two historical records from the Delhi area, from 1879 and October 1969. A couple of mid-winter and one late April reports from Okhla and the Sultanpur National Park neighbourhood in 2002–03, of single birds or two or three together, and one recent record of two individuals at the Chandu wetlands near Sultanpur in the third week of September 2022, are the only later records. Separation of first-year birds from similarly aged migratory Little Ringed Plovers *C. d. curonicus* can be difficult, leading to potential confusion in identification.

Little Ringed Plover *Charadrius dubius*

Breeding Resident and Winter Visitor Common

Widespread residents, subject to migratory movements in part, across Eurasia, North Africa, the Middle East, the Indian subcontinent and Southeast Asia. Three subspecies are recognized, two of which occur in India: wintering *curonicus* of northern Eurasia, whose black head markings and breastband turn brownish in winter, and the resident *jerdoni* of southern Asia, which lacks any markedly distinct non-breeding plumage.

Little Ringed Plovers are resident in the area.

Both races of Little Ringed Plovers are recorded about Delhi, *curonicus* as winter visitors and *jerdoni* as nesting residents. The river sandbanks are a preferred habitat, but open expanses of bare mud or short grass in the floodplains and borders of wetlands are equally patronized. They breed in the summer months before the monsoon breaks, but newly fledged chicks have been recorded as early as March. They normally keep in pairs, but small parties may gather after breeding in June. Increased numbers by the Yamuna in September and March are presumably migrant *curonicus*.

CHARADRIIFORMES: WADERS, GULLS AND ALLIES

ROSTRATULIDAE: PAINTED-SNIPES

The order Charadriiformes is remarkable for the number of families and species that exhibit unusual breeding economies. The painted-snipes, a small family of three species of which one occurs in India, have a breeding biology based on polyandry; the females mate sequentially with several males during a season. They are very different from snipes, in spite of their name, and are more akin to rails in behaviour and flight.

India: 1 species, Delhi: 1

▲ *The Painted-snipe, a female of this polyandrous species*

Greater Painted-snipe *Rostratula benghalensis*

Mainly Summer Breeding Visitor · Fairly Common

Resident, but subject to marked local movements over large parts of temperate and tropical Africa and Asia. Monotypic.

Greater Painted-snipes, exceptional birds both in their appearance and lifestyles, are most in evidence about Delhi in summer, from April to August and until October, with very few winter reports. Their preferred habitat is a mosaic of wet grass, reed beds and open pools, but overgrown seepage marshes, even sewage overflows will do as well, and they are well distributed though never very common at any given site.[92] They are most active in early summer in the morning and evening, feeding in the open, flying about and calling as the females establish territories. Females are sequentially polyandrous, mating with a number of males within their territory in succession. The nests, built by both parents, are shallow cups of grass concealed in marsh vegetation. Once the eggs are laid, usually in clutches of four, males alone take up the responsibilities of incubation and parental care and have been observed accompanying well-grown chicks in May, June and even September.

CHARADRIIFORMES: WADERS, GULLS AND ALLIES
JACANIDAE: JACANAS

Jacanas are a remarkable pan-tropical family of which India hosts two species, both of which occur in the Delhi area. They require shallow freshwater wetland habitats with abundant floating vegetation, on which they can balance and walk with their very long, slender toes. Female jacanas practise simultaneous polyandry: each female holds a large territorial space that overlaps a few males' individual territories, and she mates with the males of her 'harem' promiscuously. Both sexes may join in nest building as part of the courtship, with the nests being floating pads of aquatic vegetable material on water lily leaves. However, only males incubate the eggs, usually four in a clutch, and look after the chicks. They are also well known for their remarkable habit of shifting their eggs, one by one, to a new nest, if the original nest becomes vulnerable due to disturbances or changing water levels. Even the precocial chicks are carried away if danger threatens, grasped between the adult's wing and body.

India: 2 species, Delhi: 2

▲ *A Pheasant-tailed Jacana in flight, showing very long toes*

Pheasant-tailed Jacana *Hydrophasianus chirurgus*

Mainly Summer Breeding Visitor, a few remain in winter · Fairly Common

Resident, but with marked seasonal movements in parts of its range from the Indian subcontinent eastwards across the Oriental region. Monotypic.

These superb jacanas are rather common summer visitors to the Delhi area from April to end September; they are very striking birds in their resplendent breeding dress at this season, but most leave the area after nesting, only small numbers staying on through the winter months when they are less conspicuous. They affect ponds and marshes with floating vegetation such as water lilies, lotus, water chestnut, water hyacinth, etc. and can be quite common in any wetland that meets this primary requirement. Nesting in the area appears to be mainly between June and August, the birds being very active and noisy all through May and June as they set up territories in the marshlands.

Bronze-winged Jacana

Bronze-winged Jacana *Metopidius indicus*

Breeding Resident Fairly Common

Resident of the Indian subcontinent, east to Myanmar and Indochina, overlapping with Pheasant-tailed Jacana *H. chirurgus* over much of this range and very similar to that species both in choice of habitat and a polyandrous breeding biology. Unlike that species, however, Bronze-winged Jacanas are relatively poor on the wing, usually low over the floating weeds with a flapping flight, their long toes trailing. Monotypic.

Bronze-winged Jacanas were very scarce about Delhi half a century ago but have been expanding westwards from their Gangetic floodplain strongholds and are far more widespread today. Ganguli, writing in the 1960s, lists them as vagrants, mentioning only three records, two in 1953 in east Delhi and one in 1966 in north Delhi. By the 1980s, these jacanas had already established themselves as a local breeding species, present all year round at Okhla. They are presently reported from and are quite common in many wetlands of the Delhi area, wherever floating aquatic vegetation provides the surface for them to walk upon. Their westward expansion into the Delhi area seems to have been paralleled elsewhere at about the same time (1980s) into parts of Rajasthan and even across the Indian border into Sindh.[93]

CHARADRIIFORMES: WADERS, GULLS AND ALLIES

SCOLOPACIDAE: SANDPIPERS AND ALLIES

This is a cosmopolitan family of birds whose diversity reflects the adaptations of its members to the range of the ecological niches they occupy. It shows up mainly in appearance (the size, forms, lengths and shapes of bills) and feeding techniques that enable a number of species to occupy the same habitat without directly competing for food. The family comprises five major lineages that are widely regarded as subfamilies: curlews Numeninae, godwits Limosinae, turnstones and stints Arenariinae, dowitchers and snipes Scolopacinae, and sandpipers and phalaropes Tringinae.

Mating systems in the family are varied. Most species are monogamous, but others exhibit specialized breeding behaviours, including 'lekking', polygyny and polyandry in various forms. Most nest on the ground, though some in the genus *Tringa* appropriate the abandoned nests of other birds in trees. Only 5 of the 42 species on the Indian list are known to breed within Indian limits and none in the Delhi area. Most nest in the high Arctic and migrate across great distances to winter in the tropics and the Southern Hemisphere. The Phalaropes *Phalaropus* winter on the open ocean. In some sandpipers, adults form flocks and migrate south first, immediately after the chicks fledge, while the juveniles follow a month or two later on their own. Many arrive as early as July–August, often still in breeding dress. Before they leave, many are again in fresh breeding plumage with showy orange and chestnut highlights. In their native marshes and tundra, they advertise their nesting territories with aerial displays and haunting song flights.

Wader migration can be strikingly evident along the river sandbanks and wetlands in April and May when large numbers of diverse species, in their breeding finery, gather in the evenings, calling excitedly as flocks take off, one after another, northwards for the night's staging hop.

India: 42 species, Delhi: 26

▲ *Ruffs throng the wetlands at Basai near Sultanpur.*

Whimbrel *Numenius phaeopus*

Passage Migrant · Rare

Nest in the circumpolar subarctic tundra and winter in mudflats, tidal marshes and beaches along tropical and southern coasts. Four subspecies are recognized; most birds wintering in India are of the nominate race that breeds in northern Europe and western Siberia and winters along the coasts of Africa and the Indian Ocean; although the eastern Palaearctic race *variegatus*, that winters from Southeast Asia eastwards to Australasia, is also recorded, rarely, from the eastern coast. Whimbrels are primarily coastal during migration and in their wintering areas, and although they do overfly land masses in non-stop flights, they are unusual inland.

Sporadic records of Whimbrels from the wetlands near Delhi are probably attributable to birds compelled to descend due to weather or other circumstances during autumn passage – thus, they were observed in the Rohtak district of Haryana on 25 August 2012, 7 August 2016 and 25 September 2020; and at inundations along the Najafgarh canal, Haryana, on 4 September 2016 and 14–15 October 2017. An apparently overwintering individual was seen at the Hastinapur Wildlife Sanctuary on 2 February 2019.

Eurasian Curlew *Numenius arquata*

Winter Visitor · Uncommon

Widespread as a breeding species across temperate Eurasia, wintering further south in Europe, Africa and southern Asia. Two subspecies have been recognized: the nominate nesting in Europe and wintering further south, and the paler, Siberian breeding *orientalis* that winters in Africa and Asia.

These large waders visit the Delhi area in small numbers, presumed to be of the eastern subspecies *orientalis*, mainly between September and April with late records till 19 May, affecting *jheels*, open marshes, flooded grounds and river sandbanks. Single birds or small parties of up to a dozen or so are the norm in the area. Ganguli mentions 120 birds collecting on the Yamuna sandbanks in the evening, but does not mention the date; it is unclear whether this could have been a migrating party in transit. Such numbers have not been reported in recent decades.

An Eurasian Curlew with a snail at the Chandu Budhera marshes

Bar-tailed Godwit *Limosa lapponica*

Vagrant

Nesting across the Eurasian Arctic and Alaska, and essentially maritime in winter along the coasts of the eastern Atlantic, Indian ocean and southwest Pacific ocean. Four subspecies are recognized: the Alaskan breeding *baueri* is a long-distance migrant with perhaps the longest known non-stop flight of any bird, covering 11,000 km between Alaska and New Zealand without resting for over a week. Birds of the races *taymyrensis* and *yamalensis* that breed in northern Siberia overfly the land mass of Asia during migration to their wintering areas, from Africa to the South Asian coasts. Regional specimens have been identified as *yamalensis* and are only accidental inland with a few odd reports from the northern Indian plains and the western peninsula.

There are two records of accidental vagrants from the Delhi area, one from the Dighal wetlands, Rohtak district, on 17 September 2015 and the second from the Chandu wetlands in the Sultanpur National Park neighbourhood, which remained at the site from at least 10 to 18 September 2022, studied and photographed by many observers.

Black-tailed Godwit *Limosa limosa*

Winter Visitor Common

Widespread breeders at mid- to high latitudes across Eurasia, wintering from the Mediterranean, sub-Saharan Africa, the Middle East and across South Asia to Australasia. Three subspecies have been recognized, and the western nominate race that nests in Europe and Siberia, and the smaller, darker *melanuroides* of the Russian Far East, are both recorded wintering within Indian limits.

A Black-tailed Godwit in winter plumage

These godwits, presumed race *limosa*, are common in the winter in the Delhi area, arriving in end July early August, while most have departed by late April–May. A fair number of non-breeding birds linger through the summer, and small parties may be seen even in late June. They affect the larger waterbodies, *jheels* and village ponds, flooded fields and grasslands, where typical site counts may touch 100 or more. The numbers vary from year to year, and up to 2,000 have been recorded from wetlands, like Okhla or Dhanauri. Flocks often fly high, their contrasting wing patterns and graceful flight against a gloriously blue winter sky presenting an unforgettable picture of avian beauty.

Ruddy Turnstone *Arenaria interpres*

Vagrant

Among the most northernly breeding waders of tundra habitats across the Holarctic, Turnstones are ubiquitous in winter along coasts worldwide, but only accidental inland on the subcontinent during passages. Two subspecies are recognized; the nominate visits the Indian coasts in winter.

Accidentals have been recorded in the Delhi area during both the spring and autumn passages. These rare records include, historically, one at the Najafgarh Lake in mid-May 1957 with Curlew Sandpipers *Calidris ferruginea* and Lesser Sand Plovers *C. mongolus*.[94] There have been six reports in subsequent years: two birds in late August 2012 and one in early November 2014, at the Dighal wetlands, Rohtak district, and one at the Najafgarh wetlands on 18 August 2021.[95] In 2022, and this is unusual, there have been at least three records of Ruddy Turnstones on autumn passage through the Delhi area: three birds at the Mandothi Jheel, Jhajjar district, on 15 August 2022, and singles at the Chandu wetlands near the Sultanpur National Park, on 5–7 August 2022 and again on 17 September 2022.

Ruff *Calidris pugnax*

Winter Visitor Common

Ruffs, well known for their fascinating polygynous breeding economy, breed across the Palaearctic from northern Europe east across Siberia and winter from Mediterranean Europe, sub-Saharan Africa and the Middle East to India and Southeast Asia. Monotypic.

Ruff are amongst the more common migrant waders about Delhi, affecting inundated fields, shallow margins of lakes and river sandbanks across the area. They arrive early, often by early July, but their peak movement in autumn is in August–September when flocks of 1,000 or more may be seen at favoured locations. On their outward passage, March to May, the numbers increase again with 10,000 (!) being estimated once in April at the Basai wetlands, Haryana.[96] Delayed migrant males in June are sometimes seen in nearly complete breeding plumage.

Ruff, in winter plumage

Broad-billed Sandpiper *Calidris falcinellus*

Vagrant

Breeds in Scandinavia and subarctic northern Russia, and winters in maritime habitats along the Asian coasts east to Australia. They are rare inland on passage. Two subspecies are recognized, of which the nominate *falcinellus* winters along South Asian coasts.

Historically, there is just one report of two Broad-billed Sandpipers seen in Delhi by Alexander before 1963, with no further details except that the birds were in breeding dress, therefore presumably, this record was spring, during northward passage. There are only two later reports: April 1999 from Okhla and May 2003 from the Basai marshes.[97]

Sharp-tailed Sandpiper *Calidris acuminata*

Vagrant

An East Asian calidris, breeding in northeastern Siberia and wintering in Australasia, but prone to wander, with vagrancy records scattered as far as North America, Africa and Europe. There are very few records of such vagrants from the subcontinent during the autumn passage. Specimens were collected on 1 August 1880 at Gilgit and 18 September 1958 in Sri Lanka, and one was observed in the Kashmir Valley on 10 August 2021. There are a handful of reports from Sri Lanka and peninsular India as well, probably correct but often insufficiently documented. Monotypic.

One was photographed at Najafgarh, Haryana, on 24 September 2019.[98]

Curlew Sandpiper *Calidris ferruginea*

Mainly Passage Migrant Uncommon

Widespread, nesting at high latitudes across Eurasia, and wintering along the coasts from Africa, West and South Asia to Australasia. Monotypic.

Curlew Sandpipers are regularly recorded during both passages through the Delhi area, between late July and November in autumn and May–June in summer, usually in small parties of up to 10–20 birds by the river and the shallow margins of lakes. However, flocks of 60–80 birds have been counted at Okhla on peak migration days in May and early June. They seem less numerous in autumn, and a smaller number of birds remain all winter. When they arrive, and before they leave in May, most birds are in their chestnut breeding plumage. Later, in autumn, juveniles predominate in September-October, with distinctive pinkish-buff hues on the sides of the neck and breast.

An elegant Curlew Sandpiper to the right; Dunlin in typical hunched posture to the left

Temminck's Stint *Calidris temminckii*

Winter Visitor Common

Nest in the Siberian far north and winter in Africa and southern Asia. Monotypic. In their native tundra, Temminck's Stints follow a breeding economy based on 'successive bigamy', a two-clutch system in which each female lays a first clutch incubated by the first male, followed by a second clutch fertilized by a second male, which she incubates herself. The first male, having fulfilled his first round of parental duties, then partners with a second female to fertilize another clutch of eggs and occasionally even a third clutch.

These small waders are common in winter about Delhi, arriving from about the second week of July and leaving latest by mid-May, and they are generally more widespread and frequently met with than the Little Stint *C. minuta* at inland muddy pools and marshes, irrigated fields, village ponds and sewage farms. Temminck's Stints keep singly or in small parties in winter, but flocks of 20–25 may gather at wetlands and by the Ganga and Yamuna sandbanks for the spring migration in April, outnumbered then by Little Stints.

Dunlin *Calidris alpina*

Winter Visitor Scarce

Dunlin, in winter plumage

Widespread, breeding in the circumpolar Arctic, and wintering south to central America, the Mediterranean region, Arabia and the Persian Gulf to northern India and in East Asia. Nine subspecies are recognized, of which the nominate, *C.a. alpina* that nests in northern Russia, is believed to winter on the subcontinent.

Dunlins are scarce winter visitors to the Delhi area, where they have been reported from several wetlands where waders congregate – by the Yamuna, shallow flooded fields and lake margins. They are more regularly reported, albeit still in small numbers, during the northern passage between mid-April and early June, when many individuals may be moulting into their distinctive breeding plumage, and during autumn passage in September–October. Winter numbers rarely exceed a dozen at a site, and it may be scarcer here now than earlier, going by former reports.

Little Stint *Calidris minuta*

Winter Visitor · Common

Breeds in the Siberian tundra and winters from Africa eastwards to India. Often monogamous, but like Temminck's Stints *C. temminckii*, Little Stints can also be 'successively bigamous' in their breeding economy. Monotypic.

Little Stints are recorded commonly as winter visitors to the Delhi area between the third week of July and late April–May on river sandbanks and muddy inlets, the margins of lakes and village ponds. They usually keep in small parties, but larger numbers totalling a hundred or more may gather on the river sandbanks in April and early May in bright breeding plumage colours before emigration.

Little Stint

Asian Dowitcher *Limnodromus semipalmatus*

Vagrant

Breeds on the Asian steppes and winters mainly from Southeast Asia to northern Australia; rare on the Bay of Bengal coasts and merely vagrant elsewhere in India. Dowitchers have a distinctive 'sewing machine' feeding technique that distinguishes them from other members of the family. Monotypic.

An Asian Dowitcher adult in breeding plumage was recorded at the Basai wetlands, Haryana, in July 2002 with a migrant flock of Marsh Sandpipers *Tringa stagnatilis*.[99]

Long-billed Dowitcher *Limnodromus scolopaceus*

Vagrant

Long-billed Dowitchers are known global wanderers from their usual range in northeastern Russia and western North America; vagrants have been recorded on rare occasions in the Indian subcontinent as well. Monotypic.

There are two records from the area: two birds at the Sultanpur National Park, Haryana, that stayed through most of March 2013, one in breeding plumage at Sonipat between 17–26 July 2020.[100]

Jack Snipe *Lymnocryptes minimus*

Winter Visitor · Scarce

Widespread breeder in the boreal zones of Scandinavia across Russia to eastern Siberia, wintering in Europe, northern Africa and the Middle East, and the subcontinent to Southeast Asia. Monotypic.

This little snipe is a scarce winter visitor to the Delhi area between October and March, reported from swampy ground with low herbage on the Yamuna floodplain and the margins of wetlands and *jheels*, with a number of recent records from the Sultanpur National Park neighbourhood.

A Common Snipe in the marshes

Common Snipe *Gallinago gallinago*

Winter Visitor Common

Widespread breeders in marshlands and swampy ground across Eurasia, as close to the Delhi area as Kashmir, wintering in Europe, Africa, the Middle East and across much of temperate and southern Asia. Two subspecies are recognized, of which the nominate race *gallinago* winters in India.

These snipes are common visitors to the Delhi area from early August to March–April. They affect shallow marshes and inundated fields with soft mud where they can probe for food with their sensitive bills, usually singly or in wisps of two to ten. But they can be numerous on occasion, with a 100 or more seen scattered over the marshes, feeding in the bright winter daylight.

Pin-tailed Snipe *Gallinago stenura*

Winter Visitor Rare

Breeds in the boreal wetlands and meadows of eastern Siberia and winters in South and Southeast Asia; common visitor to the eastern and southern parts of the subcontinent, entering from the east, but much scarcer towards the northwest. Monotypic.

There are very few records of Pin-tailed Snipes from the Delhi neighbourhood, but they have probably been overlooked. Ganguli, writing about the 1960s, mentions old hunting records from Delhi. One bird was flushed at Basai, Haryana, in March 2003; it 'was identified in flight from Common by dark underwings and lack of white trailing edge. I was very familiar with the species after four years in Madras [Chennai] and three in Dhaka.'[101] A couple of later reports are plausible, but cannot be verified.

Terek Sandpiper *Xenus cinereus*

Vagrant

Breeds in the boreal taiga zone across Siberia and winters along the coasts from East Africa, the Middle East and the subcontinent as far to the east as Australia; only occasional inland on migration. Monotypic.

There are a few scattered records of these sandpipers from the area during both the spring and autumn passages, and these were probably migrants forced to land due to adverse weather conditions, or by getting mixed up in flocks of other waders. They have thus been recorded on the river sandbanks at Okhla in May 1968, May 1979, August 1998 and August 2002[102] and from wetlands in Haryana, such as at the Bhindawas Bird Sanctuary, in September 2000; from Dighal, Rohtak district, where a small migrating party was seen on 25 August 2012; a single bird at Kurana, Panipat district, on 8 September 2020; and another at the Chandu wetlands in the Sultanpur National Park environs, on 24 July 2022. The Dighal record was at the same time that two other mainly coastal waders rarely recorded inland, the Whimbrel and Ruddy Turnstone, were found at the same site, all probably compelled to descend during migration by stormy monsoon weather.

Red-necked Phalarope *Phalaropus lobatus*

Vagrant

Breeds in the Holarctic far north and winters pelagically in the Arabian Sea and the tropical Pacific. Usually monogamous, though polyandry is also recorded; the brighter coloured females defend territories and court males, who then undertake incubation duties and raise the chicks alone. They winter offshore along the coasts of the subcontinent, most commonly off the Pakistan and Gujarat coastlines, and less regularly and in smaller numbers along the entire Indian coastline. They are recorded sporadically from a number of inland localities during the spring and autumn passages, usually in very small numbers but rather more regularly and even in larger parties from inland lakes on the north-western subcontinent.[103] Monotypic.

They are accidental on migration so far inland, with only a few records of single birds from the area: north Delhi in early June 1961; Sultanpur in December 1980;[104] the Najafgarh wetlands in early February 2015; near the Dighal wetlands, Rohtak district, on 11–12 October 2018; and on two occasions at the Chandu wetlands near the Sultanpur National Park on 3 April and 22 September 2022.

Common Sandpiper *Actitis hypoleucos*

Winter Visitor Common

With a wide breeding catchment that extends across temperate Europe and Asia and as far south through Central Asia as Kashmir, these distinctive sandpipers winter in Africa, southern Asia and Australia. Monotypic.

Widespread winter visitors about Delhi between the third week of July and mid-May. They avoid squelchy marshes and swamps and are usually seen teetering about man-made structures, weirs, canal locks and bunds, etc. They are the least numerous of the regularly occurring sandpipers, territorial even in winter, though they may collect in small groups on migration.

Common Sandpiper

Green Sandpiper *Tringa ochropus*

Winter Visitor Common

Breed across northern Eurasia and winter in the Mediterranean basin, Africa and southern Asia. On their breeding grounds, Green Sandpipers are associated with swampy, boreal forests where they lay in rather unusual locations for a wader – the abandoned nests of forest birds or similar sites in trees – though sometimes they may nest in the more usual ground sites as well. Monotypic.

Green Sandpipers arrive by the first week of July in the Delhi area and may stay till mid-May with small numbers over-summering. Never in flocks, they avoid the marshes beloved by Wood Sandpipers and are widespread in village ponds, streams, sewage trickles and even artificial waterbodies in city parks. Small groups of three to five may collect on migration.

Spotted Redshank, moulting into breeding plumage

Spotted Redshank *Tringa erythropus*

Winter Visitor · Fairly Common

Breeds in subarctic regions across Eurasia, and winters from the Mediterranean and tropical Africa, east to the subcontinent and Southeast Asia. Monogamous in their breeding economy as a rule, but they are apparently sometimes also polyandrous; females may leave soon after laying their eggs to begin their migration southwards, while the males undertake most of the incubation duties and tend to the fledglings largely alone. Monotypic.

Spotted Redshanks winter in the area from mid-July to May, but there is a marked spring passage in end April and May with most birds in their handsome dark breeding plumage. In winter, they are usually seen singly or in small parties, but during the spring migration, may collect in monospecific flocks of up to 150 birds at some of the larger lakes. Spotted Redshanks prefer the open shores of ponds and lakes but sometimes deeper waters as well, where they feed by swimming and picking food from the surface or even upending like ducks.

Common Greenshank *Tringa nebularia*

Winter Visitor · Common

Breeds across northern Europe and Siberia, wintering from the Mediterranean and Africa east to India, Southeast Asia and Australasia. Monotypic.

Greenshanks are common visitors to the Delhi area, arriving early in July and remaining till May, with records in June as well of over-summering birds or very early arrivals. Single birds or sometimes two or three together may be found at a variety of wetland habitats, where their loud calls sound the alarm at any danger. They may collect, however, in parties of 10–12 birds on river sandbanks and other wetlands in late April before taking off in the evenings for the night migration.

Marsh Sandpiper *Tringa stagnatilis*

Winter Visitor · Fairly Common

Breeds in a broad belt across northern Europe and Asia, migrating in winter to Africa, southern Asia and Australia. Monotypic.

Marsh Sandpipers, the most elegant and finely built of the *Tringas*, visit the Delhi area in comparatively small numbers from early July till April–May, but they can be much more common on passage in autumn, late July through September and then again in April on the outward passage. At these times, up to a 100 or more birds may be present compared to winter numbers which rarely exceed 10 at a site. They prefer open areas of soft, wet mud and shallow pools and seepage marshes where they can dip and probe with their long, fine bills for food under the water surface.

Wood Sandpiper *Tringa glareola*

Winter Visitor Common

Widespread breeder across Eurasia, wintering to Africa, India and Australia. Like Green Sandpipers, Wood Sandpipers are associated with swampy, boreal forests in the breeding season, where they may nest in trees, in the old nests of thrushes and other passerine species, though also commonly in the more usual ground sites. Monotypic.

Wood Sandpipers are perhaps the commonest and most generally distributed of the waders in winter, coming in early by the first week of July and staying till May. A few linger through the summer with records in end June. They keep in parties of 4–12 birds, sometimes larger, preferring swamps and flooded grounds, seepage marshes, paddy fields, etc.

Common Redshank *Tringa totanus*

Winter Visitor Common

Widespread breeding bird of temperate Europe and Asia, wintering to Africa, the Middle East, India and south China. Six subspecies are recognized, of which both the nominate race of Europe and western Siberia, and *eurhina*, which nests in Central Asia, Ladakh and Tibet, winter in India.

Redshanks are common migrants in the area, some arriving as early as the first week of July and leaving in April–May. Records in late June may be over-summering individuals or early migrants. They keep singly or in twos and threes, sometimes in scattered groups of up to 15, and their liquid trisyllabic calls are a familiar sound of the wetlands.

Wood Sandpiper (on the left) with Common Redshank, in winter plumage

CHARADRIIFORMES: WADERS, GULLS AND ALLIES
TURNICIDAE: BUTTONQUAILS

Small Buttonquails live in remnant grasslands, such as Jhanjraula

The buttonquails of Africa, Asia and Australasia are quail-like birds of grassland and open scrub country, but have only three toes, lacking a hallux. They, too, exhibit a sequentially polyandrous breeding economy with a reversal of sexual roles; females being the more attractively coloured, mate successively with several males in their respective territories, who then assume all parental duties. They nest in a pad of grass in a scrape on the ground, usually sheltered by overhanging grass, and in most cases, only the male incubates the eggs and raises the chicks, which are extremely precocial. Three species represent the family in India, and all three have been recorded in the Delhi area.

As a family, they are notoriously elusive; often nomadic or locally migratory in many areas and dependent on the seasonal creation of a suitable habitat – all of which make them difficult to study and determine their exact status.

India: 3 species, Delhi: 3

Small Buttonquail *Turnix sylvaticus*

Summer Breeding Visitor · Uncommon

A widespread species, mainly resident in its range but subject to rainfall-related local movements, in Africa and in South and Southeast Asia. Nine subspecies are recognized; the Indian race *dussumier* represents the species about Delhi.

These elusive birds appear to be regular but localized summer visitors, nesting locally in uncultivated grassland patches in the Delhi area. Historically, they were considered scarce, with a few scattered reports in April and May from near the Yamuna. However, in recent years and especially since 2016, Small Buttonquails are being recorded annually between May and end October, especially from the grassy flats near the Sultanpur National Park which probably reflects the intensive ornithological coverage of this area more than their actual distribution about Delhi. These records typically involve single birds or small parties, with recorded maxima of eight together in first week August 2017, and 11, including juveniles, in June–July 2018. One photographed on 30 March 2020 in the same general area may be an early returning migrant. Another recorded in the scrub jungle at Maroudi Jatan, Rohtak district, is the only recent record away from the Sultanpur vicinity.

Yellow-legged Buttonquail *Turnix tanki*

Summer Breeding? Visitor · Data Deficient

A South and East Asian species, nomadic and subject to marked seasonal movements. Two subspecies are recognized, with nominate *tanki* on the Indian mainland.

This pretty buttonquail is as elusive and difficult to see as most of its congeners. The paucity of records make it impossible to assess these buttonquails' status about Delhi; they may turn out to be scarce, possibly erratic, monsoon visitors to suitable grassland habitats in the area. Ganguli, in her *Guide* published in 1975, mentions pre-1950 reports in May and August, and there are only three verifiable reports from later: a just identifiable photograph of a male crossing a sandy track at Jhanjraula in the Sultanpur National Park neighbourhood on 23 July 2016; one at Manethi, Rewari district, on 27 June 2021; and another photographed near Sultanpur (apparently at the same grassland area at Jhanjraula as the 2016 record) on 19 June 2022.[105] The dates fall within the breeding season of the species.

Barred Buttonquail *Turnix suscitator*

Breeding Resident · Fairly Common

Resident across much of the Oriental region. Seventeen subspecies are recognized over its range, with three in India: *plumbipes* in the Gangetic Plain and the north, *taigoor* in central and southern India and *benghalensis* in lower Bengal.

This buttonquail is the most widespread of the group in India and a fairly common resident about Delhi, presumably of the race *taigoor*, in the grass and scrub country on the Ridge and in rural areas, in large gardens as long as there is some cover into which it can retreat on the approach of danger. They nest mainly in the monsoon months, laying a clutch of four eggs in a scrape on the ground in grass or under a scrubby thicket, though a male leading three well-grown chicks was seen as early as late May on the New Delhi Ridge. The females' advertising call, a low, sustained 'drumming', likened by Sálim Ali to the beat of a distant two-stroke motorcycle, is heard till October.

Barred Buttonquail, female, most widespread of the genus in the area

CHARADRIIFORMES: WADERS, GULLS AND ALLIES

GLAREOLIDAE: COURSERS AND PRATINCOLES

The family, with two subfamilies, coursers Cursoriinae and pratincoles Glareolinae, is widespread in southern Europe, subtropical and tropical Asia and Africa. Both the long-legged, plover-like coursers of arid habitats and the long-winged, tern-like pratincoles of wetland margins display a tendency towards nomadism, and they can appear irregularly in unexpected places. They feed on insect life, the coursers running and dipping, plover-style, to pick food from the ground, while pratincoles usually hunt flying insects in aerial pursuit. Pratincoles are gregarious, strong fliers that gather at times in very large flocks that tend to be most active in the early mornings and evenings, hawking swarming insects over marshes and cultivated fields. These flocks spend the hot hours of the day on the ground, standing scattered over fallow fields, pasture lands or sandbanks. Coursers breed as solitary pairs, pratincoles often in loose colonies, and both lay their eggs, usually two or sometimes three in a clutch, in a simple scrape on the ground. Both parents incubate and look after the precocial chicks, which are superbly camouflaged to merge with their habitat. Like some of the smaller plovers, pratincoles may perform 'broken wing' distraction displays to lure predators away from their nests and chicks.

India: 6 species, Delhi: 4

▲ *An Oriental Pratincole in the Sultanpur neighbourhood*

Cream-coloured Courser *Cursorius cursor*

Vagrant

A resident of the great desert belt that stretches across North Africa eastwards to Rajasthan in India, nominally resident but with marked nomadic tendencies. May associate with parties of Indian Coursers *C. coromandelicus* where their ranges and habitats overlap. Three subspecies are recognized, amongst which the eastern race *bogolubovi* extends from Iran to northwestern India.

One was observed near the Sultanpur National Park in late January 2002, apparently having arrived with a party of Indian Coursers *C. coromandelicus* that may also have wintered in Rajasthan, and remained in the area for a couple of weeks till they were driven off by the latter species as they settled down to breed.

Indian Courser *Cursorius coromandelicus*

Mainly Breeding Pre-Monsoon Visitor, also partly resident Uncommon

An Indian subcontinental endemic. A terrestrial species of dry open country, usually in pairs and small parties that run about on the ground, feeding like plovers on overgrazed grazing grounds and fallow fields. Strong fliers that show some local, seasonal movements. Monotypic.

A locally distributed and declining breeding species about Delhi, this elegant courser has seen its short-grass plains habitat becoming increasingly fragmented and lost to agricultural and urban expansion. Ganguli deemed it a 'not-too-common' resident in the 1960s, in the bare open country in southwestern Delhi, near Tughlaqabad and across the Yamuna. In recent decades, most records are from degraded fallow land in the environs of the Sultanpur National Park and elsewhere as well as in Haryana, where it seems to be mainly a dry-season visitor, arriving in small parties from about November, pairing off to nest by March and leaving before the monsoon breaks in June, probably for drier habitats in Rajasthan. Some birds remain in barer patches through the year.

An Indian Courser at the Sultanpur flats

Oriental Pratincole *Glareola maldivarum*

Summer Breeding Visitor Fairly Common

Breeds locally in suitable habitats in southern and eastern Asia, spreading more widely in the non-breeding season across India and Southeast Asia to Australia. Monotypic.

These pratincoles are summer visitors to the area, mainly between March and October. Parties of 20–25 birds have been observed moving northwards along the Yamuna in April and early May and southwards again in the third week of October. They breed locally in May and June in small, scattered colonies, with nesting records from the river floodplain and fallow ground in Haryana, including the Sultanpur National Park and Najafgarh neighbourhoods, in Rohtak and Sonipat districts and very likely at other localities east of the Yamuna, such as near the Dhanauri wetlands in the GautamBuddha Nagar district. A marked post-breeding influx is noticeable from about June, when larger numbers, including many juveniles of the year, congregate and hunt over the marshes and irrigated fields. Both Hutson, in the 1940s, and Ganguli, in the 1960s, had recorded such flocks of 200–300 birds in June–July and recent reports, mainly July–September, typically involve 50–300 birds hawking over grassland or marshland in the evenings.

Small Pratincole *Glareola lactea*

Summer Breeding Visitor Fairly Common

Endemic to southern Asia from the subcontinent east to Cambodia, mainly resident but with marked local movements. Monotypic.

Small Pratincoles are local summer visitors to the Delhi area; though they have been recorded in most months, they are most visible and numerous between February and March and especially during the monsoons and post-monsoon months, from late June to September–October, when large flocks – of hundreds at times – swarm like swallows at dusk over the Yamuna River at Wazirabad or Okhla and the Ganga at the Haiderpur barrage, about wetlands such as Sultanpur and Najafgarh, and possibly elsewhere. These probably represent pre-breeding and post-breeding movements of this species, the details of which remain unclear. They have been recorded nesting in the Ganga *khadar* at the Hastinapur Wildlife Sanctuary, and Ganguli[107] had considered them 'not-too-common' residents in Delhi, breeding on the Yamuna sandbanks in April and May in small colonies with River Terns *Sterna aurantia*. There is circumstantial evidence of breeding, including distraction displays or mobbing behaviour, which suggests that some pairs may still attempt to breed on sandbars of the Yamuna, though sand mining downstream and barrage operations probably destroy many nests.

Small Pratincole, in flight

CHARADRIIFORMES: WADERS, GULLS AND ALLIES
LARIDAE: GULLS, TERNS AND SKIMMERS

This family includes three lineages that are recognized as separate subfamilies, all of which are represented in India and the Delhi area: gulls Larinae, terns Sterninae and skimmers Rynchopinae. Gulls are cosmopolitan, aggressive, scavenging and commensal birds that have greatly benefited from the refuse and garbage spread through human agency. Terns, also geographically widespread like gulls, differ in their agile and elegant flight, but rather than benefiting from human proximity, several tern species have suffered from the unrelenting disturbance to and loss of their breeding grounds. The skimmers form a small, distinctive clade within the family, with one species each in South Asia, Africa and America.

Most Indian breeding terns and the skimmer nest colonially or semi-colonially, laying their clutches of two or three camouflaged eggs in scrapes on bare sand, islands, coastal habitats and inland along the braided, sandy courses of the larger rivers. In northern India, as about Delhi, they nest during the dry season, when the river levels are low and the sandbanks are exposed, before the advancing monsoon rains raise water levels and flood the channels. Marsh terns, however, like the Whiskered Tern *Chlidonias hydrida*, build their nests on floating vegetation in freshwater wetlands.

India: 38 species, Delhi: 18

▲ *River Terns*

Slender-billed Gull *Chroicocephalus genei*

Vagrant

Slender-billed Gulls nest in coastal lagoons and inland saline lakes in widely scattered locations from West Africa, the Mediterranean basin, the Middle East and in Central Asian steppes; winter visitors to the Arabian Sea coasts of the subcontinent with occasional stragglers recorded inland as well. Monotypic.

There are only five verified reports of Slender-billed Gulls from the Delhi area to date, usually of single birds in flocks of Black-headed Gulls *C. ridibundus* on the Yamuna, mainly in mid-winter but with one record in October: from the Okhla area on 21 January 1990, in early January and again in late December 2018; at Chilla *khadar*, Mayur Vihar, east Delhi, on 10 October 2019; and from Wazirabad, north Delhi, on 31 December 2020.

Black-headed Gull *Chroicocephalus ridibundus*

Winter Visitor Common

Widespread and familiar gulls, nesting across northern and temperate Eurasia and wintering southwards of that region. Monotypic.

Common through the winter, between September and April, on the rivers and less frequently on lakes and reservoirs of the area, these gulls are often in mixed flocks with Brown-headed Gulls *C. brunnicephalus*. Ganguli remarked[108] that these species have greatly increased about Delhi since the early 1960s, and large gatherings of these two smaller gulls on the river and at roosts on the Yamuna at Wazirabad and Okhla may number in the thousands – up to 10,000 or more (as recorded by Harvey in the *Atlas*). However, this huge increase may only reflect the greater availability of food, organic waste and garbage in the river.

Brown-headed Gull *Chroicocephalus brunnicephalus*

Winter Visitor Common

Purely Asian gulls which breed on the islands and marshes of inland lakes at high altitudes in Central Asia, Ladakh and the Tibetan Plateau and winter on the Indian subcontinent. Monotypic.

A Brown-headed Gull in breeding plumage

Brown-headed Gulls are common on the rivers and larger waterbodies of the Delhi area between September and March–April but tend to be outnumbered by Black-headed Gulls *C. ridibundus*. Both gulls have similar habits while they are here: flocks frequent fishing villages and ghats, living off handouts or scavenging on refuse, food scraps, etc. carried by the current. They are also seen at lakes and fish ponds, shuttling between them and the Yamuna. The increase in their numbers, noted as early as the 1970s, has visibly continued over recent decades, and the gull flocks beating up and down the river all day are quite a sight during the winter.

Little Gull *Hydrocoloeus minutus*

Vagrant

Breeds in northeastern Europe and across Siberia and normally winters to the Mediterranean Sea and Caspian Sea; recorded as a vagrant to the subcontinent, but with very few verifiable reports. Monotypic.

One individual of this smallest of the gulls was reported from Okhla in end December 1992 (by Paul Holt), but no details are available and later references have conflicting information.[109] However, another first winter Little Gull was photographed in a flock of c. 300 Black-headed Gulls *C. ridibundus* on 17 December 2014, also at Okhla, with the observer's comment: 'Though photographs are not sharp, the comparative size in relation with the Black-headed Gulls, black transverse band on upper wings, more round wings, black terminal band, and white on secondaries separate this from other gulls, including first winter Black-legged Kittiwake *Rissa tridactyla*'.[110]

Pallas's Gull *Ichthyaetus ichthyaetus*

Winter Visitor Uncommon

Breeds in Central Asia, wintering from the eastern Mediterranean to the Middle East and South Asia. Monotypic.

These splendid gulls are present in the area between August and April, usually singly or in small groups of four to six, but are more common during the spring passage in February–April. They frequent the Yamuna as well as the larger lakes, such as Sultanpur and Bhindawas in Haryana. They tend to keep to themselves and not mix with the smaller gulls, except at roosts. Pallas's Gulls are readily identifiable by their sheer size at any stage of plumage, and a bird in full breeding dress in spring is a truly magnificent creature.

A Pallas's Gull in breeding plumage in March

Mew Gull *Larus canus*

Vagrant

Widespread, breeding in northwestern North America and across Eurasia and wintering in Europe, the Middle East and coastal East Asia. Four subspecies are recognized; *heinei*, the race breeding in central Russia, has occurred as a rare vagrant to the subcontinent.

A first winter bird was recorded at Okhla on 19 January 1992 and a second year bird on the Ganga at Gajraula, UP, on 5 March 1993.[111]

Caspian Gull *Larus cachinnans*

Winter Visitor Uncommon (Taxonomy dependent)

Breeds in central Eurasia and winters from Europe, the Middle East and the Persian Gulf to western and northern India, with recent confirmed records from as close as Pong Dam, Himachal Pradesh, in December 2017. Monotypic.

A few birds showing characters of the Caspian Gull – slender longish bills, sloping foreheads and pale grey upper parts – have been reported and photographed in the Delhi area amongst the small numbers of Large White-headed Gulls (LWHGs) that winter here. Yet, these also often show some features typical of the Steppe Gull *L. f. barabensis* – deeply coloured legs and bills, more solidly black wingtips and primary patterns. Any Caspian that winters in Delhi would most likely be from the eastern part of its breeding range, about Kazakhstan, which abuts the range of *barabensis*; these 'eastern Caspians' begin to share features of *barabensis* and vice versa and could in fact be Caspian/Steppe hybrids resulting from some genetic exchange between the two. It is moot whether they should be labelled 'eastern' *cachinnans* or the (normally) darker backed *barabensis*.

The 'Herring Gull' complex – LWHGs

Three forms in this complex that have been variously referred to as Caspian Gull *cachinnans*, Steppe Gull *barabensis* and Heuglin's Gull *heuglini* have been reported from the Delhi area. Phylogenetic relationships amongst the various forms are complicated by hybridization within the complex, and the resultant taxonomic and systematic uncertainties, phenotypic variation and the fact that the larger gulls can take up to four years to reach full adult plumage, has confounded gull identification in past years. Following a revision of species limits based on genetic studies, both *heuglini* and *barabensis* are now considered subspecies of the Lesser Black-backed Gull *L. fuscus*. Rasmussen & Anderton, in *Birds of South Asia*, 2012 (henceforth Rasmussen & Anderton), consider Caspian hypothetical in South Asia and confirm only the taxa *heuglini* and *barabensis* on the Indian list. The Caspian Gull is, however, included in the *Checklist of the Birds of India*, v7.0, which forms the taxonomic basis for this book.

Lesser Black-backed Gull *Larus fuscus*

Winter Visitor Uncommon

Two of the five recognized subspecies of the Lesser Black-backed Gull have been recorded wintering in South Asia: Steppe Gulls *L. f. barabensis* that breed in the Central Asian steppes eastwards to the Altai and winter from Arabia to India, and their genetically poorly differentiated northern neighbours Heuglin's Gulls *L. f. heuglini* of northern Russia that winter mainly along the Indian Ocean coasts of East Africa and the Arabian Sea to western India, and also in eastern China.

Some amongst the adult LWHGs that winter about Delhi between September and March show light grey mantles, matching the Caspian's tone. However, rather darker-backed, yellow-legged gulls also occur, and these are surely *barabensis*. The darkest, slate-grey-backed *heuglini* are the scarcest, mere stragglers on passage as they are mainly coastal in winter. Their numbers are small, usually under 10, and they are seen mainly along the rivers but may wander on the larger lakes of the area occasionally.

Little Tern *Sternula albifrons*

Summer Breeding Visitor Uncommon

Breeds locally on the coasts and inland in parts of Europe, Africa and Asia. Six subspecies are recognized, two of which occur in India: the nominate as a winter visitor to the west coast and *pusilla* breeding locally by the lakes and rivers of the Indo-Gangetic Plain and in Sri Lanka.

Small numbers of the north Indian breeding race *pusilla* are regular on river sandbanks in the area, both on the Yamuna and Ganga, in summer from late February through the monsoon months till October. Ganguli had recorded a few pairs nesting in April in the 1950s and 1960s on the Yamuna sandbanks and at Najafgarh, but pollution, changes in the river's hydrology and sand mining in the *khadar* downstream of Okhla render breeding here very unlikely now. However, if protection can be ensured, Little Terns may still find conditions suitable for nesting on any undisturbed sandbanks of the Yamuna above Wazirabad, north Delhi, where courtship feeding by a pair was observed in May 2017. They nest in small colonies along the Ganga *khadar* in the Hastinapur Wildlife Sanctuary as well. There is a post-breeding influx into the Delhi area from presumably not-too-distant nesting colonies, with both adults and juveniles – up to 25 at a site – observed during the monsoon months between June and September at Okhla and Wazirabad, on the Yamuna, when they also often appear at flooded areas away from the large rivers.

Gull-billed Tern *Gelochelidon nilotica*

Winter Visitor Uncommon

A Gull-billed Tern in flight, non-breeding plumage

Widely distributed terns, breeding in coastal colonies and inland at scattered sites across Eurasia, Australia and the Americas, wintering to Africa and southern Asia. Six subspecies are recognized; the nominate race *nilotica* represents the species on the subcontinent, nesting on the Makran coast and Indus sandbanks in Pakistan and passage migrant or winter visitor to wetlands, marshes and coastal mudflats elsewhere in the region.

The status of these terns about Delhi has been changing for reasons that are unclear. Ganguli had first noted a change around 1960; they were earlier considered 'uncommon winter migrant(s) seen singly in January, April and May', but then began to be recorded all through the year in the 1960s, occasionally in 'large flocks of more than 100' between May and September with birds in breeding plumage from March. A decade later, the numbers had dropped again, and yet parties of 20–25 birds with many in breeding plumage continued to be recorded between March and June in the 1970s and 1980s, but with no evidence of nesting. Their status seems to have changed yet again in recent years, with fewer winter sightings and rarely more than 10 at a site; mainly between September and May, with increased numbers, possibly passage migrants, from March.[112] They do not plunge to feed as many tern species do and may hawk over marshes and even adjacent fields, dipping down in flight to pick small vertebrate and other food items.

Caspian Tern, a vagrant photographed near Delhi, in breeding plumage

Caspian Tern *Hydroprogne caspia*

Vagrant

Widely distributed globally, breeding populations are scattered, both along the coastlines and inland in North America and Eurasia, including around the Black Sea, Caspian Sea and Aral Sea in Central Asia, eastern China, coastal Pakistan and Sri Lanka and locally elsewhere in Africa and Australasia. The Central Asian breeders winter from the Persian Gulf to the subcontinent, where this is mainly a coastal species in India and is much scarcer inland.

There is one historical record of this large tern from the Yamuna in November 1963 and only three more records since then, all of which are accidentals in the area on spring passage. One was reported from the Sultanpur National Park on 2 February 1984, one from the Yamuna *khadar* in March 2003, and two adults in breeding plumage were photographed at the Chandu wetlands in the Sultanpur National Park neighbourhood on 2 April 2022.[113]

Black Tern *Chlidonias niger*

Vagrant / Reconfirmation desirable

Black Terns breed, in two subspecies, in temperate latitudes in both Eurasia and North America, the nominate race *niger* wintering mainly in coastal West Africa. The occurrence of this tern in India has been a contentious issue mainly due to field identification problems.

The first report of this marsh tern in the Indian subcontinent was a sight record by H. Alexander in October 1949 on the Yamuna in Delhi. Two birds were reported in Okhla in late September 1998, described in detailed fieldnotes, and one in July 1999. However, the identification of Black Terns in winter plumage, based solely on plumage, can be problematic and structural features and behaviour are critical in isolating this species from other marsh terns. Rasmussen & Anderton accept that this tern could be a very rare visitor to the Indian subcontinent but treat it as hypothetical for want of proper corroboration. There are, however, subsequent validated records from Gujarat, Karnataka and Kerala, and its status here is retained as a vagrant, requiring reconfirmation.[114]

White-winged Tern *Chlidonias leucopterus*

Passage Migrant · Rare

Breeds from central Europe across Eurasia, wintering in Africa and southern Asia to Australia.

These lightly built, elegant marsh terns are scarce and irregular passage migrants through the area. The few spring records of birds in their distinctive breeding plumage are all between late April and the second week of June and are from Najafgarh in 1968, Okhla in 1980, 1986 and 2000, and the Ganga *khadar* in 2018. In autumn, when most birds are in their juvenile or moulting into their first winter plumage, there are scattered reports of singles or parties of up to four birds together from Okhla, the Sultanpur neighbourhood at the Basai and Chandu wetlands and the Dighal wetlands between late July and October, at times even remaining at a favoured site for up to two weeks during passage on occasion.

Whiskered Tern *Chlidonias hybrida*

Summer Breeding? Visitor · Common

Breeds in Eastern Europe, Asia and in South and East Africa, widespread in the tropics in winter. Three subspecies are recognized, with race *hybrida* in its Eurasian range. In India, they nest in the lakes of Kashmir and very locally in the Gangetic Plain and Assam – these plains colonies, however, tend to be unstable, dependent on monsoon conditions and disturbance levels. A large colony of over 100 pairs was recorded nesting for the first time in Gujarat, at the Nalsarovar Sanctuary, in 2021 and again in 2022, indicative of such opportunistic establishment of breeding colonies of this tern if circumstances are favourable.

Whiskered Terns may be seen all year in the Delhi area but are most abundant during passage from early March to mid-May, when counts – with most birds in full breeding plumage – may exceed a thousand or more on the Yamuna and at other wetlands and marshes. Ganguli, in her *Guide* published in 1975, believed they had increased since the 1950s and had located a breeding colony at the Najafgarh Jheel that was active between July and September 1962. Since then, however, there has been no reconfirmation of nesting, though birds in breeding dress are seen throughout the summer at Okhla and other wetlands where conditions seem suitable. As with other terns, there is a post-breeding influx of birds, and the numbers visibly increase in September–October and are spread more widely over the area.

Common Tern *Sterna hirundo*

Vagrant

Widespread Holarctic birds that winter in tropical maritime habitats and rarely recorded on passage inland on the subcontinent. Three subspecies are recognized, and the birds recorded about Delhi are probably race *tibetana*.

Ganguli, writing in the early 1970s, considered Common Terns to be uncommon, irregular passage migrants and noted records in September, March and April in the 1960s from north Delhi and Najafgarh. There have been three verifiable reports of straggling passage migrants from the Delhi area since then: an adult on the Yamuna above Wazirabad, north Delhi on 18 June 2016; a first-year immature at the Mandothi Jheel, Jhajjar district, in late August–early September 2021; and two immature individuals at the Chandu wetlands in the Sultanpur neighbourhood on 30–31 July 2022.

Black-bellied Tern *Sterna acuticauda*

Former Breeding Resident · Scarce

A resident river tern of the subcontinent, now endangered with a declining population, with disturbance and destruction of breeding habitat being a major threat. Monotypic.

A delicate, elegant species, the Black-bellied Tern's declining population is reflected in its status in the Delhi area as well. These terns bred on the Yamuna sandbanks at least until the 1960s, when Ganguli considered them 'fairly common, but not as numerous as River Tern'. There has been no report of these terns nesting in recent decades but till the early 2000s, up to 25 birds, including many juveniles, could still be recorded at Okhla during the annual post-breeding influx of river-nesting tern species that occurred between about the third week of June till mid-October. A reduced post-breeding influx is still noticeable; however, with reports of small numbers at Wazirabad and Okhla between June and September–early October, the recent maxima of 14 birds at Okhla on 2 October 2016 and 44 at Wazirabad on 5 July 2018[115] being pleasantly high. They are reported in small numbers from the Ganga River as well, but all the records are from the non-breeding season, from mid-July to January, with no evidence of nesting.

River Tern *Sterna aurantia*

Former Breeding Resident · Fairly Common

Resident across the subcontinent and Southeast Asia, where it is now rare. Monotypic.

Small colonies of this tern used to breed between late March and June on the Yamuna sandbanks till the 1960s, but there has been no reconfirmation of nesting about Delhi over the last three decades at least. The dismal state of the riverine habitat is no doubt responsible. They probably breed in some not-too-distant stretch of the Yamuna as there is a noticeable post-breeding influx of adults and fledged juveniles from about the third week of June, when fairly large numbers, up to 100 or more, appear along the river, notably upstream of Wazirabad and below Okhla. By end September, these flocks begin to disperse, numbers drop again and only a few remain in winter. Away from the Yamuna, they occasionally wander to village ponds and other wetlands and have even been seen hawking over artificial waterbodies in the Central Vista in New Delhi. They are regularly recorded all through the year on the Ganga at the Hastinapur Wildlife Sanctuary and Haiderpur Wetland just upstream, but with no confirmation of nesting.

River Tern, non-breeding plumage

Indian Skimmer *Rynchops albicollis*

Former Summer Breeding Visitor · Rare

Ranges across much of the northern and central subcontinent and Myanmar, but populations have declined due to widespread degradation, pollution and disturbance of their riverine habitat. Largely restricted to major rivers, particularly along calm stretches with sandbars; has wandered, albeit rarely, to inland lakes and inshore coastal waters. Monotypic.

Ganguli[116] mentions breeding records from the 1940s and 1950s for April–May on the Yamuna sandbars, adding that skimmers were 'not too common' but more frequently seen in summer in small groups, which probably represented some post-breeding dispersal. Skimmers continued to be decreasingly reported from the Yamuna till the mid-1980s but thereafter, records have become very scarce, with barely more than a dozen reports over the last three decades: from the river at Sonipat, from Wazirabad and Okhla and, exceptionally, away from the river as well, with immature individuals recorded at Najafgarh in mid-March 2021 and another at the Chandu wetlands in the last week of July 2022, both sites in the Sultanpur neighbourhood. Skimmers are also recorded from the Ganga as occasional visitors both upstream and downstream of the Hastinapur Wildlife Sanctuary in the east of the area. Most recorded occurrences in recent years involve small numbers, in single digits, usually just one or two individuals. The records cluster in January–March and late May–August with one record from Okhla in end October as well, which suggests that the skimmers may be moving to and from some nesting sites somewhere either upstream or downstream of Delhi. The fact remains, though, that the near disappearance of this charismatic species from our riverine landscapes has been tragic.

Two Indian Skimmers on a sandbank of the Chambal river

CICONIIFORMES: STORKS

CICONIIDAE: STORKS

Delhi is rich in stork species. Adjutants *Leptoptilos* are restricted to Asia and Africa, with two and one species respectively. The Painted Stork is one of four species in the genus *Mycteria* that, between them, inhabit the tropical wetlands of the world. The Openbill *Anastomus* and Black-necked Stork *Ephippiorhynchus* each have a congener in Africa. Two storks, in the widespread genus *Ciconia*, are migrants from Eurasia and one is a resident of the Indian and African tropics. Our storks' affinities, thus, are overwhelmingly African.

Some stork species are colonial breeders, often in mixed heronries near wetlands; others are solitary nesters. Two to four eggs, as a norm, are laid in bulky nests of sticks built in large trees or sometimes on artificial structures, such as buildings or pylons. These nests then create a resource that attracts reuse by other species, notably some raptors and owls, after the young of the original nest builder have fledged.

On the Delhi list are some of the most spectacular, as well as threatened, members of the family. We can justifiably be proud that India's capital offers refuge to such wonderful creatures, but such wealth also imposes a special responsibility on the governments and citizens of Delhi and the NCR to conserve these remarkable birds and their wetland habitats.

India: 8 species, Delhi: 7

▲ *Asian Woolly-necked Stork*

Asian Openbill *Anastomus oscitans*

Breeding Resident · Fairly Common

Resident across India and Sri Lanka to Thailand and Indochina. Monotypic.

Delhi is towards the western limits of the Openbill's range, possibly accounting for its relatively reduced abundance here than elsewhere in India. Compared to the 1940s, when Frome found them 'in irrigated and flooded areas particularly in north and east Delhi',[117] they are now more widely spread in wetlands but still not in usually large numbers. A maximum of 30–40 birds may be present at favoured sites, mainly between March and August, when they are in their white breeding plumage; by late August most are in grey out-of-season dress. Their numbers decrease in mid-winter, suggesting that some move out of the Delhi region as it gets colder (possibly because the populations of their primary food, snails, decrease?). Openbills nest between May and August in heronries at the Damdama Lake (12–15 pairs in a single species colony), the Sultanpur National Park, Surajpur lake, along canals in the Sonipat district of Haryana, in the Hastinapur Wildlife Sanctuary and Haiderpur Wetland in UP, and probably elsewhere in the area as well.

Asian Openbill

Black Stork *Ciconia nigra*

Winter Visitor · Scarce

Breeds across the temperate Palaearctic from Europe to north-eastern China and in South Africa, wintering more widely in Africa and southern Asia. Monotypic.

Apparently mainly passage migrants through the area, usually in ones or twos on northward passage in late February–April and in autumn in October–November; sometimes in small parties with a maximum of about 40 birds over south Delhi in mid-October 2017. There was just one historical record till the 1970s, from Surajkund in March; later records cluster in the two decades after the year 2000, which may reflect either better ornithological coverage or a real increase in population. Individuals may winter, as suggested by a few December reports from waterbodies in Haryana and from the Asola–Bhatti Wildlife Sanctuary.

Asian Woolly-necked Stork *Ciconia episcopus*

Breeding Resident · Fairly Common

Resident across much of southern Asia; Monotypic.

These striking and conspicuous storks are fairly common residents throughout the area in fallows and cultivation usually, but not invariably, near wetlands, where they feed on the land, not by wading in water. They usually keep in pairs, but gatherings of 15–20 or more have been recorded as well. Pairs nest singly during the monsoon on tall, mature trees standing in farmlands or along irrigation canals, not in heronries like many other storks. A recent study in Jhajjar and Rohtak districts in Haryana[118] has recorded Dusky Eagle Owls *Ketupa coromandus* selectively reusing the nests of this stork, that inhabits and nests in the same habitat mosaic, in a novel commensal relationship.

White Stork *Ciconia ciconia*

Winter Visitor · Rare

Breeds in Europe, North Africa and West Asia to Turkestan, wintering in Africa and India. Two subspecies are recognized: the nominate nesting in the western Palaearctic, *asiatica* in Central Asia.

White Storks used to be considered irregular winter visitors, but their rare appearances in the Delhi area now hardly warrant even this status. The only recent reports include two at Okhla in March 1992 and one in February 2005; two at Palwal in Haryana in mid-winter 2004–05; and one observed from the Yamuna Expressway south of Jewar, UP, in March 2019.

A female Black-necked Stork with yellow iris

Black-necked Stork *Ephippiorhynchus asiaticus*

Breeding Resident · Scarce

Range from India to northern Australia, though sparsely distributed over much of this range. Two subspecies are recognized: the Asian population, race *asiaticus*, is declining and may not exceed 1,000 individuals, and India, particularly the south-western Gangetic plain adjoining the Delhi area, may presently hold the largest breeding numbers of these magnificent storks in Asia.

Ganguli, while writing of the 1960s, had considered these storks 'fairly common' with occasional reports of 'more than 40 birds' in summer on the Najafgarh lakes. Their decline, since then, is strikingly evident and only single-digit numbers are now reported from the larger wetland complexes in both Haryana and UP, with occasional records on the Yamuna River in Delhi, mainly from Okhla and the Ganga (at the Hastinapur Wildlife Sanctuary). These storks breed after the monsoon, October to December, with at least four or five known nesting sites in the area.

Greater Adjutant *Leptoptilos dubius*

Former Non-breeding Summer Visitor · No recent records

Globally endangered with a breeding range now limited to a few localities in Bihar and Assam and a small population in Cambodia. This population collapse is reflected in their status in the Delhi area as well. Monotypic.

Greater Adjutants were historically considered irregular visitors, in small numbers, to lakes and rubbish dumps, but Ganguli had recorded flocks of up to 150–250 birds – unimaginable today – between May and August in the 1960s at the Najafgarh lakes.[119] These were presumed migratory gatherings. The only later records are all from the 1980s: one by the Yamuna River in July 1981, five at Sonipat in July 1982, followed by three records in Okhla – five in July 1985, three in July 1986 and five again in August 1989.[120]

Painted Stork *Mycteria leucocephala*

Breeding Resident · Common

Bulk of the global population found mainly in the Indian subcontinent, holds the bulk of the global population of these colourful storks, though their range extends east to Cambodia. Monotypic.

Painted Storks are still common in our wetlands, nesting in the heronries at the Delhi Zoo and Sultanpur National Park between August and December. The Delhi Zoo colony has been in existence since 1960, and while numbers do fluctuate year to year, with a maxima of about 300 nests, they may be less affected by water levels and other factors as the zoo's pools are artificially stocked and food shortage should not be an inhibiting factor. The Sultanpur heronry was established in the early 1990s after the creation of islands in the lake planted with acacia as part of the sanctuary's management plan in the 1980s, changing the character of the lake from a seasonal saline inundation to a wooded wetland. Nesting has been annual since the year 2000, with an estimated maximum of 150 pairs.[121] After breeding, the storks disperse to wetlands in the area. Okhla was a regular foraging site for the Delhi Zoo colony, with up to 200 birds recorded from there in the 1980s, but it is too polluted now to attract such numbers.

Painted Stork

SULIFORMES: DARTERS, CORMORANTS, ETC.

ANHINGIDAE: DARTERS

This unique family is represented in the Delhi area by one species, the Oriental representative of four in the genus *Anhinga* that, between them, inhabit the tropical and subtropical regions of the world. Like cormorants, darters hunt underwater but lack waterproof plumage; with their hunger satiated, they need to perch with their wings open to dry. They are colloquially called 'snakebirds' because of their habit of swimming with just their thin bills and snake-like necks showing above the surface. Their legs are short, all four toes connected with webs for better propulsion as they hunt fish underwater, harpooning them with their sharp bills. Surfacing, they toss the fish into the air, catch and swallow it head first.

India: 1 species, Delhi: 1

◄ *Oriental Darter*

Oriental Darter *Anhinga melanogaster*

Breeding Resident · Fairly Common

Widely distributed across the Oriental Region. Monotypic.

Darters are fairly common residents in the Delhi area, though numbers may have declined as Ganguli had considered them 'common' some five or six decades ago, 'around any large sheet of water deep enough for it to swim and dive in for fishing'. More individualistic than cormorants, they are usually seen singly or in single-digit numbers about Delhi except when nesting. They are colonial breeders, either in small groups of their own species as at the Yamuna Biodiversity Park in north Delhi and the Okhla Bird Sanctuary, or in the company of other large waterbirds at the Surajpur lake, the Sultanpur National Park, the Najafgarh drain heronry, the Bhindawas Bird Sanctuary, the Hastinapur Wildlife Sanctuary, along the Yamuna canal systems in Haryana and probably elsewhere as well.

SULIFORMES: DARTERS, CORMORANTS, ETC.

PHALACROCORACIDAE: CORMORANTS

This widespread family is represented in the Delhi area by three species, two that are largely restricted to southern Asia and one well known and widely distributed species. They are fish eaters that dive to hunt underwater, with longish hooked beaks and short legs with all four toes webbed, enabling them to swim and chase prey underwater easily. Their plumage is not waterproof, and when resting on a perch or a sandbar after feeding, cormorants often hold their wings open to dry.

All three cormorants on the Delhi list are colonial breeders, building pads of sticks and reeds that serve as their nests, commonly in heronries with other waterbirds, in trees over water, bushes or – less often in the Delhi area – reed beds.

India: 3 species, Delhi: 3

▲ *A group of Great Cormorants, both adults and juveniles*

Little Cormorant *Microcarbo niger*

Breeding Resident · Common

Resident across the Oriental Region. Monotypic.

These common Asian cormorants are widespread about Delhi and may be found easily in any wetland area, large or small, and even in artificial waterbodies in city gardens. They occupy their own ecological niche, keeping singly or in small parties, and fishing alone and not in coordinated groups as Indian or Great Cormorants often do. They breed during the monsoon, from June to September, often in small colonies with Cattle Egrets *Bubulcus ibis* and Pond Herons *Ardeola grayii* beside temple ponds and the like, or in the large mixed heronries at the Delhi Zoo and the larger lakes.

Great Cormorant *Phalacrocorax carbo*

Breeding Resident and Winter Visitor · Common

The most widespread species of the family, distributed across the Old World and Australasia. Largely resident, but Central Asian populations winter to southern Asia. Six subspecies are recognized, and Indian birds, both residents and winter visitors, belong to the Asian race *sinensis* in which the head and neck develop extensive white plumes in breeding plumage and often appear solid white.

A resident breeding population of these impressive cormorants in the Delhi area is augmented in winter by migrants from beyond our borders. The cormorants disperse in small parties to feed during the day but gather in numbers to roost; large evening flights have been recorded up and down the river, and roosts of over 200 birds have been reported on pylons at Okhla and above Wazirabad in late winter and spring. They breed locally in heronries, as at the Bhindawas Lake in Haryana, between September and December, later than their smaller congeners. Many are in their distinctive breeding plumage by April–May, and these may be birds preparing to emigrate, since the nesting season in Delhi does not begin until after the monsoon.

Indian Cormorant

Phalacrocorax fuscicollis

Breeding Resident

Fairly Common

Resident from the subcontinent eastwards to Cambodia. Monotypic.

Indian Cormorants do not seem to have been recorded from the Delhi area prior to 1950, and Ganguli has recorded that they were uncommon even in the 1960s, though they had begun breeding at the Delhi Zoo heronry by then. They are now much more widespread than these accounts suggest, and they can be seen in fair-sized groups of 20–100 birds in the larger wetlands, which often fish cooperatively in flocks like the pelicans. They breed during the monsoon between July and October in large heronries and with cattle and other egrets in smaller colonies as well.

Indian Cormorant, non-breeding plumage

PELECANIFORMES: PELICANS, HERONS AND IBISES

PELECANIDAE: PELICANS

Two species of pelicans visit the Delhi area in winter from their Central Asian breeding grounds, while a third, the Spot-billed Pelican *Pelecanus philippinus*, a regional endemic that breeds in eastern and southern India, has been recorded historically. Pelicans feed cooperatively and nest gregariously on the ground or on trees, often as part of large, mixed heronries, and their large wings enable flocks to soar effortlessly on thermal currents with synchronized movements. They swim buoyantly, with their short legs and all four toes connected by webs, and use their long, flexible mandibles with expandable, bare, gular pouches as nets to capture fish.

A few free-flying pairs of the Great White Pelican *P. onocrotalus* breed in the Delhi Zoo, a pile of twigs and sticks on the bunds surrounding the ponds serving as a nest. This and the Dalmatian Pelican *P. crispus* are ground nesting species, often in colonies in reed beds in their native breeding grounds, unlike the Spot-billed Pelican *P. philippinus* which nests in trees.

India: 3 species, Delhi: 2

▲ *The Great White Pelican develops a typical rosy tint by March.*

Great White Pelican *Pelecanus onocrotalus*

Winter Visitor · Fairly Common

Great White, or Rosy Pelicans, breed discontinuously in southeastern Europe and Central Asia, also locally in sub-Saharan Africa, wintering south to Africa and the Middle East; eastern populations winter in the northern subcontinent. Monotypic.

These impressive birds are regular visitors, typically in small flocks of less than 50 or 60 birds now, to the larger wetlands in the Delhi area. Historically, pelicans, not specifically identified, were reported off and on from the Yamuna, but large wintering flocks, occasionally of as many as 2,000 birds, began to be seen from the 1960s at the Najafgarh and Sultanpur lakes. These spectacular winter flocks continued through the 1980s,[122] though numbers tended to fluctuate year to year. A peak count of about 1,000 at Sultanpur in the late 1970s, for example, dropped to less than 100 in the following years, then rose again to 125–200 by the late 1980s, possibly reflecting abundance of some specific food item. Only smaller numbers have been reported from the Sultanpur neighbourhood and various other wetlands in the area in recent decades, with a maximum of 100–120 birds in late winter, in January/February. The ecological character of the Sultanpur lake itself has changed over the years, and these pelicans, formerly regular here, visit it only occasionally now. However, flocks of up to 200–800 pelicans have been reported on eBird checklists from just beyond the southern borders of the Delhi area, for example, at the Soor Sarovar Bird Sanctuary near Mathura in UP, and so larger numbers could conceivably occur within our limits as well. A free-flying colony established in the 1980s at the Delhi Zoo subsists on fish supplied by the Zoo authorities and breeds regularly, its numbers hovering at 30–35 birds. These birds often visit the impoundment at Okhla as well.

Dalmatian Pelican in breeding plumage at Nalsarovar, Gujarat

Dalmatian Pelican *Pelecanus crispus*

Winter Visitor · Rare

A southeast European and Central Asian nesting species, wintering from the eastern Mediterranean region and the Middle East to India. Monotypic.

Dalmatian Pelicans used to visit Delhi's wetlands fairly regularly in past decades, but in far smaller numbers than the Great White. There are historical records, from the 1960s, when they were considered rare. Singles or small parties were reported from Sultanpur and Okhla in the late 1970s and 1980s, with a maximum of 17 birds in March 1979 feeding on fish trapped at the sluice gates of the old Okhla weir on the Yamuna; three separate groups, totalling 15, were counted near Sultanpur in March 1986. Three were reported again at Okhla in late January 1990.[123] There have been very few reports since then, usually of single birds with Great White Pelicans.

PELECANIFORMES: PELICANS, HERONS AND IBISES
ARDEIDAE: BITTERNS AND HERONS

Bitterns, a group of small- to mid-sized, secretive herons represent a distinct evolutionary clade within the family, subfamily Botaurinae. The Delhi area hosts one widely distributed migrant bittern and four smaller species that nest here. The remaining herons form another clade, subfamily Ardeinae, and are either widely distributed species themselves or have close relatives that range across all continents.

Most herons on the Delhi list require wetlands for foraging and nesting. Their carnivorous diet includes varied aquatic life, vertebrates such as fish, amphibians, reptiles, etc., and various invertebrates, hunted either actively in the marshes or through a wait-and-watch technique; some species are even known to lure prey within reach with bits of food. Cattle Egrets *Bubulcus ibis* follow large grazing mammals that flush insect prey from the grass. Most herons are colonial nesters, building large platforms of sticks in trees, usually in the open and often mixed with other large waterbirds such as storks, ibises or cormorants. Bitterns nest solitarily, in the safety of cover in dense reed beds. Clutches of three to five eggs appear to be the norm amongst the Indian species.

Herons in the Delhi region seem to be maintaining their populations satisfactorily, unlike most storks whose numbers have declined.

India: 23 species, Delhi: 15

▲ *Purple Heron, an immature bird*

Eurasian Bittern *Botaurus stellaris*

Winter Visitor · Uncommon

Breeds in a broad belt across Eurasia from Britain to Japan, wintering in southern Europe, sub-Saharan Africa and southern Asia. Two subspecies are recognized: nominate *stellaris* is widespread, while a second race, *capensis*, is resident in South Africa.

This Old World bittern is a scarce winter visitor to the Delhi area. The species is not listed in Ganguli's *Guide to the Birds of the Delhi Area*, published in 1975, but there are a growing number of records in recent decades, between November to as late as May, from various wetlands, typically mosaics of reed beds and flooded channels with floating vegetation. Most reports are of single birds but up to seven have been recorded at a site. A couple of records in midsummer in July, from Najafgarh in 2013 and the Yamuna *khadar* in 2018,[124] may be early returning migrants or over-summering birds – the latter was actually heard 'booming', its advertising call during the breeding season.

Yellow Bittern *Ixobrychus sinensis*

Mainly Summer Breeding Visitor · Fairly Common

Widespread and largely resident from the subcontinent to China, Japan and Southeast Asia, but with many populations in colder regions migrating southwards in winter. Monotypic.

Yellow Bitterns are breeding visitors to the area between May and September, though some linger through mild winters. They were considered uncommon till the 1980s, when the impoundment of the (then) newly constructed barrage below Okhla created the extensive reed beds that are now a stronghold of this species in the Yamuna floodplain. Yellow Bitterns are also recorded regularly from several other wetlands about Delhi and have become more common and widely distributed in the area in recent decades. In May, soon after they arrive, pairs may be seen in slow, flapping courtship flights over the reed beds and are aerially active through much of the day in the monsoon when feeding young. The *Atlas* records that 98 birds were counted at Okhla, flying to and fro over three hours of observation in July 2003, a remarkable figure for a single site.

A Yellow Bittern in flight

Little Bittern *Ixobrychus minutus*

Summer Breeding Visitor · Rare, Sporadic

A male Little Bittern at Mandothi

Widespread, a summer nesting visitor in Europe and Central Asia, southwards to Kashmir where they breed commonly in the valley lakes. Adventitious breeding of Little Bitterns south of Kashmir in the subcontinent has been recorded from lower Sindh, Pakistan and, on the last decade, from Gujarat.[125] Based on earlier records of breeding in Delhi and their experience in Gujarat, these observers conclude that 'just six records of the species in Gujarat in past eighteen years, with a single instance of breeding (in 2017), suggests that it may be nomadic and breeding occasionally at suitable habitats ... quite far from its known breeding range in Baluchistan or Kashmir'. The entire population is believed to winter in Africa. Three subspecies are recognized: nominate *minutus* in the Palaearctic and two others in sub-Saharan Africa and Madagascar.

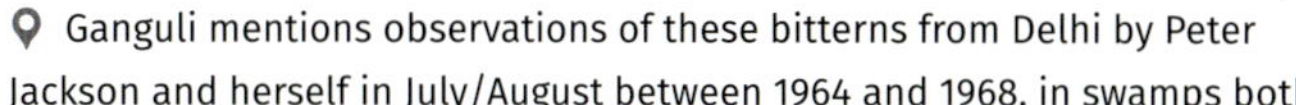

Ganguli mentions observations of these bitterns from Delhi by Peter Jackson and herself in July/August between 1964 and 1968, in swamps both south of Okhla and in north Delhi, with the birds nesting at both these sites. After a long hiatus, one was reported at Badli in north Delhi in 1988. After another, longer gap of over 30 years, two males and a female were recorded at Okhla on 26 June 2021, but could not be located at the site thereafter despite an intensive search but a number of individuals – males, females and flying juveniles as well – were recorded later in the same season at the Mandothi wetlands, Jhajjar district, from late August through September 2021.[126] At least one pair, the male in peak breeding dress with a bright red flush to the bill, was recorded at the same site the following year as well, from end July through at least September 2022. The records of the 1960s from the Yamuna floodplains and now from Mandothi in 2021 and 2022 suggest episodic breeding in the Delhi area, perhaps temporary colonies established at specific local sites where some undefined factor may be conducive to breeding in certain years.

Cinnamon Bittern *Ixobrychus cinnamomeus*

Mainly Summer Breeding Visitor · Fairly Common

Widespread resident with local movements across the Oriental region from the Indian subcontinent to China and Southeast Asia, with populations in colder regions moving further south in the cold season. Monotypic.

These bright little bitterns are summer breeding visitors to the Delhi area, appearing from early May to September, with very few records in winter. They are widespread in the Delhi area, common at favoured sites, such as Okhla, but may nest even in small reedy marshes in the midst of cultivation. Normally crepuscular like Yellow Bitterns, they have similar courtship flights early in the breeding season; pairs fly together over the reeds in the evenings with slow deep wingbeats, with the male often calling in a long series of low hoots in flight. Yellow Bitterns, on the other hand, are usually silent. Cinnamon Bitterns may be active all day during the monsoon – with a record at Okhla of 40 counted in three hours in July – and they are often in the air along with Yellow and a few Black Bitterns. Fledged juveniles appear by late August. They were believed more numerous than Yellow Bitterns till the 1990s, but the position may have reversed.

A Black Bittern with prey

Black Bittern *Ixobrychus flavicollis*

Mainly Summer Breeding Visitor

Uncommon

Largely resident, with local movements, from the subcontinent to China, Southeast Asia and Australasia. Black Bitterns are placed by some authorities in the monospecific genus *Dupetor*; three subspecies are recognized: the nominate in mainland Asia and two others extend the species to Australasia.

A breeding summer visitor from early May to October, they appear in smaller numbers compared to the other small bitterns, with occasional records in winter as well. Black Bitterns were first recorded in May 1962 from the marshes along the Agra canal south of Okhla, and this remained the only record from the Delhi area till the 1980s when they began to be recorded in increasing numbers every year from the Okhla reed beds. They have since been reported much more widely from the larger, well-vegetated wetlands across the area. Normally crepuscular, they do occasionally expose themselves – possibly more often than their smaller cousins – to feed in shallow open water, but never far from the sheltering reeds.

Grey Heron

Grey Heron *Ardea cinerea*

Breeding Resident Fairly Common

Very widely distributed across the Palaearctic, Oriental and African regions. Four subspecies are recognized, with nominate *cinerea* breeding over most of the Palaearctic, Africa and South Asia. The northern Asian populations are migratory, wintering in the Middle East and southern Asia.

Fairly common and widespread at all the larger wetlands of the area, they affect open waters including the sandbars and braided channels of the Yamuna and Ganga Rivers. These herons usually keep singly or in small numbers when hunting, but sometimes in loose parties of 20 or more, which gather, when satiated, to loaf on sandbanks, tall trees, or even pylons. Grey Herons breed locally in trees, up to 10–20 pairs in the mixed heronries in the Delhi area, including the Delhi Zoo, or sometimes in single species groups as at the Bhindawas Bird Sanctuary, where they nest in bushes amongst tall grass.

Purple Heron *Ardea purpurea*

Breeding Resident · Common

These lanky, colourful herons have a largely overlapping, but more southerly, distribution than the Grey. Palaearctic populations move south to winter in Africa and southwest Asia. Four subspecies are recognized, of which the species is represented in India by the Asian race *manilensis*.

Their numbers appear to have increased considerably in the Delhi area over the last several decades. Ganguli considered them only 'fairly common' in the 1960s, though more common than the Grey Heron even then. Today, they are found in most wetlands in the area with bordering reed beds and vegetation-choked waters. They nest freely in such habitats in large, dispersed colonies of their own species, between May and September, with just fledged young observed in nests in July and August. Nesting in reed beds is the norm in the area, but unusually, some 25 pairs have nested on thorn trees in the main heronry at the Bhindawas Bird Sanctuary.[127]

Great Egret *Ardea alba*

Breeding Resident · Common

A cosmopolitan heron, whose four subspecies range from central and southeastern Europe across Eurasia (nominate race *alba*), across southern Asia and Australia (race *modesta*), sub-Saharan Africa (race *melanorhynchos*) and the Americas (race *egretta*). Birds breeding in higher latitudes move to the south of the breeding range in winter.

A common resident of wetlands in the area, these stately herons have maintained their status and abundance in the Delhi area over past decades. Typically, only a few birds are present at a site, but up to 40 have been recorded huddled with other herons and storks, hunting aquatic life stranded in a desiccating pond. Great Egrets breed in all the large heronries in the Delhi area, where they make a striking picture in the nesting season with their diaphanous plumes, that are dramatically erected during courtship displays, their black bills, turquoise green lores and red tibia.

Great Egret in non-breeding plumage

Intermediate Egret *Ardea intermedia*

Breeding Resident · Fairly Common

Widespread resident from sub-Saharan Africa across South and Southeast Asia to Australia. Three distinct subspecies are recognized, with nominate *intermedia* in Asia. Some authorities accord species rank to the populations in Africa *brachyrhyncha* and Australia *plumifera*, which have differing bare part colours.

Widespread about Delhi, Intermediate Egrets have maintained their numbers over the past decades and are common in most waterbodies and wet cultivation in the area. They breed colonially in mixed species heronries, including at the Delhi Zoo. Breeding birds have black bills, bright yellow lores and very long and graceful breast and dorsal plumes.

Great, Intermediate and Little Egrets in breeding colours

Little Egret *Egretta garzetta*

Breeding Resident · Common

Widespread resident across the Palaearctic, Oriental, African and Australasian regions. Three subspecies are recognized, with nominate *garzetta* in India.

These elegant egrets are generally common at wetlands around Delhi in parties and small flocks that mix readily with other egrets and large wading birds of their habitat. They breed widely across the area, both in small colonies with Cattle Egrets, Pond and Night Herons, and in large, mixed heronries. Breeding birds look particularly lovely with their crest plumes and ornamental, diaphanous feathers on their breasts and backs, while their lores, which are usually yellowish, turn pink-purple in peak breeding condition. An egret, showing the structural characters and bare parts colours of a Little Egret but in a patchy lavender grey plumage phase has been documented at least once in the area at the Bhindawas Wildlife Sanctuary in late 2016, early 2017.[128] Koparde & Yesou[129] argue that inland records of putative 'dark phase' Little Egrets in South Asia show mixed characters and represent hybrids with vagrant individuals of the closely related Western Reef Egret.

Western Reef Egret *Egretta gularis*

Vagrant

Western Reef Egrets, resident herons of maritime habitats from West Africa to India, are prone to vagrancy with claimed records from far inland, both on the subcontinent and elsewhere. This species and Little Egrets have been considered conspecific in the past, although they differ in their ecological niches, and assortative breeding is the norm where ranges overlap. However, Koparde & Yesou (see above) point out that mixed pairs with *Egretta garzetta* are not unknown, especially inland where vagrant Western Reefs may interbreed with local *garzetta*. Two subspecies are recognized, of which *schistacea* represents the species in coastal East Africa, the Red Sea, the Persian Gulf and western India.

Western Reef Egret, at Pulicat lagoon, AP

There are two historical records of grey phase birds from Delhi in November 1963 and July 1969, and three white phase birds were seen in March 2003 at the Bhindawas Wildlife Sanctuary, with Little Egrets[130] enabling a comparative study of bare part colours and morphology. One in the grey morph was recorded in the Jhajjar district of Haryana in August–September 2014 and two near the Sultanpur National Park in mid-August 2021.

Cattle Egret *Bubulcus ibis*

Breeding Resident · Common

A cosmopolitan heron with two subspecies: the western (African) nominate *ibis* and eastern (Asian) *coromandus*. Some authorities give specific rank to the two, which differ especially in the extent and depth of the golden-orange colours of their breeding dress. A commensal relationship with humans and domestic cattle has helped them extend their ranges since the early twentieth century – the eastern birds to Australia and the western to the Americas – in one of the most extraordinary natural range expansions of any bird, at a speed and scale unmatched by any other avian species. Yet, Asian-African plumage differences have persisted; they have not invaded each other's ranges or interbred.

Cattle Egrets are ubiquitous, attending grazing cattle, feeding in rubbish dumps, frequenting rural markets for scraps, or hunting insects in irrigated fields. They roost communally in trees near villages or temples, where they are unmolested and breed in small colonies by themselves, along with Pond Herons and Little Egrets, or as part of the large heronries in the area.

Indian Pond Heron *Ardeola grayii*

Breeding Resident · Common

The subcontinental representative of a group of five small Old World herons of the genus *Ardeola*; also found westwards to the northern Persian Gulf and eastwards to Myanmar. Monotypic.

A familiar bird in ponds and marshes of all kinds, large or small, the Indian Pond Heron may be seen stalking about and hunting food even in the artificial waterbodies and irrigated lawns of the Central Vista and New Delhi gardens. Pond Herons breed on the margins of the larger heronries in the Delhi area, in small colonies of their own or mixed with egrets and Night Herons, often near villages.

Striated Heron *Butorides striata*

Breeding Resident · Scarce

Widely distributed in both fresh water and saltwater habitats, from South America, across Africa and southern Asia to Australasia. Up to 26 subspecies are recognized, of which race *chloriceps* represents this heron in the Indian mainland.

These small herons prefer the firmer banks of rivers, canals and irrigation impoundments in the area, especially where these are shaded by overhanging vegetation, to reed beds or squelchy ground. They are scarce but quite widely distributed about Delhi, with historical nesting records in July from the canals at Okhla, and more recently at the Delhi Zoo and the Sultanpur area.

Black-crowned Night Heron *Nycticorax nycticorax*

Breeding Resident · Fairly Common

The most widespread heron in the world, its global range spanning all continents except Australia. Four subspecies are recognized, with the Old World nominate race representing it in South Asia.

A fairly common breeding resident about Delhi. Breeding colonies, too, are well distributed, either of Night Herons alone or as part of the large mixed heronries as at the Delhi Zoo and doubtless elsewhere.

PELECANIFORMES: PELICANS, HERONS AND IBISES
THRESKIORNITHIDAE: IBISES AND SPOONBILLS

A cosmopolitan group of medium to large birds, formerly classified with storks in the Order Ciconiiformes, but phylogenetic studies demonstrate that ibises and spoonbills are evolutionarily closer to the herons and pelicans and are now placed in the Order Pelecaniformes.

Despite morphological differences between ibises and spoonbills, they are similar in many respects. Both are carnivorous with a range of prey items, including aquatic invertebrates, fish, amphibians, reptiles and even small mammals. Ibises probe the mud and water with their long, decurved bills, while spoonbills sweep their spatulate bills in the water from side to side to locate and snap up their food. Most species breed in colonies with other large wading birds, while others are solitary nesters. Nests are typically large platforms constructed of sticks, and clutches of three or four eggs appear to be the norm in the family.

India: 4 species, Delhi: 4

▲ *Eurasian Spoonbill taking flight*

Glossy Ibis *Plegadis falcinellus*

Non-breeding, Year-round Visitor · Fairly Common

Widespread, with localized populations in all continents. Migratory, nomadic and dispersive, with apparently random fluctuations in breeding populations. Monotypic.

Their status in the Delhi area appears to have fluctuated over the years, and Glossy Ibises have become visibly more common in recent decades. Ganguli considered them irregular visitors in the 1960s with maxima of 150 in January and 100 in May. There were few reports of this ibis in subsequent years till the 1980s, when numbers began to build up again. They are now regular at most wetlands through the year, but there is no confirmation of breeding about Delhi. Their numbers increase from September to November when counts of 120–150 birds at wetlands in Haryana may represent post-breeding local dispersal, or winter arrivals of migrants from beyond our borders. The numbers increase again from March but not to such levels.

Black-headed Ibis *Threskiornis melanocephalus*

Breeding Resident · Fairly Common

The Asian representative of a group of four very closely related ibis species that extend from Africa to Australia. Ranges across much of South Asia and very locally eastwards into Southeast Asia. Monotypic.

A common breeding resident, the Black-headed Ibis has, by and large, maintained this status in the area over the past decades. They remain widespread, but numbers rarely exceed 10–25 at a site; post-breeding flocks may be larger, and 75–100 birds may be counted between September and November at the Sultanpur–Basai–Najafgarh wetlands complex. Nesting colonies often also tend to be small, with 20–30 pairs in the large heronries at the Delhi Zoo; Surajpur, UP; Sultanpur, Najafgarh and Bhindawas, Haryana, and smaller ones at favoured village ponds elsewhere.

Red-naped Ibis *Pseudibis papillosa*

Breeding Resident · Fairly Common

Also called Indian Black Ibis and is a subcontinental endemic. This is one of the ibises that have shifted to a mostly terrestrial existence away from water. Monotypic.

These are common breeding residents of the Delhi area. Singles, pairs or parties of up to 10–15 may be seen in fallow cultivation, village precincts and damp ground, often in the vicinity of wetlands though they do not actually feed in water. In recent years, they seem to have adapted well to human-modified habitats, resting or feeding on lawns in central Delhi, roosting on the bare branches of tall trees in larger public gardens, even perching on roadside electricity poles, and almost certainly nesting within city precincts. They breed from April to August in trees, appropriating the old nests of kites or constructing one themselves, and increasingly on electricity pylons too.

Red-naped Ibis

Eurasian Spoonbill *Platalea leucorodia*

Non-breeding, Year-round Visitor · Fairly Common

Distributed as a nesting species from southern Europe across to Central and East Asia and south to Arabia, India and Sri Lanka, subject to migratory movements. Three subspecies are recognized; birds of the race *leucorodia* of Eurasia are both breeding residents in India and winter visitors from Central Asia.

Spoonbills are regularly recorded all through the year from the Delhi area, but there is no evidence of nesting to date. Larger wetlands usually hold up to 20–25 birds, but the numbers tend to increase between December and May–June, when flocks of up to 200 have been recorded at *jheels* such as at Sultanpur, Najafgarh and Bhindawas. Such large assemblages have been noted from as far back as the 1960s[131] and may represent nomadic movements of birds that have bred locally within subcontinental limits or winter migrants from Central Asia. Post-monsoon flocks include many immatures, while by April, most are in breeding plumage with well-developed crests and yellow-gold chest bands.

ACCIPITRIFORMES: BIRDS OF PREY

PANDIONIDAE: OSPREYS

The Order Accipitriformes is segregated into three families, of which two concern us: Pandionidae, Ospreys, with one cosmopolitan species, and Accipitridae, that includes all our other birds of prey. Ospreys are in their own family because of their phylogenetic distance from other raptors and their structural adaptations to a piscivorous lifestyle – a reversible outer toe, soles with prickly projecting scales or spicules and strong claws to enable a firmer hold on slippery prey.

Ospreys characteristically hunt by hovering over the waterspread, searching intently for swimming prey. On spying a fish, the bird plunges downwards with wings closed and legs extended, often immersing itself in the water with a splash, then emerging with strong flaps of its wings carrying the prey in its talons, to be consumed at some favoured perch.

India: 1 species, Delhi: 1

◄ *An Osprey with fish*

Osprey *Pandion haliaetus*

Winter Visitor · Scarce

Nest widely in North America and across the Palaearctic from temperate Europe and northern Asia, locally south to Arabia and Australia. Four subspecies are recognized; the nominate Palaearctic race *haliaetus* is migratory, wintering widely in Africa and South and Southeast Asia.

Ospreys are winter visitors in small numbers to the rivers and large *jheels* of the area, mainly from October to March. They have rarely been reported from the Yamuna River itself in recent decades other than for an occasional passage bird at the cleaner stretches above Wazirabad and at Okhla, but they are more frequently reported on the Ganga. Most records are of single birds in winter from the larger lakes of the area, but up to five birds have been seen together on rare occasions, which are probably loose migratory groups.

ACCIPITRIFORMES: BIRDS OF PREY

ACCIPITRIDAE: KITES, HAWKS AND EAGLES

A cosmopolitan family, raptors as a group are unmistakable with their piercing eyes, strongly hooked bills and powerful talons. Yet, they are highly diversified in size, morphology, flight styles and hunting techniques and live in varied habitats, from the arctic to the equatorial and in tundra, grasslands, deserts and forests. Many are long-distance diurnal migrants, often in spectacular numbers.

In many birds of prey, females are larger than the males, often markedly so, and both sexes share parental duties, often with the female incubating the eggs while the male brings her food. Nests are typically platforms of sticks with a central depression lined with finer material. They are built on trees or ledges of cliffs, sometimes even artificial structures including old forts and monuments, pylons and the like. Traditional sites are often reused year after year by the same pair, and with new material added each season, their nests may develop to enormous sizes. Some species alternate between two preferred sites every other year. Clutch sizes range from one to four, rarely five, eggs, and the larger species, such as vultures and eagles, usually lay just one or two eggs. Chicks hatch asynchronously, and brood reduction through fratricide is widespread .

The systematics of the Accipitridae have been variously treated by taxonomists, but for the purposes of this overview, the family may be considered to comprise, broadly, three lineages that are often designated as subfamilies: the elanine kites in the subfamily Elaninae, the 'African vultures' (including *Gypaetus*, *Neophron* and the pernine kites, *Pernis*) in the Gypaetinae and all the remaining members of the family in the subfamily Accipitrinae.

Birds of prey have fascinated and excited the imagination of human beings throughout history. They have been storied and symbolized by many cultures, trained to hunt wild quarry for food and sport since ancient times. But as apex predators, they need space and territory. They are very vulnerable to the pollutants and poisons that we spread over our land. Delhi boasts a rich and diversified assemblage of these lords of the air, but here too, as elsewhere, they are under severe stress, and it is incumbent on us to do what we can to look after these fabulous creatures.

India: 61 species, Delhi: 36

▲ *A handsome male Marsh Harrier*

Black-winged Kite *Elanus caeruleus*

Breeding Resident · Fairly Common

Widespread from southwestern Europe, Africa and across South and Southeast Asia. Four subspecies are recognized, of which *vociferus* inhabits much of mainland southern Asia.

This graceful species is the only Indian representative of the small clade of elanine kites. It is a common breeding resident, but with some local movements, in all types of lightly wooded country and cultivation. The open scrub landscapes of the Aravallis are a favoured habitat. It appears to breed mainly in the late summer and monsoon in the area, building a small nest of sticks in the upper branches of a small tree, as juveniles of the year are regularly seen in winter. It has maintained its status and numbers in the Delhi area over the last several decades, wherever spreading urbanization has not destroyed its habitat.

Egyptian Vulture *Neophron percnopterus*

Breeding Resident and Winter Visitor · Now Uncommon

Widely distributed from southern Europe and Africa to India; a representative of the African Vultures in the Delhi area (DNA studies indicate that these form a separate clade from the Old World Vultures). Two subspecies occur in India: the widespread nominate of Eurasia and Africa and the subcontinental endemic *ginginianus*.

Egyptian Vultures of the nominate race *percnopterus*, with dark beaks, are winter visitors to the area, presumably from the Central Asian breeding population, while those of the Indian race *ginginianus*, with yellow bills, are resident and breed between January and April on old monuments and ruins such as Tughlaqabad, rocky cliffs in the Aravalli landscape, and occasionally on trees. Outside the breeding season, they collect at communal roosts on trees or pylons from where birds may be seen flying out, one by one, to refuse dumps and other feeding grounds every morning. They inhabit open country, the neighbourhoods of villages, tanneries and slaughterhouses and are regular visitors to municipal landfills along with the multitudes of Black Kites. Described as 'common' in the past, a decline in their numbers was becoming evident by the late 1990s, and the *Atlas*, in 2006, referred to them as 'a widespread breeding resident in small numbers'. It is unclear whether they show the same physiological intolerance to the drug Diclofenac as the *Gyps* vultures, but recent evidence suggests that they may – estimates showed an 80% decline in their numbers between 1991 and 2003 and also that the decline slowed and even reversed marginally as the ban on veterinary Diclofenac took effect, thus pointing to a direct linkage.[133] Other likely factors responsible include pesticide poisoning and reduction in availability of domestic cattle carcasses. Nevertheless, the species has fared better than the *Gyps* species.

Egyptian Vulture, an adult individual

Oriental Honey Buzzard

Pernis ptilorhynchus

Breeding Resident · Fairly Common

Oriental Honey Buzzard, female

Distributed across Asia in two disjunct breeding populations: a northern population in Siberia and northeast Asia, which is migratory and winters in southern Asia, and a southern group of races resident in South and Southeast Asia. Six subspecies are recognized: *orientalis* in North Asia and five others in the southern part of their range. Two occur in India: *orientalis* as winter visitor and the Indian subspecies *ruficollis* as resident.

Honey Buzzards are widespread residents about Delhi, breeding freely in urban parks and gardens, in woodland patches on the Ridge and in the countryside. In the breeding season, in March–April, males can be vocal and may display in undulating flights high in the air, characteristically raising and fluttering their wings at the peak of each ascent. They frequent areas where the large combs of the rock bee *Apis dorsata* are found under the great arches of monuments such as Humayun's Tomb, or on large trees. Ganguli referred to them 'not too common' in the 1960s, though today they are one of the more frequently observed raptors in urban areas after the ubiquitous Black Kite *Milvus migrans* and the Shikra *Accipiter badius*.

Vultures on the Brink

The once-thriving populations of the Old World Vultures have been decimated by a veterinary drug, Diclofenac, which causes renal failure in the birds if they ingest it even in small quantities from domestic cattle carcasses. The catastrophic decline in the populations of three of India's commonest vultures – including two on the Delhi list: the White-rumped and the Indian – by 99% in just a decade and a half, between 1990 and 2005, is perhaps the most rapid and precipitous collapse of any bird species in recorded history and has been a huge environmental disaster. Vultures are long-lived, but slow breeders; they mature when they are five years old and lay just one egg per year. They are all now ranked Critically Endangered.

With growing public awareness of their ecological value, government support and a ban on veterinary Diclofenac imposed since 2006, there is some evidence that the declines may be slowing, and remnant populations are stabilizing at very low levels as the ban takes effect. Under a joint initiative by the BNHS and the Central and state governments, nationwide surveys of vulture populations are being held every four years, the last in 2022 (delayed because of the pandemic), for which the data is being analysed and is yet to be published.

The most recent assessment, published in December 2017 following the 2015 survey, suggests – to quote from a study by Dr Vibhu Prakash Mathur, Deputy Director of the BNHS and head of the vulture conservation programme – that the previously rapid decline of the White-rumped Vulture may have reversed since the ban, but 'only a few thousand of this species, possibly up to the low tens of thousands, remained in India in 2015'. The Indian Vulture continues to decline, 'though probably at a much slower rate than in the 1990s. This remains the most numerous ... with about 12,000 individuals in 2015 ...' It is just permissible to hope that these long-lived, but slow-breeding, birds that were such a feature of our skies may slowly recover.

Red headed Vulture, over the Southern Ridge

Red-headed Vulture *Sarcogyps calvus*

Former? Breeding Resident

Formerly Fairly Common, now Rare

Naturally sparsely distributed, not commensal with humans like the other dominant vulture species and may have retreated because of competitive exclusion from carcasses as the latter proliferated beyond a point. As a result, Red-headed Vultures suffered severe declines over their entire range from India to Cambodia much before the collapse of vulture populations in South Asia. It is unclear whether they are as intolerant to Diclofenac as the *Gyps* vultures, but evidence suggests that they may be: they suffered a 91% decline between 1991 and 2003, but then this appears to have slowed or even marginally reversed as the ban on veterinary Diclofenac took effect, which suggests a linkage.[135] Monotypic.

Red-headed Vultures nest on trees and have historically been recorded breeding between January and April even within New Delhi city limits.[136] Ganguli described them as 'fairly common' in the 1960s, but these vultures seem to have suffered a decline soon thereafter as there have been few records since then. Since December 2010, however, after nearly two decades, fairly regular but rare sightings of singles or two birds together have been reported from Southern Ridge habitats over most months of the year, and a bird on a low tree nest has been seen in the Asola–Bhatti Wildlife Sanctuary but with no other sign of breeding activity.[137] There are a few records from the Sultanpur National Park as well, in October.

White-rumped Vulture *Gyps bengalensis*

Former Breeding Resident Formerly Common, now Rare

Formerly ranged widely across southern Asia from Iran to Cambodia and largely commensal with humans everywhere in the Indian subcontinent. Even before the Diclofenac crisis, they had been all but extirpated from Southeast Asia because of persecution, pesticides and changed animal husbandry practices. With a countrywide decline of over 99.9% between 1992 and 2007, the White-rumped has been the most seriously impacted of the Indian vultures, and 'only a few thousand of this species, possibly up to the low tens of thousands, remained in India in 2015'.[138] Monotypic.

White-rumped Vultures had proliferated to unnaturally high numbers about Delhi and their nests were a common sight during the winter in trees along roads and parks in New Delhi and the surrounding countryside, with a nesting density as high as 2.7 nests/sq km within the city itself.[139] Colonial roosts on trees, or on domes of tombs, held over 100 birds at times. This situation continued till the early 1990s, when population declines began to be reported, with increasing alarm, through the second half of that decade in the Delhi area, as indeed over the rest of the subcontinent as well. Small numbers still bred within New Delhi city limits till 2001 but appear to have been extirpated in the Delhi area by 2005. The only records since then are an adult photographed on 23 March 2014 at Khidwali near Rohtak, and an immature *Gyps* at the Asola Wildlife Sanctuary in early January 2017 was identified as of this species.

Indian Vulture *Gyps indicus*

Non-breeding Visitor · Now Rare

Cliff nesting vultures, unlike the White-rumped which nests on trees, never as common about Delhi as the latter. This species is endemic to peninsular India with a breeding range that extends northwards along the Aravallis at least as far as Sariska and Alwar. Vultures range far in search of food and can appear in areas distant from their breeding colonies. Monotypic.

This vulture had not been recorded from the Delhi area till the early 1960s, possibly overlooked, and only rarely in the following years. There were reports, however, in the 1980s from the Southern Ridge around the Sohna and Damdama lakes, and three were recorded at Sultanpur in early 1998.[140] Indian Vulture populations were decimated by the Diclofenac crisis, dropping by 97% between 1990 and 2005 when the veterinary drug had its most disastrous impact, and recent assessments suggest it continues to decline, even after the ban on veterinary Diclofenac, though 'probably' at a much slower rate than over the decade before the ban was imposed.[141] The only recent records from the Delhi area are one at Sonipat on 23 May 2009[142] and another from the Sultanpur National Park neighbourhood on 4 May 2019.

Himalayan Vulture

Gyps himalayensis

Winter Visitor · Rare

A vulture of the higher Himalayas and Tibet, north to the Pamirs and central China, a few juveniles of which have taken to straggling southwards across the northern Indian plains in winter (especially towards Rajasthan and Assam, but occasionally reaching southern India) since the last couple of decades – possibly because of reduced competition after the collapse of resident vulture populations. Monotypic.

Himalayan Vulture, a juvenile from Sultanpur

Records from the Delhi area are all of juvenile birds. One at Okhla on 27 April 2014 was the first confirmed occurrence in the Delhi area. A group of 15 were observed roosting at the Aravalli Biodiversity Park, Gurugram, in mid-November 2019, but dispersed thereafter over the next couple of days.[143] There have been a few reports since from Southern Ridge landscapes, including from the Asola Wildlife Sanctuary in January 2020; from near Sohna, Haryana, in late May (an over-summering juvenile?); one over the JNU campus, south Delhi, in January 2021; and another at the Bhondsi Nature Park, Gurugram district, in early April 2022; with records from the GautamBuddha Nagar district in January and February 2022 as well. Diclofenac has been shown to be toxic to Himalayan Vultures[144] but which largely escaped the 1990s crisis because of their high mountain habitat, and hopefully the ban on veterinary Diclofenac will allow these vultures to return safely from their wintering sojourns in the plains.

Griffon Vulture *Gyps fulvus*

Winter Visitor · Rare

Range from southern Europe and North Africa to Central and South Asia, spreading in small numbers across northern and central India in winter. Two subspecies are recognized; the eastern race *fulvescens* that breeds in the northwestern parts of the subcontinent winters in India.

Scarce winter visitors, recorded from the Delhi area since the 1940s, but always considered uncommon. Later reports include six at Sultanpur in February 1977 and a few records of single birds from the 1980s, soaring or perched with other vultures thus enabling comparison of size and plumage; and a clutch of reports over the past decade, including a dead juvenile collected near Delhi Airport by Wildlife Rescue on 24 April 2014, a distant, 'possible Griffon' photographed at Okhla in mid-March 2016, one over the JNU campus, south Delhi, in early December 2019, another at the Dhanauri wetlands in mid-December 2019, two at Greater Noida in the first week of November 2020 and one over the Aravalli outspurs at Rewari, Haryana, on 1 May 2022.

Crested Serpent Eagle *Spilornis cheela*

Non-breeding? Visitor · Uncommon

Two genera of Serpent Eagles occur in India: the purely Asian, forest haunting *Spilornis* and the mainly African, open country *Circaetus*. Crested Serpent Eagles are widely distributed across South Asia eastwards to southern China and Southeast Asia. No less than 21 subspecies are recognized, many of these being populations isolated on forested small islands. Two races in India, *cheela* of northern India and *melanotis* in the peninsula.

These handsome raptors of well-wooded, well-watered country are uncommon in the Delhi area (race *cheela*) and may appear in any wooded patch, even in city parks or the Ridge, but have an affinity for wooded wetlands such as at Okhla, Sultanpur or Bhindawas, Haryana. Records from the area, usually of single birds, are scattered over the year, mainly between September and May, but a few in June and July as well. Juveniles have been recorded in April, May and September. Rishad Naoroji in his *Birds of Prey of the Indian Subcontinent* notes that they breed between late February and June–July, and while there is no report of nesting from Delhi's immediate neighbourhood, the pattern of records suggests dispersal from not-too-distant breeding areas, such as the Hastinapur Wildlife Sanctuary, where they are listed as breeding residents.

Crested Serpent Eagle, adult

Short-toed Snake Eagle *Circaetus gallicus*

Breeding Resident Uncommon

Short-toed Snake Eagle, in flight

Ranges from southern Europe, North Africa and the Mediterranean region, to Central Asia and India. Subcontinental populations are resident, while northern, European and Central Asian breeders are migratory, the latter wintering in South and Southeast Asia. Two subspecies are recognized, with only the nominate race occurring in India.

These impressive raptors of open habitats have bred near Delhi in the past, and Malcolm Macdonald's *Birds in My Indian Garden* mentions a nest with a half-grown chick, photographed near Dadri, UP in June 1959. Recent records come from open rural countryside, often from grassy, uncultivated patches near wetlands, and from valleys and plateaux of the Southern Ridge south of the city. The *Atlas* considers them scarce winter visitors to the Delhi area, but the several reports throughout the year suggest they may be scarce residents. Their breeding season in India is from January through June/July, and a pair showing territorial behaviour against a solitary conspecific at Dighal, Haryana, in early December 2016[145] suggests the possibility that it still breeds here.

Black Eagle *Ictinaetus malaiensis*

Winter Visitor Rare

Black Eagle, showing its distinctive flight silhouette

The Black Eagle is in a group of large birds of prey that most fit the popular conception of an 'Eagle', four genera of which, *Ictinaetus*, *Clanga*, *Aquila* and *Hieraaetus*, occur in the Delhi area. Black Eagles are widespread residents of forested hill ranges of South and Southeast Asia. Two subspecies are recognized, and subcontinental birds are all referable to the race *perniger*.

Delhi's landscapes are not these eagles' typical habitat; thus a solitary historical record[146] could be discounted were it not for a rather unexpected series of records since 2012, all between November and March: first from the Aravalli Biodiversity Park, Delhi, in February 2012, and one or two reports almost annually since then from forested Southern Ridge landscapes with a couple of reports from wooded wetlands in the river floodplain as well.[147] In their Himalayan range, Black Eagles are believed to breed between about November and March, and it is interesting that the occurrences in the Delhi area are around the same time period. They are not listed as residents of the Aravalli Range, yet these records are mirrored by recent winter reports from the Ranthambore and Sariska National Parks in Rajasthan. In the light of this evidence, they can no longer be considered as mere vagrants and are probably better categorized as rare, irregular visitors to the area, presumably non-breeders dispersing from the normally resident Himalayan foothills population. There seems to have been a definite change in the status and movements of this eagle over the past decade or so, reasons for which are unclear.

Indian Spotted Eagle *Clanga hastata*

Breeding Resident · Uncommon

Locally resident in the subcontinent with relict populations in Myanmar and Cambodia. Indian Spotted Eagles are tree-nesting raptors of open woodland, often near water. Monotypic.

Historical records (as Lesser Spotted Eagle, formerly considered conspecific) are vitiated by identification issues, but there are old reports from Najafgarh. With identification criteria now clearer, there have been several records in recent years from the Yamuna *khadar* in north Delhi, the Sultanpur neighbourhood; confirmed breeding in eucalyptus along canals in Sonipat and from the Rohtak and Jhajjar districts,[148] with at least 10 nests reported from Rohtak district (1,668 sq km) and two in a 15 km transect in Sonipat; that Indian Spotted eagles 'can breed in reasonably high densities in small, conducive, open agricultural habitats with tall trees and high rodent densities (six nests in 10 sq km)' has been observed elsewhere.[149] They probably nest about Dhanauri wetlands, GautamBuddha Nagar district, from where they are reported through the year. The only resident eagle here with which confusion is possible is Tawny, which is now evidently scarcer than Indian Spotted; any eagle of the *Clanga/Aquila* group in summer in the area is quite likely to be of this species and should be thoroughly scrutinized and documented.

Greater Spotted Eagle *Clanga clanga*

Winter Visitor · Fairly Common

Greater Spotted Eagles breed in Eastern Europe, Russia and China, wintering to northeast Africa, the Middle East, the northern subcontinent and Southeast Asia. Associated with wetlands. Monotypic.

These are now perhaps the most frequently observed eagles near lakes and wetlands of the Delhi area in winter, from mid-September to April. One or two may be present at any of our larger *jheels*, and small parties sometimes gather before migration in spring. Ganguli, in the 1960s, considered them irregular visitors but not infrequent in some years, which seems too parsimonious a description to be applied to this eagle today. Piracy by larger eagles that share its habitat has been observed by Steppe *A. nipalensis*, Imperial *A. heliaca* and Pallas's Fishing Eagles *Haliaeetus leucoryphus* (the last, formerly, when they were still around!).

Booted Eagle *Hieraaetus pennatus*

Winter Visitor · Fairly Common

Fierce eagles that breed from southern Europe and North Africa across to Central Asia and the western Himalayas; winter visitors to sub-Saharan Africa and the Indian subcontinent. Monotypic.

Booted Eagles are regular winter migrants to the area from September to March, often near wetlands but also elsewhere, including over suburban Delhi. They are active predators, usually seen singly or in pairs (often in March, before they emigrate) when they may hunt cooperatively, one bird flushing prey while the other, with agile twists and turns, pursues and cuts off its escape. The colour phases of this eagle are unrelated to age or sex, and though there is no solid evidence of geographical variation, there seems to be a tendency towards a higher ratio of dark phase birds in the eastern parts of its range; as about Delhi, perhaps in the proportion 3:2.

Tawny Eagle *Aquila rapax*

Formerly Fairly Common Breeding Resident · Now Scarce

Widespread Afro-Asian eagle ranging from northwestern and sub-Saharan Africa and Arabia to Pakistan and much of India. Three subspecies are recognized, of which *A. r. vindhiana* represents the species in India.

The *Handbook of the Birds of India and Pakistan*[150] had called it 'our commonest and most widely distributed eagle', and this was indeed the situation in the Delhi area half a century ago. Ganguli, writing of the 1950s and 1960s, considered them fairly common in open country and they bred in and around Delhi then; there are historical records of them nesting in gardens in Old Delhi and near the Red Fort in July 1959.[151] Tawnies were regularly observed in the then open country across the Yamuna from Delhi till the late 1970s, but a rapid decline in numbers appears to have set in by the following decade. There have been no reports of nesting from the Delhi area since then, and records in recent years are scattered and sparse, mainly between October and March, from the river *khadar* and the Dhanauri wetlands neighbourhood on the UP side, the Southern Ridge and the Sultanpur National Park environs in Haryana. Tawny Eagles appear to have lost ground in large swathes of their Indian range for reasons that are not entirely clear, and their scarcity in the Delhi area today parallels their decline in the riverine landscapes of the Gangetic Plain where they used to be quite common.

Steppe Eagle *Aquila nipalensis*

Winter Visitor · Fairly Common

Steppe Eagle, an immature in flight

Breeds widely across Central Eurasia eastwards to Mongolia and winters in East Africa, Arabia and southern Asia. Though seriously declining in the western part of the range, by over 80% in Europe, overall status largely determined by trends in its Asian heartland, where it remains the commonest eagle of its size. Two subspecies, of which the Central Asian breeding *nipalensis* winters in India.

These large eagles are still fairly common migrants about Delhi between late October and March, but numbers are declining across their range. They were regularly recorded in the Yamuna floodplain all through the 1980s, when 10–15 birds, or more, would join larger numbers of kites and vultures at landfills and gathered to roost on pylons at Okhla. Wintering numbers have dropped since the 1990s, but they remain the eagles most often seen soaring over the countryside in winter, and especially around wetlands, where ones and twos may typically be found alongside similar numbers of Greater Spotted Eagles. Up to 10 birds have been seen in recent years, congregating with kites and other raptors at animal carcass dumps. As with some other *Aquilas*, there are several reports of these eagles gathering in loose parties during migration, especially in March–April.

An immature, but very impressive, Eastern Imperial Eagle

Eastern Imperial Eagle

Aquila heliaca

Winter Visitor · Uncommon

These impressive eagles nest at relatively low densities from Central and east Europe across Eurasia to Mongolia and northwestern China, and winter in East Africa, Arabia and the northern Indian subcontinent. Monotypic.

A winter visitor about Delhi, in small numbers, from November to March. There are historical records going as far back as the 1940s from refuse dumps and the river floodplain, while Ganguli considered it an irregular visitor to the wetlands of the Delhi area. The Yamuna *khadar* at Okhla remained a favoured area until the 1980s with one or more recorded annually before urban expansion isolated these sites from the surrounding country. Most records now are from the teeming wetlands at Dhanauri, GautamBuddha Nagar district, the Sultanpur National Park neighbourhood and other lakes in Haryana; usually one or two at a site, but four together have been seen in March before emigration.[152] They are sluggish eagles, usually seen perched on a tree or low eminence on the ground for hours.

Bonelli's Eagle *Aquila fasciata*

Breeding Resident · Uncommon

Powerful eagles distributed from northwest Africa, the Mediterranean to southern Asia. They are resident over much of the subcontinent, nesting on cliffs, ancient forts and large trees between November and March; non-breeders tend to disperse to wetlands where wintering waterfowl, waders and other species congregate. Two subspecies are recognized: nominate *fasciata* in the greater part of its range, a second isolated in Indonesia.

Most records in the Delhi area are of birds frequenting wetlands in winter where they live off the teeming waterfowl flocks, or soaring over Aravalli landscapes of the Southern Ridge. They were known to breed in the Delhi neighbourhood historically, at Tughlaqabad and elsewhere on the Southern Ridge, and after a hiatus of over half a century, nesting has been reconfirmed in recent years. A pair nested on a power pylon (an unusual site) in farmland near Sohna, Gurugram district. Pairing activity was first observed in November 2017, nest building in end December and January with the female incubating by late January, feeding two chicks by late February and flying juveniles by mid-April 2018 though they remained in the proximity of the nest site for the rest of the month. A second pair built on a large eucalyptus tree the same year, in November 2017, at the Sultanpur National Park,[153] but though the pair continued to patrol the nest site till late March 2018, it is unknown whether the chicks fledged successfully.

White-eyed Buzzard *Butastur teesa*

Breeding Resident · Fairly Common

Endemic to South Asia and Myanmar, this raptor of lightly wooded grasslands is a widespread resident in India, with local movements depending on seasonal abundance of food items. Monotypic.

Ganguli considered this species to be a fairly common breeding resident in the 1960s and 1970s, nesting during the monsoon months between July and early September. It is no longer as common in the area as this would suggest, as much suitable habitat, typically areas of rough grass and open woodland, has been lost to cultivation and urbanization. Recent breeding records are mainly from the Sonipat and Rohtak districts, Haryana, and the Asola Wildlife Sanctuary on the Southern Ridge. There is, however, a very noticeable post-breeding influx of White-eyed Buzzards around Delhi between late July and October, when single birds or small parties comprising both adults and juveniles appear in areas where they are not usually seen during the rest of the year – presumably birds that have bred outside our limits as breeding records here are too few to account for the numbers involved. Similar seasonal movements and concentrations have been observed at Tal Chappar and Sariska in Rajasthan, the buzzards arriving soon after the rains and staying until end September.[154] The post-monsoon profusion of grasshoppers and mole crickets is exploited during these influxes.

Western Marsh Harrier *Circus aeruginosus*

Winter Visitor · Common

These well-known harriers breed widely across Europe eastwards to Central Asia and winter in Africa, the Middle East and the Indian subcontinent. Two subspecies are recognized; the nominate race *aeruginosus* is a winter visitor to India.

Common in most wetlands of the area, arriving as early as mid-August and staying till mid-May, stragglers remaining as late as June. The approach of a Marsh Harrier causes great commotion in a raft of duck and coot on a *jheel*, the birds rising en masse with a roar of beating wings to escape its attentions. The larger wetlands may host several individuals each winter, and up to 15–20 may gather at dusk at communal roosting sites, such as at Okhla.

Hen Harrier *Circus cyaneus*

Winter Visitor · Rare

Hen Harriers breed across the northern Palaearctic in Europe and northern Asia and winter in Europe, the Middle East and much of temperate Asia; wintering birds in the subcontinent are largely confined to its northern half, rarely extending further south as its congenerics. Monotypic.

Occasional records from the area between September–October and March, from cultivation and open country in general, often near *jheels*, including the Sultanpur National Park–Najafgarh complex of wetlands and grassy meadows, the Dhanauri wetlands in the GautamBuddha Nagar district and from scrubland on the Southern Ridge, Haryana.

A male Hen Harrier hunting over fields at Haiderpur

Pallid Harrier *Circus macrourus*

Winter Visitor Scarce

Pallid Harrier

These harriers breed in eastern Europe and the Central Asian steppes and winter in East Africa and West Asia, eastwards to the Indian subcontinent. The species is suffering a population decline in Europe, but numbers believed more stable in their Asiatic strongholds. Monotypic.

Ganguli, in her *Guide* published in 1975, had listed these harriers as uncommon winter visitors in the 1960s, and though scarcer now, they are still the most likely of the four 'white, grey and black' harriers to be recorded on passage, mainly September–October and February–April. A very few overwinter. Pallid Harriers seem to prefer more arid, open ground rather than cultivation, and most reports are from such barer, short grassland patches and fallow land in the river floodplain and open landscapes of the Southern Ridge.

Pied Harrier *Circus melanoleucos*

Vagrant

An East Asian harrier that nests in southern Siberia and the Amur region and winters in eastern and southern India, southern China and Southeast Asia. Monotypic.

These very handsome harriers are only stragglers as far west as Delhi with just three recent records from the area, all from sites in Haryana: Hodal in February 2006; Hailey Mandi, near Farrukhnagar, in February–March 2008; and Chandu, Gurugram, on 24 March 2021.

Montagu's Harrier *Circus pygargus*

Winter Visitor Rare

Breeds in temperate Europe and Central Asia, wintering south to sub-Saharan Africa and the Indian subcontinent, where they are apparently commoner towards its western and southern parts. Monotypic.

There are a few reports from the Delhi area in winter, but mostly of birds on passage between October and November and in March.

Shikra *Accipiter badius*

Breeding Resident · Common

Widespread in lightly wooded country in sub-Saharan Africa, Central Asia (where mainly summer visitor) and throughout southern Asia. Six subspecies are recognized, with two resident races in India: *dussumieri* in the north and *badius* in the south.

Shikras are common residents of the Delhi area in both urban and rural habitats. They breed between March and June, raising a family of usually three chicks in a crow-like nest in a medium-sized tree, and flying juveniles are regularly reported in June and July. As the breeding season approaches, pairs perform sky dances in courtship displays with slow, deep, wingbeats and are particularly noisy in this season. Shikras are quintessentially bird hunters, though they will catch and eat any small living creature they can lay their talons on and have even been observed making repeated sallies from a branch to capture pipistrelles *Pipistrellus* at dusk.

Besra *Accipiter virgatus*

Winter Visitor · Rare

Largely resident from the hilly parts of the subcontinent to China and locally in Southeast Asia. Many Besras that breed in the Himalayan mid-hills descend to the foothills in winter, at times drifting further southwards into the northern plains during cold waves. Eleven subspecies are recognized, including several populations isolated on Southeast Asian islands, with three in India: *affnis* in the Himalayan foothills, *besra* in south India and Sri Lanka and *abdulalii* in the Andamans.

Stragglers have been recorded off and on, presumed race *affnis*, with scattered winter reports from wooded patches in the area.[155]

Eurasian Sparrowhawk *Accipiter nisus*

Winter Visitor · Uncommon

A widely distributed sparrowhawk, breeding from Europe across temperate Asia to eastern Siberia, China and Japan, and south to the Himalayas. The northern populations are migratory, wintering in northeast Africa, the Middle East and southern Asia. Six subspecies are recognized, with two in India wintering Siberian *nisosimilis* and the largely resident, Himalayan nesting, *melaschistos*.

Most Eurasian Sparrowhawks that winter in the Delhi area, mainly between October and April, appear to be of the paler Siberian breeding subspecies *nisosimilis* rather than the well-differentiated *melaschistos* of the Himalayas that also descends to the foothills in winter. They are regularly reported in these months from the larger city gardens, the Ridge habitats, at Okhla and the better wooded parts of the surrounding countryside.

A Eurasian Sparrowhawk in flight

Northern Goshawk *Accipiter gentilis*

Vagrant

Largest of the genus, a resident of boreal and temperate forests across the Holarctic, a partial migrant that at times also displays dispersive movements possibly relating to prey availability during winter. In Eurasia, the Northern Goshawk's range extends as far south as the Himalayas; it presumably nests sparingly in the western and central Himalayas above 2,400 m, but is better known as a winter visitor to the range, down to the foothills. A few wander further south into Pakistan and north India, and such vagrants have been reported from the peninsula south to Gujarat, Karnataka and Mysore.[156] Nine subspecies are recognized, of which the eastern race *schvedowi* breeds from southeast Russia to China and the Himalayan ranges.

Vagrant Northern Goshawks have been recorded as close to our area as the Ranthambore National Park in the Aravallis. Ganguli, in her book published in 1975, mentions a 'doubtful' sight record by H. Alexander, 'a large noisy hawk chased by crows at Faridabad, February 1950.' A record as vague as this is unacceptable, but its rare occurrence in the area stands confirmed by one photographed during a raptor bird count at the Asola Wildlife Sanctuary on 26 February 2022.[157]

A Black Kite of the resident Indian subspecies

Black Kite *Milvus migrans*

Breeding Resident and Winter Visitor Common

A widespread Old World raptor, commensal and adaptable to human-modified habitats, and one of the commonest of the family. Seven subspecies are recognized, which fall in three subspecies groups: the Black Kite group (*migrans*, *govinda*, *affinis*) from Europe, South Asia and Australasia; the Black-eared Kite group (*lineatus* and *formosanus*) from East Asia; and the Yellow-billed Kite group (*aegyptius* and *parasitus*) from Africa. These groups are sometimes awarded separate species status, but intergradation between them suggests they are probably best retained as a single species. Black Kites of the resident race *govinda* are widespread across the subcontinent; Black-eared Kites, race *lineatus*, breed widely across the Chinese mainland and East Asia south to the Himalayas, and migrate through the Karakoram and Himalayan ranges along the major river valleys to winter in the plains of the northern subcontinent.

The Indian race *govinda* is a common and familiar resident of urban Delhi and its satellite towns, nesting in winter between November and April, when pairs perform sky dances and chases near their nest sites with much screaming. Black Kites gather at communal roosts outside the breeding season and their daily commutes, a continuous stream of birds strung out singly or in twos and threes, are familiar city sights at dawn and dusk. The migrant race *lineatus*, Black-eared Kites, are common winter visitors to the area from September to April, at the same time when their resident cousins are nesting. They are much less dependent on urban habitations but are common by the Yamuna River where they live off floating refuse and offal. Thousands gather at garbage dumps on Delhi's outskirts.

Brahminy Kite *Haliastur indus*

Non-breeding Visitor Scarce

The two species of *Haliastur* kites have feet that are densely covered with spicules for grasping slippery prey, an example of convergent adaptation between these kites, fish eagles and the Osprey for a fish diet. Brahminy Kites are resident raptors, with some local movements related to monsoon rains, from the Indian subcontinent to Indochina and Australasia. Four subspecies are recognized, of which nominate *indus* inhabits the subcontinent and eastwards across mainland Southeast Asia.

The status of this water-loving kite, otherwise common and widespread across much of its Indian range, seems to have changed in the area and it is nowhere as regular about Delhi as formerly. Ganguli, in the 1960s, had recorded 'small numbers throughout the year, but more frequently during the rainy season', with as many as 50 together in mid-October 1961. Records from about Delhi in recent decades are sparse, mostly in the late monsoon, with regular but scattered reports in other months as well. They are rather more frequent in the Ganga *khadar* at the Hastinapur Wildlife Sanctuary and Haiderpur Wetland just upstream, where they are reported between late June and October and between mid-January and March. Similar changes are evident across its northern Indian range; an influx of Brahminy Kites, juveniles and adults, was a feature of the monsoon along the rivers of UP till the late 1960s, after which a similar decline seems to have set in there too. Although juvenile and immature kites are regularly recorded, the first and only report to date of breeding in recent years from the Delhi neighbourhood was a video taken of an adult feeding two juveniles on 21 June 2022 near Karnal, Haryana, which is, however, just outside the area covered by this book.[158]

White-tailed Eagle *Haliaeetus albicilla*

Winter Visitor Rare

White-tailed Sea Eagles are largely resident across northern and central Eurasia, but a few move south in winter to the Mediterranean, the Middle East, the northern and western parts of the Indian subcontinent and southeastern China. Monotypic.

These eagles had not been recorded from the Delhi area prior to 2007, when a juvenile was first photographed at the Sultanpur National Park in early December of that year. There have been a few more records, all of juveniles, in recent years since then, including from the Bhindawas Lake, Haryana, in early February 2016; from the Ganga at the Hastinapur Wildlife Sanctuary and Haiderpur Wetland (Bijnor barrage) between November and April; and by the Yamuna in late February 2017 and December 2020. Records of a juvenile bird by the river, in north Delhi in January 2021 and at Okhla from late February till mid-March 2021, probably pertain to the same individual that appears to have spent the winter here;[159] and yet another remained at the Sultanpur National Park from mid-December 2022 through the winter of that year.

An immature White-tailed Eagle, at Sultanpur, December 2022

An immature Pallas's Fish Eagle photographed at the Haiderpur Wetland, March 2022

Pallas's Fish Eagle

Haliaeetus leucoryphus

Former Breeding Resident

Now Rare

Pallas's Fish Eagles were earlier believed to comprise two sub-populations, one migratory, breeding north of the Himalayas in Central Asia, Mongolia and Tibet and moving south to winter in the northern Indian subcontinent alongside a second, resident sub-population there. Recent evidence and re-evaluation of historical data has revealed that, in actual fact, there is only a single population that breeds in northern India and disperses north of the Himalayas in its non-breeding season (May to September), performing an extensive migration of over 4,000 km from India to Mongolia.[160] These eagles used to be conspicuous denizens of the rivers and lakes of the Gangetic Plain but have seriously declined over the last few decades, for which a combination of factors are probably responsible: habitat loss with pollution and reduction of river flows, loss of and disturbance at nesting sites and accumulation of pesticide residues in freshwater fish. The species is now virtually extirpated from much of its range in western and northern India, with remnant breeding populations in Uttarakhand and Assam in India, in northeastern Bangladesh, Bhutan and Myanmar. In Nepal and the Upper Gangetic Plain, it is only a rare migrant now. Monotypic.

The extirpation of this magnificent eagle from the Delhi area has been a tragic loss. They bred in Delhi historically, from October through February, with a traditional nest site at Okhla and possibly another by the Yamuna in north Delhi till the mid-1950s.[161] Records continued till the 1980s: single immatures at Okhla in the winters of 1978–79 and 1985–86, an adult pair at Sultanpur in winter 1982–83 and at Bhindawas in 1987.[162] Thereafter, there have only been rare reports of immatures in recent years from the Ganga *khadar* between November and April, but none from the immediate vicinity of Delhi until October 2020, when one bird (named Ider), radio-tagged in Mongolia,[163] was tracked entering India from eastern Nepal and moving westwards across UP to cross Delhi on 27 October 2020 towards the Jhajjar thermal power plant, Haryana, where it remained till about mid-November 2020 before moving out of the area (and ultimately returning to Mongolia by April 2021, to the same location where originally tagged. The following year, 2021, Ider again entered India from the east but remained for the 2021–22 winter in the Kaziranga National Park in Assam and its environs).

Lesser Fish Eagle *Haliaeetus humilis*

Winter Visitor Rare and Sporadic

An immature Lesser Fish Eagle, at Okhla

Smallest of the *Haliaeetus* genus, distributed across southern Asia from India to Indonesia in two subspecies, of which *plumbea*, of southern Asia, breeds along forested, fast flowing rivers in the Himalayan foothills from Himachal Pradesh eastwards, and locally in the southern peninsula where it has adapted to dammed rivers. In the Himalayas, it is a scarce resident, but while most birds remain around the breeding grounds all year, a few straggle into the northern plains in winter. There are apparent ecological differences between this species and Grey-headed Fish Eagle; Lesser prefers swift flowing rivers and streams rather than still waters, the Ramganga and Kosi rivers in the Corbett National Park, for example, being excellent habitats for this species.

A rare, sporadic visitor in winter, this fish eagle had not been recorded historically from the area; the first report is of one at Okhla in early October 2014, identified as this species though without supporting detail.[164] One was photographed in flight over south Delhi on 8 October 2019, and two well-studied individuals were present at Okhla in late winter and spring 2021 – a first-year immature that remained at the site for more than a month, from late February to the second week of April, and an adult that stayed for a week from 1 to 6 March.[165] An adult bird was reported again from Okhla in the third week of November 2022. These records suggest that Lesser Fish Eagles should preferably be considered rare, sporadic visitors in the Delhi area, mainly on passage, rather than mere out-of-range vagrants.

Grey-headed Fish Eagle *Haliaeetus ichthyaetus*

Winter Visitor Rare and Sporadic

Until recently, the scattered records of 'grey-headed' fish eagles from the Delhi area were all believed to be of this species, *H. ichthyaetus*, which is widely but only locally distributed over much of the Indian subcontinent and across Southeast Asia – it gets progressively more frequent eastwards towards Bengal and Assam and in peninsular India. Westwards it is much scarcer, with rare records from eastern Rajasthan, Haryana and Delhi. Monotypic.

Ganguli mentions historical records of Grey-headed Fish Eagles between November and April from the Yamuna River and the Sultanpur lake; later post-1975 records include one in December 1984 at Rithal, Gohana, and two or three at Barhwasani, Sonipat, Haryana, in January 2001,[166] an adult at the Asola Wildlife Sanctuary, Delhi, in October 2015, and a first winter bird at the Bhindawas Wildlife Sanctuary, Haryana, in late January 2017 that stayed through February 2017.[167] However, with the Lesser Fish Eagle also confirmed from Delhi recently, and in the absence of detailed descriptions, it is permissible to question whether some historical observations were actually Grey-headed, as reported, or Lesser; but have been accepted as this species as the observers had presumably observed the distinctive tail pattern and were confident of the identification.

Common Buzzard *Buteo buteo*

Winter Visitor Scarce

A common species in Europe to Central Asia, largely resident in the west, but with the northern and eastern populations wintering mainly in Africa and also in South Asia. Six subspecies are recognized, but only race *vulpinus* (Steppe Buzzard), the migratory eastern subspecies, concerns us here. The bulk of the *vulpinus* population winters in Africa, but birds from the easternmost parts of its range (Eastern Steppe Buzzard) winter in India too. Many wintering birds in India are, however, believed to be hybrid *rufinus* x *vulpinus*.[168] In addition, reliable data about their winter distribution in India is lacking, vitiated by taxonomic confusion and identification problems, but they are most regularly reported from southwestern India, Punjab, Rajasthan and Gujarat. These latter records complement the situation in Pakistan, where *vulpinus* is recorded mainly from areas adjoining the eastern borders with India.[169]
Common Buzzards have been identified from the Delhi area only in recent decades, having been possibly overlooked earlier, and are irregular and rare visitors here at best. Confusion with Long-legged is widespread, but the few convincing records are from the Greater Noida area and the Sultanpur/ Najafgarh neighbourhood, Haryana, mainly in mid-winter.

Long-legged Buzzard *Buteo rufinus*

Winter Visitor Fairly Common

Long-legged Buzzard

Long-legged Buzzards breed from North Africa to Central Asia, including in the western Himalayas. Two subspecies are recognized; the continental race *rufinus* is a winter visitor to northern India.
Recorded between September and April in the Delhi area, in cultivated areas, often in the vicinity of wetlands or in rocky scrub country of the Aravalli landscapes. Though well distributed in the Delhi area, wintering numbers tend to fluctuate, but with a gradually declining trend over recent decades.

STRIGIFORMES: OWLS
TYTONIDAE: BARN OWLS

Barn Owls have a distinctive morphology, with a heart-shaped face and forward-facing eyes that provide excellent binocular vision in low light, and a facial disc which can be adjusted to focus sounds on to the owls' asymmetrically placed ear cavities that enable them to pinpoint the exact location of the slightest sound in the darkness.

The Tytonidae comprise two subfamilies: Tytoninae (Barn and Grass Owls) and Phodilinae (Bay Owls). Both are represented in India, but only one species of Barn Owl occurs in the Delhi area.

India: 5 species, Delhi: 1

Common Barn Owl *Tyto alba*

Breeding Resident Fairly Common

Perhaps the most widespread of all owls, occurring in every continent except Antarctica. Up to 28 races have been named over this vast range, many restricted to islands, with three or more subspecies groups sometimes recognized within the complex that have been given species rank in some taxonomies. Indian birds belong to race *stertens*, which extends eastwards to Indochina as well.

Barn Owls are widespread about Delhi, well adapted to human-modified habitats, but with strictly nocturnal habits which renders them inconspicuous. They are most often noticed flying overhead at night with their white underparts illuminated in the city lights. They live around old as well as modern buildings in urban centres, in ancient monuments and garden tomb complexes, where their screeches attract attention to them in the dark, and they are also reported from grain storage depots in rural areas. They breed over much of the year, perhaps raising two broods, and commonly in holes and sheltered corners of historical and even occupied buildings and other structures, laying as many as four to six eggs in a clutch. Incubation usually begins with the laying of the first egg, which are laid at about two-day intervals, so they hatch asynchronously and the young of a single brood may show marked disparities in growth and size.

Barn Owl

STRIGIFORMES: OWLS

STRIGIDAE: OWLS

The owls of the Delhi area fall into three clades or subfamilies: Striginae (Typical Owls, including *Otus*, *Bubo*, *Ketupa* and *Asio*), Surniinae (Owlets, including *Athene*) and Ninoxinae (Hawk Owls, *Ninox*). Owls are small- to large-sized, mostly nocturnal birds of prey and, like their Barn Owl cousins, have large, forward-facing eyes that give them binocular vision over a large arc, facial discs surrounding asymmetric ears that help to locate prey by sound alone, with strong legs and sharp claws to capture it. Some have upstanding feather tufts on their heads which make them look 'eared' but are just adornments. Owls' flight feathers have a velvety structure and fringed edges that enable them to fly silently in the dark.

Most owls conceal themselves well during the day, and their calls at night are usually the first indication of their presence. The small scops owls, in particular, are extremely secretive and easily overlooked when not calling. Others, like Short-eared Owls *Asio flammeus*, may hunt by day, searching for prey by sight in low flight. Undigestible parts of their food are regurgitated as 'owl pellets'.

Owls, in general, do not build their own nests, laying on the ground in a hole or depression in an earth bluff or cliff ledge, in a tree cavity, or by appropriating another bird's nest after the young of the original nest builder have fledged. A recent study has thrown up evidence of one owl species, Dusky Eagle Owl *Ketupa coromandus*, becoming preferentially reliant on, and reusing, nests of Woolly-necked Storks *Ciconia episcopus*, in the Delhi area at least; such reliance would suggest the owls' ability to track the nesting habits of a single species.[170]

The smaller resident species of the Delhi area, including scops owls *Otus* and owlets *Athene*, usually lay three or four, rarely five eggs in a clutch; larger ones such as some Eagle Owls *Bubo* and Fish Owls *Ketupa* may lay just one, or occasionally two, eggs. Indian Eagle Owls *Bubo benghalensis*, however, lay a clutch of three or four eggs, which hatch asynchronously and the chicks usually show marked disparities in growth within a single brood.

India: 31 species, Delhi: 10

▲ *A family of Spotted Owlets*

Indian Scops Owl

Otus bakkamoena

Breeding Resident

Fairly Common

Delhi's parks offer refuge to birds, such as this Indian Scops Owl.

Subcontinental endemic, used to be part of the Collared Scops Owl complex in its widest sense, now split into four species, of which two occur in India – Collared Scops Owl *O. lettia* in the Himalayas and northeastern India, and *O. bakkamoena* in the rest of the country. Four subspecies are recognized; race *gangeticus* of northern India represents the species about Delhi.

Indian Scops Owls are frequent residents of well-wooded localities and gardens, in the quieter tree-lined side lanes of residential New Delhi, the Aravalli landscapes of the Ridge and woodland patches and mango groves in the countryside. Nests in a hole or hollow in a tree trunk, mainly March to May in the Delhi area when birds are most vocal, but their soft notes can be easily missed amidst the noises of the city.

Pallid Scops Owl *Otus brucei*

Winter Visitor Rare

Owls of dry, sparsely wooded, country from West and Central Asia ranging east to Pakistan, that winter in western and north-western India in dry scrub and woodland habitats, wandering as far south as Kerala.[171] Birds wintering in India probably nest in Central Asia where they are summer visitors, since the populations further to the south are said to be mainly resident. Four subspecies are described, but their validity remains unsubstantiated.

Pallid Scops Owls have been reported from the Delhi area on a few occasions in recent years in winter, between late November and January, first recorded at Tughlaqabad in December 1981, but there have been a number of later observations from woodland patches since 2015.[172] They possibly winter about the Delhi area more regularly than these few records suggest, albeit in small numbers, and have been overlooked in the past.

Oriental Scops Owl *Otus sunia*

Passage Migrant · Rare

Breeds in the lower Himalayas and peninsular hills of India, in Sri Lanka and across much of East Asia from northeastern China, Korea and Japan, southwards to Myanmar, Thailand and Indochina. Northern populations in China and Japan are migratory, and they are also believed to be mainly summer visitors in the western Himalayas. Nine subspecies are recognized, of which three occur in India: *sunia* in the Himalayan foothills and northeastern India, *rufipennis* in southern India and *nicobaricus* in the Nicobar Islands.

There are just five records of this scops owl from the Delhi area, presumably of race *sunia*: two specimens collected on 26 February and 22 March 1925, the latter in breeding condition; one trapped in a mist net on the New Delhi Ridge on 8 February 1973 and listed as *O. scops*, of which *sunia* was then treated as a race. There are two more recent reports: one photographed at New Palam Vihar, Gurugram, on 5 March 2011 and a pair, one in rufous and the other in grey phase, that remained at the Asola Wildlife Sanctuary from 22 February till the second week of March 2018.[173] The dates show a consistency that suggests this owl is probably best considered a rare and probably overlooked passage migrant in spring, in February and March.

Indian Eagle Owl *Bubo bengalensis*

Breeding Resident · Uncommon

Indian, or Rock Eagle Owls as they are also called, are subcontinental endemics, resident from Pakistan and Nepal across most of India except in the northeast. This species is part of the *B. bubo* species group and has been considered conspecific with Eurasian Eagle Owl *B. bubo* in the past: the two differ, however, in DNA and voice. Monotypic.

Local breeding residents about Delhi, in eroded ravines, abandoned stone quarries and thorny, rocky scrubland, they are characteristic of the Aravalli landscapes and the Southern Ridge, old forts and ruins such as Tughlaqabad and (formerly) Purana Qila and Humayun's Tomb. They nest in such terrain in holes and crevices in cliffs and rocks, or even in a depression in the ground shaded by bushes. They also affect plantations alongside canals at Najafgarh and Sonipat, large trees surrounding wetlands as at the Dighal and Bhindawas lakes, Haryana, and in cultivated landscapes as well, where excavations, embankments and earth bluffs also offer resting and nesting sites.

Indian Eagle Owl

Brown Fish Owl *Ketupa zeylonensis*

Breeding Resident Scarce, Localized

Distributed from the subcontinent across to Indochina and very locally as far west as Turkey. Four subspecies are recognized, of which *K. z. leschenaulti* inhabits India and Myanmar, with other races in Turkey, Sri Lanka and Southeast Asia.

Historically, Hutson in his *Birds about Delhi* (1954) recorded 'one or more of these owls' from the Yamuna at Okhla, presumably in the 1940s, but they have only been reported since then from two other locations: from the Surajpur wetlands, Greater Noida, where one was photographed in September 2014, and from tree-lined canals at Rithal, near Gohana, and at Sonipat, Haryana, where these owls have been recorded nesting, usually in a cradle-like cavity created by a branching trunk of a large, mature tree, or occasionally reusing nests of another large raptor.[176]

Dusky Eagle Owl *Ketupa coromandus*

Breeding Resident Uncommon, Localized

Breeding residents across the northern subcontinent, inhabiting wooded plains country and preferring the proximity of water, can be common at some sites and scarce elsewhere for reasons that are unclear. Two subspecies are recognized, the nominate in the subcontinent, while *K. c. klossi* extends into Myanmar and southern Thailand.

These large eagle owls appear to have been much more widespread in the area in the past. Norman F. Frome recorded them in the 1940s 'in old established parks and gardens ... along canals and in riverain tracts', breeding in mid-winter; Ganguli found them '...in small numbers' in the 1960s and believed that they were declining even at that time. Dusky Eagle Owls seem to have been last recorded from the city itself in the early 1970s,[174] but are still regularly reported from tree-bordered wetlands, often eucalyptus plantations, in Haryana near Jhajjar (Bhindawas Wildlife Sanctuary) and Rohtak (Dighal and Baland), where they are known to nest regularly. A recent study in these districts from 2011 to 2020 has recorded these owls selectively reusing nests of Woolly-necked Storks *Ciconia episcopus* that inhabit and nest in the same habitat mosaic of farmland, irrigation canals, groves and old roadside avenue trees, in a novel commensal relationship. Dusky Eagle Owls appeared to preferentially reuse stork nests despite the area having other bird species, eagles, ibises, etc. that build large nests – Indian Spotted Eagles *Aquila hastata* and Red-naped Ibis *Pseudibis papillosa* nests were also reused, but only in two cases of the former and one of the latter, as against twenty stork nests, especially those close to irrigation canals.[175] A pair was resident at the Sultanpur National Park between late 2002 and 2005 but with no evidence of breeding. These are vocal owls, unlikely to be overlooked where present.

Dusky Eagle Owl, photographed near Rohtak, Haryana

Spotted Owlet *Athene brama*

Breeding Resident · Common

A Spotted Owlet peers, wide-eyed, from a hole in a tree trunk.

Common residents over much of southern Asia, from southern Iran and the subcontinent to Myanmar. Four subspecies are recognized, with two within Indian limits: *A. b. indica* in northern and central India and nominate *brama* in the south.

A familiar resident owl about Delhi, in wooded parks and city gardens, temple groves and large trees about villages as also old buildings and ruins where they nest in holes in the ancient walls. Pairs and small family parties live in traditional trees and may be found there day after day.

Long-eared Owl *Asio otus*

Vagrant

A widespread owl of the temperate Northern Hemisphere, with two historical records of breeding in the Indian subcontinent, in Kashmir and Baluchistan. It is an uncommon winter visitor, often in small parties, to the northwestern parts of the subcontinent, with isolated records south to Kachchh and eastwards along the Himalayan foothills as well. Three subspecies are recognized; birds recorded in India are of the nominate race *otus*, of Eurasia.

A rare visitor to the area. Ali & Ripley, in their *Handbook*, Vol. 3 (published 1969), included 'Delhi District' in a list of localities from where it has been recorded in the past. There were no reports since then, until two were seen and photographed by many observers in a wooded patch at Maroudi Jatan, Rohtak, Haryana, in February 2020, and reported again from the same site on 10 November and 1 December 2020.[177] It is uncertain whether the two birds remained the entire year at the site or returned to the same location the following winter. A Long-eared Owl was received by the Wildlife Rescue Centre, Wazirabad, Delhi, on 15 December 2021 from the Jain Bird Hospital, Old Delhi (presumably rescued from the immediate neighbourhood of the city, but no details available), and treated there for wing muscle injuries, and yet another photographed at the Jamia Millia Islamia University campus, New Delhi, on 12 November 2022.

Short-eared Owl *Asio flammeus*

Winter Visitor Uncommon

One of the most widely distributed owls of open grasslands breeding in northern latitudes over much of North America, Europe and Asia and also South America. Winters further south, including to India. Ten subspecies are recognized, including several island populations; the owls wintering in India belong to the widespread race *flammeus* of Eurasia and North America.

Short-eared Owls are most frequently reported as a passage migrant in October–November and in March/early April, in reed beds and grassland areas which are not too heavily overgrazed and bare, where single birds or small parties may often be seen hunting by day. Many birds, however, remain all winter. With suitable habitat increasingly being lost to agriculture and urban expansion, these owls are now localized and most recent records come from uncultivated pockets in the vicinity of the larger wetlands such as Sultanpur and open grassy ground in the Ganga and Yamuna floodplains.

A Short-eared Owl out in the daytime

Brown Hawk Owl *Ninox scutulata*

Non-breeding? Visitor Scarce

Resident in much of India, Sri Lanka and across Southeast Asia. Nine subspecies are recognized, of which five are recorded within Indian limits, and *N. s. lugubris* represents the species in northern and central India.

Brown Hawk Owls are breeding residents in the Himalayan foothills and more sparsely in the Gangetic Plain. The Delhi area is towards the western limits of their range in the plains, and they have been recorded, variously as visitors or sparse residents, from as close to Delhi as Meerut, Agra, Bharatpur and Ranthambore. Their occurrence here is therefore not unexpected, yet Brown Hawk Owls were first reported from the area relatively recently, from Meerut in March 1984, and from Kosli, Rewari district, and Matanhale Bani, near the Bhindawas Sanctuary in Haryana, in April and September 2012. In the last few years, however, there have been several records in the winter months, from woodland patches at the JNU campus in late December 2016; at the Asola Wildlife Sanctuary in mid-November 2017 and February to mid-March 2018 (possibly the same individual); at the Surajpur lake, GautamBuddha Nagar district, in February 2018; at the Sultanpur National Park from November 2020 through most of January 2021 and again in December 2021; in January–February and October 2022 at the Hastinapur Wildlife Sanctuary near Meerut towards the east. There are two records in summer as well: a pair, one of which was heard calling at Mangar Bani, Faridabad district, in late July–early August 2020, and one at Sultanpur, Haryana, in second-week August 2021. The records fall broadly outside the breeding season, which is believed to be between March and June/July in northern India. They have been categorized here as scarce visitors, but may well be confirmed as breeding residents sooner rather than later.

BUCEROTIFORMES: HOOPOES AND HORNBILLS

UPUPIDAE: HOOPOES

This unique Old World family is represented across Europe, Africa and Asia by two unmistakable and well-known species, with striking plumage of orange, black and white, and conspicuous crests. One of these is widespread as a breeding resident in the subcontinent, and a winter visitor as well. They require rather bare ground or short-grass areas such as lawns and overgrazed village precincts, where they forage by walking about and probing the soil for insects with their long slender bills. They nest in cavities, both natural and in man-made structures, in which five to seven eggs may be laid.

India: 1 species, Delhi: 1

▲ *Common Hoopoe*

Eurasian Hoopoe *Upupa epops*

Breeding Resident and Winter Visitor · Fairly Common

Widely distributed across the temperate Palaearctic, African and Oriental regions, the northern populations being migratory and the African and southern Asian populations resident. Eight subspecies are recognized, of which two, *ceylonensis* and *longirostris*, are resident across India while two others, *epops* and *saturata*, visit the western and northeastern parts of the country in winter.

Hoopoes are well distributed in the Delhi area, with a resident race *ceylonensis* that breeds in March–June in holes in walls and buildings, and wintering nominate *epops* that has been reported in small numbers between September and April. Ganguli, writing in the 1970s, referred to them as common. A drop in the numbers of breeding pairs may have taken place with the creeping loss of open spaces in urban and suburban areas in recent decades. Hoopoes prefer heavily grazed, bald or short-grass areas; rural dirt tracks, and village precincts and lawns are popular feeding grounds.

BUCEROTIFORMES: HOOPOES AND HORNBILLS

BUCEROTIDAE: HORNBILLS

This spectacular family of birds, with long, decurved beaks and elaborate casques, inhabits a range of wooded habitats in sub-Saharan Africa and southern Asia. Remarkable for their unique breeding biology, where the female seals herself with mud into a nest cavity in a tree trunk within which she incubates and broods the chicks, while the male of the pair provides food for her through a small slit in the mud wall of the opening. She emerges after the chicks have hatched and partly developed, having undergone a rapid moult during this period, and then both birds feed and rear the chicks till they fledge.

This family is represented in the Delhi area by one subcontinental endemic as a resident, and another widespread forest species as a rare, casual visitor.

India: 9 species, Delhi: 2

Indian Grey Hornbill *Ocyceros birostris*

Breeding Resident · Fairly Common

Endemic to the subcontinent, in India and adjoining northeast Pakistan. Monotypic.

These relatively small hornbills are residents of urban parks and gardens, roadside trees and village groves throughout the Delhi area. There are, however, signs of local declines in the numbers of this conspicuous, noisy species since about 2015–2016, with hornbills no longer in evidence in parts of urban New Delhi where they were regularly seen. Their primary requirement is large trees with cavities and holes for nesting, in which they breed from February to June, and family parties are commonly seen in late-March and April.

Oriental Pied Hornbill *Anthracoceros albirostris*

Non-breeding Visitor · Rare, also probable escapes

A widespread and common hornbill from northern and eastern India to most of Southeast Asia. Two subspecies are recognized; nominate *albirostris* in India.

Delhi is not in the usual residential range of this hornbill, and records from the area could be attributed to vagrant individuals from the Himalayan foothills, given that this species is known to wander southwards into the Indo-Gangetic Plain occasionally. Historically, single birds or pairs had been recorded between mid-November and mid-February in the 1940s and on the New Delhi Ridge in February and March 1973.[178] There were no reports from the area over the next four decades; a rather enigmatic flurry of recent reports since 2013, however, are spread over most months of the year.[179] Single birds were recorded in early May 2013 at the Aravalli Biodiversity Park, Delhi; in April 2016 at Okhla; a female, first reported in late July 2016 in a south Delhi garden, showed clear behavioural signs of having escaped from captivity and was seen off and on in south Delhi over the next four years till at least 2020; and a presumed pair were observed repeatedly in a wooded campus in Ghaziabad for over a year from May–June 2020 to at least July 2021.[180] It is unclear at this stage whether this pair of hornbills could have been escapes from some local aviary or had taken up residence naturally in a conducive well-wooded area and may even have nested there.

CORACIIFORMES: KINGFISHERS, BEE-EATERS AND ROLLERS

ALCEDINIDAE: KINGFISHERS

Conspicuous birds with distinctive, pointed beaks and boldly patterned, often colourful plumage. Many kingfisher species, true to their name, are expert fishers, capturing fish underwater by diving, for which their eyes are specially adapted; others are primarily terrestrial and capture their prey, both vertebrate and invertebrate, on land. All nest in burrows in earth banks which they excavate themselves, in which typically five or six white eggs are laid in an unlined chamber at the end of the tunnel.

Three species of kingfishers are recorded from the Delhi area, each with its own different hunting technique, and representing the three separate clades or subfamilies (Alcedininae, Halcyoninae and Cerylinae) of these well-known birds.

India: 12 species, Delhi: 3

▲ *A White-throated Kingfisher with fish*

Common Kingfisher *Alcedo atthis*

Breeding Resident · Uncommon

Widespread in the Palaearctic and Oriental regions, mostly resident across its range but central Eurasian populations tend to disperse further south in Asia. Seven subspecies are recognized, of which three, *atthis*, *bengalensis* and *taprobana*, are resident in northwestern, northern and southern India respectively.

In spite of their name, these kingfishers are not particularly 'common' birds about Delhi. Single birds or pairs, presumed race *benghalensis*, may be seen in waterbodies, both natural and human modified, where they breed along lakes, canals, weirs and fish ponds. They are scarce by the Yamuna River itself, though they occur at Okhla where there is overhanging vegetation along the canals, which provides them the low perches where they wait for small aquatic life to come within range, or briefly hover to locate prey in the water. They breed between March and June in holes they excavate in earthen banks.

White-throated Kingfisher *Halcyon smyrnensis*

Breeding Resident · Common

Ranges from the Middle East, across the Indian subcontinent to Southeast Asia. Five subspecies are recognized, three of which occur within Indian limits: *fusca* over most of the country, *perpulchra* in eastern India and *saturatior* in the Andaman Islands.

White-throated Kingfishers are common, colourful and noisy residents of urban parks, gardens and the countryside throughout the Delhi area. Like several other kingfishers of this genus, they frequently hunt over land. They are vocal when breeding between March and June, though birds begin to pair off, 'sing', and court each other, opening their wings and showing off their brilliant contrasts in display, from February or even earlier.

Pied Kingfisher *Ceryle rudis*

Breeding Resident · Fairly Common

A widely distributed and very distinctive kingfisher, resident in sub-Saharan Africa, parts of the Middle East and southern Asia, with three subspecies are recognized, of which *leucomelanurus* inhabits most of India, except for *travancoreensis* of the southwest.

Pied Kingfishers used to be common on the Yamuna as it passes by Delhi, a most attractive adjunct to the river scenery. Since the last couple of decades, however, they have become far scarcer than formerly in the heavily polluted stretch of the river between Wazirabad and Okhla, which is now too opaque for their method of feeding, hovering expertly above the water and diving into it for fish. They breed in tunnels excavated in sand bluffs left by the receding river after the monsoons and are still fairly common residents in its cleaner stretches above Wazirabad and in the Ganga *khadar* at the Hastinapur Wildlife Sanctuary; many disperse to, and may even breed, in irrigation canals, artificial waterbodies and the larger natural wetlands across the area.

A Pied Kingfisher, hovering before the plunge

CORACIIFORMES: KINGFISHERS, BEE-EATERS AND ROLLERS
MEROPIDAE: BEE-EATERS

Bee-eaters are colourful, graceful, gregarious birds of Asia and Africa, with wonderful powers of flight that enable them to hunt aerial insects, including swift-flying dragonflies. Flying insects become scarce in the cold Delhi winter and Bee-eaters largely vacate the Delhi area for warmer climes during that season. Many are colonial breeders, excavating their own nest burrows in vertical earth banks and road cuttings, or even in the ground, in which four to six eggs are laid as a norm. Three breeding species and two vagrants figure on the Delhi area checklist.

India: 7 species, Delhi: 5

◄ *Blue-tailed Bee-eater in flight*

Asian Green Bee-eater *Merops orientalis*

Mainly Summer Breeding Visitor · Common

Familiar birds of the Indian countryside, represented by three subspecies in India: *beludschicus* in the west, the nominate in most of the country and *ferrugeiceps* in Assam and further eastwards. The Asian Green Bee-eater was formerly considered part of the very widespread Green Bee-eater, with eight subspecies in three groups – yellow-throated (*viridissimus* group) in Africa, blue-faced (*cyanophrys* group) in the Middle East and Arabia and russet-crowned (*orientalis* group) in southern Asia; these are now recognized as three separate species, the African, Arabian and Asian Green Bee-eaters respectively.
These bright cheerful birds are breeding summer visitors to Delhi (nominate subspecies *orientalis*) though a few may linger through milder winters. Numbers begin to increase in spring, when parties of up to 20 birds may also be seen in the mornings (they are day migrants) in late February and March, flying in a northwestern direction over the Delhi area. They move southwards again between late September and November, passing through in small parties and feeding as they go, and in this season may swarm in large numbers around roosting trees in the evenings. They affect a variety of habitats, cultivation, parks, wetlands, canal embankments, etc., and breed between April and June, nesting singly or in loose colonies in tunnels that they excavate themselves in an earth bank, sandy mound, or even in uneven ground.

Blue-cheeked Bee-eater

Merops persicus

Summer Breeding Visitor

Fairly Common

A Blue-cheeked Bee-eater pair, breeding summer visitors to the Delhi area

Migratory bee-eaters nesting across North Africa and West Asia to western India, the entire population wintering in Africa. Two subspecies are recognized, with the nominate in Asia.

Delhi is at the eastern edge of their range, and here they breed sympatrically with their sibling species, the Blue-tailed Bee-eater. They arrive by late April or early May and often prefer the proximity of marshes or waterbodies where flying insects, especially dragonflies, are abundant. Ganguli, in her *Guide to the Birds of the Delhi Area*, states that they were recorded breeding 'in great numbers' in the late nineteenth century at Garhi Harsaru near Sultanpur, Haryana, from mid-May to mid-July; they still nest, albeit in smaller colonies and numbers, in low earth hills and cuttings in the same general area. They are also likely to be confirmed nesting in other locations, where they are present through the summer and with flying young later in that season, as in Aravalli habitats at the Mangar Valley, Faridabad, and about Bhondsi, Gurugram, alongside the Blue-tailed Bee-eaters that also breed there. From late July to end September, very large gatherings sometimes collect near marshes and in the Yamuna floodplain preparatory to migration. By end October, all have left Delhi.

Blue-tailed Bee-eater *Merops philippinus*

Summer Breeding Visitor Fairly Common

Replaces the Blue-cheeked Bee-eaters *M. persicus* to the east of their range, from northern India to Southeast Asia. These two closely related, parapatric species occupy similar ecological niches but some barrier, morphological, behavioural or genetic, prevents inter-breeding and separates the population into two sister species. Instead of migrating south-westwards to Africa as the Blue-cheeked Bee-eaters do, Blue-tails move southwards and south-eastwards into the Indian peninsula, Sri Lanka and Southeast Asia for the winter.

They are present in the Delhi area from the third week of March onwards, but numbers visibly increase between late July and October before they leave for their winter quarters. At this time they collect about wetlands and grassy woodland on the Southern Ridge and feed on the post-monsoon profusion of dragonflies and other aerial insect life. Writing of the 1940s, Hutson considered them passage migrants, and Ganguli, of the 1960s, mainly monsoon visitors, but they breed locally in the Delhi area in colonies of 50 or more pairs, excavating their nest tunnels in earth cuttings and banks in July–August.

European Bee-eater *Merops apiaster*

Vagrant

Breeds in warm, dry and open habitats over a wide range that spans southern Europe, the Mediterranean region and Central Asia, with a disjunct resident population in southern Africa. The species has expanded its range northwards in Europe and eastwards in Asia over the last decades, especially marked since the 1990s. In Asia, its nesting range extends to Kazakhstan and Mongolia, south to Afghanistan and the northwestern subcontinent in western and northern Pakistan, and in India, in Kashmir. The bulk of the population winters in Africa, but small numbers, presumably birds that have bred towards the east of their range in Asia or in Kashmir, are non-breeding visitors in southern India as well, very locally in a belt around the mid-course of the Cauvery river in Karnataka and Tamil Nadu, and in Sri Lanka. Monotypic.

There has been just one recent observation of a small flock of these colourful bee-eaters in the area, a party of at least seven near the Najafgarh Jheel along the Delhi–Haryana border on 3 April 2022,[181] presumably a party en route from southern India to its breeding grounds further north, paralleling the very few other passage records of this species from Maharashtra and northern Karnataka in August and late March.

Chestnut-headed Bee-eater *Merops leschenaulti*

Vagrant

Resident in the Himalayan foothills and forested hill tracts in peninsular India, extending eastwards across Southeast Asia. Three subspecies recognized, two in India: *M.l. leschenaulti* ranging from India to Indochina, while *andamanensis* is isolated in the Andaman Islands.

There are only three records of these very handsome bee-eaters from the Delhi area, all within the last decade: at the Okhla Bird Sanctuary on 8 April 2017, in a New Delhi garden on 1 April 2020, and three together at Humayun's Tomb, New Delhi, on 21 July 2021.[182] They are mainly summer breeding visitors to the Himalayan foothills, and these records probably represent vagrancy during passage and, for the July record, perhaps post-breeding dispersal.

CORACIIFORMES: KINGFISHERS, BEE-EATERS AND ROLLERS
CORACIIDAE: ROLLERS

Rollers are named for their display flights in the nesting season, when they indulge in aerial acrobatics that show off the varied shades of blue in their plumage. They capture their food, invertebrate and small vertebrate life, by pouncing on them from a perch, and nest in holes in trees or earth and rock cuttings, lined with an untidy collection of grass and wood chips, on which usually three eggs are laid. Two species of rollers are on the Delhi list: one, a widespread resident whose gloriously blue-banded wings and tail are displayed to best advantage during courtship flights in spring, and the second, a passage migrant whose migration path passes by Delhi in autumn.

India: 4 species, Delhi: 2

European Roller *Coracias garrulus*

Passage Migrant · Uncommon

Widespread from Europe to Central Asia, in two subspecies, with the entire population wintering in Africa. The eastern race *C.g. semenowi* breeds from Iraq, Iran and Central Asia southwards, as close to the Delhi area as Kashmir.

Eurasian Rollers, of the eastern race, are regular migrants through the area on their southward passage between August and mid-October, when many birds are juveniles of the year. Delhi lies on the eastern fringes of their migration route in autumn, as the population breeding in Central Asia and Kashmir crosses the northwestern parts of the Indian subcontinent on the way to East Africa. They travel on a broad front and are reported from floodplain habitats, cultivated areas and especially the Southern Ridge. Numbers vary from year to year; they are usually recorded singly or in small groups in the Delhi area, but they are known to migrate in loose flocks where common, and a maximum of 60–70 birds has been reported on a late August morning at the Asola–Bhatti Wildlife Sanctuary on the Southern Ridge.

Indian Roller *Coracias benghalensis*

Breeding Resident · Fairly Common

Common residents over much of southern Asia from Iraq and the Persian Gulf to the subcontinent. Two subspecies are recognized: the nominate in north India, *indicus* in the peninsula and Sri Lanka. Formerly considered conspecific with Indochinese Roller *C. affnis*, which ranges from northeastern India to Thailand and Indochina.

These rollers are still fairly common breeding residents about Delhi, but may have lost ground in recent decades – they have declined elsewhere in northern India as well – for reasons that are not fully clear, though pesticides are a likely factor. Hutson, in his *Birds about Delhi*, had recorded pairs breeding in holes in walls of the Lodhi tombs in New Delhi in the 1940s, and till the 1980s a few still used to enter city limits in February–March, searching for nest sites and displaying over parks such as Sunder Nursery and on the Ridge. They no longer do so, but breed fairly commonly in rural areas between March and May; though normally considered sedentary, a strong post-breeding influx or passage movement is visible from late July through September when birds, including many juveniles, appear in good numbers in areas in and around the city where they are not in evidence in other months of the year.

Indian Rollers add a flash of colour to the scenery as they take wing.

PICIFORMES: BARBETS AND WOODPECKERS

MEGALAIMIDAE: ASIAN BARBETS

Barbets are small- to medium-sized, thick-billed, heavy-bodied, colourful relatives of woodpeckers with loud characteristic calls. Mainly frugivorous, they are unfailing attendees at fruiting fig trees, but will also take nectar and insect food. They excavate their own nest holes in soft wood or appropriate an already available one, the Delhi species typically laying three eggs on the bare bottom of the chamber. Two barbets are common residents of the Delhi area, their loud calls a familiar aural backdrop all summer in gardens and city parks.

India: 9 species, Delhi: 2

◄ *Coppersmith Barbet*

Coppersmith Barbet *Psilopogon haemacephalus*

Breeding Resident · Common

Very widely distributed across South and Southeast Asia to the Philippines. Nine subspecies are recognized over this range, of which race *indicus* is resident from the subcontinent to Indochina. Coppersmiths are common residents throughout the Delhi area in urban parks, gardens, roadside trees and groves in the rural countryside. Usually singly or in pairs, numbers may gather at fruiting fig trees which are an unfailing attraction, while they are also fond of nectar of silk cotton *Bombax ceiba* flowers when in bloom. They breed between March and June, though they may begin excavating their nest holes much earlier in winter and use them for roosting. Unusually for a barbet, they may sometimes collect in groups of several individuals on treetops in the evening before dispersing to roost, outside the breeding season.

Brown-headed Barbet *Psilopogon zeylanicus*

Breeding Resident · Common

Endemic to the subcontinent, a familiar bird, well adapted to human-modified habitats. Three subspecies are recognized: *caniceps* across northern India, two others in southern India and Sri Lanka. These barbets are common and widespread in Delhi in urban gardens, roadside trees and groves in the countryside. In Vol. 4 of their *Handbook*, Ali & Ripley wrote of this barbet, 'Has become conspicuously common and abundant in New Delhi during the last forty years due to increase in wooded gardens and roadside avenues with large *Ficus* and other fruit-bearing trees ... But it is not a new arrival in the area as often believed, being described as "common" in the (Old) Delhi environs even in 1893 ...'. Brown-headed Barbets begin calling in February, as soon as the weather shows signs of warming up and may be heard through till October. They breed between March and June, fearlessly contesting nest holes with parakeets and mynas.

PICIFORMES: BARBETS AND WOODPECKERS

PICIDAE: WOODPECKERS

Woodpeckers are adapted for a life on trees, climbing trunks and along branches, pecking, hammering and probing the bark to find food and excavating their own nest cavities in them – a chisel-like beak, a tongue that can be extruded a considerable distance into crevices to dislodge wood-boring grubs, ants and other insect larvae, stiff tails that prop their bodies against a tree trunk, and zygodactyl toes for a firm grip. The long tongue is curled around the back of the skull, to be extended when needed. Wrynecks, subfamily *Jynginae*, are aberrant in many respects. Woodpeckers seldom reuse the nest cavities they excavate, which then become homes for a range of other species of cavity-nesting birds, and they thus play an important ecological role in our woodlands.

India is rich in woodpecker diversity but only two or three are resident breeders in the Delhi region. Most woodpeckers, other than the distinctive wrynecks, are sedentary birds, at best dispersing to areas where there is an outbreak of the tree-destroying insects and grubs on which they feed. It is therefore surprising that one species has occurred in Delhi recently as a vagrant.

India: 33 species, Delhi: 5

Eurasian Wryneck *Jynx torquilla*

Winter Visitor · Uncommon

Breeds widely across temperate Eurasia, from Europe to northern China, wintering in Africa and southern Asia. Four subspecies are recognized, of which *J. t. himalayana* nests in the western Himalayas.

Eurasian Wrynecks are winter visitors to the Delhi area in small numbers (race *torquilla*, also *himalayana*?), commoner during passage from late August till October in autumn and mainly March in spring, when their numbers increase visibly and birds may be vocal, drawing attention to their presence. Wrynecks occur in open woodland, about river and canal embankments and open grass-and-scrub country in general, especially in the vicinity of anthills and termite nests which are their main food items.

Wryneck

Brown-capped Pygmy Woodpeckers probably nest in the Delhi area.

Brown-capped Pygmy Woodpecker

Yungipicus nanus

Breeding? Resident Scarce

A subcontinental endemic, with three subspecies, the nominate in most of India and two other races in the Western Ghats and Sri Lanka respectively.

The status of this woodpecker in the Delhi area still requires clarification. Earlier records were few and scattered; thus, the first reports appear to have been from Hastinapur, UP, in 1978 and the Southern Ridge between Surajkund and Badkhal Lake, Faridabad, in 1979,[183] and one was seen in *babool/kikar* woodland on the New Delhi Ridge in March 1996. They were then recorded through much of 2003 along *babool*-lined canals in the Sonipat district, after which this particular piece of habitat was destroyed. Since 2016, however, they have been reported, though apparently scarce, from woodland habitats of the Southern Ridge, notably from the Mangar Valley, Faridabad, through much of the year. These inconspicuous woodpeckers are resident along the Aravalli range in eastern Rajasthan as close as the Sariska National Park, and in all likelihood they are scarce residents, hitherto overlooked, in these Aravalli extensions in the Delhi area. On the other hand, they are regularly reported and evidently resident in woodlands of the Ganga *khadar* on the eastern side of the area, in the Hastinapur Wildlife Sanctuary in UP.

A Yellow-crowned Woodpecker at Bhondsi, Haryana

Yellow-crowned Woodpecker

Leiopicus mahrattensis

Breeding Resident Uncommon

Native to the subcontinent, with isolated small populations in Myanmar, Thailand and Indochina as well. Two subspecies are recognized: *L. m. pallescens* in Pakistan and northwestern India, with the nominate occupying the rest of the species' range.

Maratha Woodpeckers, as they used to be called, are uncommon breeding residents in the Delhi area in the larger city parks with remnant patches of native vegetation, in the Ridge habitats and groves in rural areas. They seem fond of *ronjh* and *babool* trees but will also be found in other woodland types. They breed from February to April, often excavating a nest hole on the underside of a branch which leads to a widened egg chamber, with three eggs forming the usual clutch.

White-naped Woodpecker *Chrysocolaptes festivus*

Vagrant

A subcontinental endemic, limited to India, the southwestern Nepal *terai* and Sri Lanka. Two subspecies are recognized, the nominate race *festivus* in India.

A female of this fine woodpecker was photographed at the Okhla Bird Sanctuary on 11 June 2017.[184] This normally sedentary species is resident along the *terai* from about Uttarakhand eastwards into Nepal and in much of peninsular India, including the Aravalli range northwards to Ranthambore and elsewhere in eastern Rajasthan (though not listed for the Sariska National Park). A stray individual could reach the Delhi area either from the Himalayan *terai* along the main rivers, or along the Aravalli range from the south; though an odd vagrant is of little significance, were this woodpecker to establish itself in the area, it would be a most welcome addition to Delhi's avifauna. There is another record of this species from just across the boundaries of the area covered by this Checklist, from the Haiderpur Wetland which forms part of the Hastinapur Wildlife Sanctuary on the Ganga, but is much closer to the *terai*, on 27 November 2020.

Black-rumped Flameback

Dinopium benghalense

Breeding Resident Common

A subcontinental endemic (as are most Delhi's woodpeckers, other than the Wryneck). Five subspecies, four in India, of which race *benghalense* is resident across northern India to Assam.

The commonest of the family about Delhi, present wherever there are large mature trees. It is immaterial whether these are in urban parks or roadside avenues, or in rural areas about villages and temple groves, their loud screaming calls immediately attract the observer's attention. They breed between April and August; their noisy flight chases, and birds searching for potential nesting sites, are a feature of its habitats from the third week of March. They excavate a round hole at mid-heights of a trunk or the underside of a branch, which leads to a widened egg chamber in which three white eggs are laid. Two broods may be raised in a season.

A Black-rumped Flameback, a conspicuous bird of gardens and avenues

FALCONIFORMES: FALCONS
FALCONIDAE: FALCONS

One of the more extraordinary results of phylogenetic DNA studies was the finding that falcons constitute their own Order, Falconiformes, which is closer to parrots Psittaciformes and passerines Passeriformes, and which has an altogether different evolutionary history from the Accipitriformes. Their similarities to the other raptors are the result of convergent evolution because of a common predatory lifestyle, not shared evolutionary history.

Falcons are hunters of open habitats; some smaller species, such as kestrels or hobbies feed mainly on insects which are captured on the ground, pounced upon from a perch or while hovering, or in the air when flying insects swarm in the mornings and evenings, but also commonly take small mammals and birds. Most larger falcons hunt birds, usually by dramatic dives or flight chases. They nest on cliff ledges, or appropriate unoccupied nests of other species such as kites or crows.

Two falcon species are residents in the Delhi area, but their numbers have been dropping and they need to be watched from a conservation point of view. Past observers had noted that both the dashing Red-necked and the larger Laggar Falcon of the hierofalcon group sometimes nested near or even in towns and villages, on minarets, old structures, etc., but there are no such recent reports from the Delhi area. On the other hand, both Laggars and Peregrines appear to have taken kindly to tall electricity pylons as hunting posts or loafing perches.

India: 12 species, Delhi: 8

▲ *A Shaheen against a blue sky*

Lesser Kestrel *Falco naumanni*

Vagrant

Breeding summer visitors from southern Europe, the Mediterranean region and West Asia eastwards to Mongolia and northern China, and winter mainly in sub-Saharan Africa. There are several mid-winter records from India too, and it seems that at least some birds from the eastern parts of their Asian range halt for the winter in the peninsula, while others carry on to Africa. Monotypic.

These brightly coloured falcons are only straggling migrants in the Delhi area: one at Sultanpur in March 1982 was the first record of the species from the Delhi area, with just two later reports, a pair at Bhindawas in November 2001 and four at Sultanpur on 30 April 2003.[185]

Common Kestrel *Falco tinnunculus*

Winter Visitor · Fairly Common

Widespread in Europe, Asia and Africa; in India, nest in the Himalayas and the Western Ghats, spreading over the rest of the country in winter. Ten subspecies are recognized over this vast range, three of which occur in the subcontinent – the nominate race of Eurasia, *F. t. tinnunculus*, breeds in the Himalayas east to Bhutan and winters in the plains; race *interstinctus* of East Asia winters in eastern India and Southeast Asia; and *objurgatus* is resident in the Western Ghats and Sri Lanka.

Winter migrants to the Delhi area between September and early May, in open grassy country of all sorts, including in cultivation, and often near archaeological ruins and monuments such as Tughlaqabad where they hunt from the ramparts. The neighbourhood of wetlands and the Aravalli landscapes with a mosaic of sparsely scrubbed ravines and cultivated patches are favoured habitats.

Red-necked Falcon *Falco chiquera*

Breeding Resident · Uncommon

Resident in generally low densities across sub-Saharan Africa and the Indian subcontinent. Three subspecies are recognized (the African races are split by some authorities); the nominate Indian race *F. c. chiquera* is wide-ranging but uncommon in lightly wooded, open landscapes, nesting often near villages.

Ganguli had considered these handsome, dashing falcons uncommon residents in the Delhi area in the 1960s, with some breeding records. They remain widespread in open country with a number of records of single birds and pairs from the river floodplains, grassy patches and the bird-rich habitats near lakes and *jheels*, such as the Sultanpur National Park and its environs, and elsewhere in the area. They nest in trees, either building their own or appropriating a kite's or crow's old nest, in which three or four eggs are laid. Pairs have been recorded nesting in palms in Delhi and at the Bhindawas Wildlife Sanctuary, Haryana, in recent decades.

The Red-necked Falcon, a dashing hunter of birds

A female Merlin, Sultanpur outskirts

Merlin *Falco columbarius*

Winter Visitor · Rare

Small falcons that breed throughout the northern forests and grasslands of North America, northern Europe and Asia, including the Central Asian steppes, from where birds migrate southwards in the cold season reaching Central America, much of Europe, the Middle East and as far as the northwestern parts of the Indian subcontinent. Nine subspecies are recognized, of which two, the Siberian breeding *insignis* and the Central Asian steppes race *pallidus*, have been recorded in winter in northwestern India.

There are scattered records of Merlins from the Delhi area since 1992, when a bird of the striking pale race *pallidus* was first reported at Sultanpur in early February. There are about a dozen later reports, mainly of single birds in lightly wooded remnant grasslands and open landscapes between November and March, including observations from the river floodplain, from Sonipat and from the Sultanpur National Park/Najafgarh wetlands neighbourhood.

Eurasian Hobby *Falco subbuteo*

Passage Migrant · Uncommon

Breeds across temperate Eurasia, including in the Himalayas, wintering mainly in Africa. Some hobbies pass the winter in the Indian subcontinent as well. Two subspecies are recognized; the nominate in India.

These elegant falcons are regular passage migrants through the Delhi area in autumn from August to mid-November, the numbers peaking in October when single birds or small parties of up to five may halt at some promising area for a few days. At this season, hobbies on passage feed on the then abundant dragonflies, sweeping over the reed beds at wetlands such as Okhla and elsewhere in the mornings and evenings, and resting in the surrounding trees at midday; they are also reported from the Aravalli landscapes of Delhi where they hunt along the ridges and dips. The northward passage, mainly in March–April, is more to the west, and they are less frequently recorded in the Delhi area at this time, but records of some very late returning migrants are noteworthy: in the second half of May over the Southern Ridge at Bhondsi and the Mangar Valley in the Gurugram and Faridabad districts, and at Surajpur, GautamBuddha Nagar district, and in the second week of June at the Basai wetlands, Gurugram; as also a couple of rare mid-winter reports.

Laggar Falcon *Falco jugger*

Former? Breeding Resident Scarce

Endemic to the subcontinent, one of a group of four closely related falcons in the subgenus *Hierofalco* – Laggar, Lanner, Saker and Gyr. The Laggar has been called 'the ordinary resident true Falcon of India' and 'the commonest and most easily identified of all our falcons'[186], descriptions that scarcely ring true today over much of its range. Rishad Naoroji, in his *Birds of Prey of the Indian Subcontinent* published in 2007, noted its decline but added that it remains common in semi-arid Gujarat and Rajasthan.

Hutson, writing of the 1940s, and Ganguli, of the 1960s, considered them breeding residents in small numbers about Delhi, and mention a pair that nested successfully on a tree near Okhla and another at on the exterior walls of Tughlaqabad. They have become very scarce in the Delhi area at least since the 1980s, paralleling their decline elsewhere. Causes are unclear but pesticides and trapping for the falconry trade may, in part, be responsible. There have been a few scattered records in recent decades: by the Yamuna River in February 2001; a pair at the Bhindawas Wildlife Sanctuary, Haryana, in September 2004; and from the Sultanpur National Park neighbourhood in August 2005, October 2019 and October–November 2020. But a number of recent reports, mainly from 2014 onwards, seem to cluster around the Greater Noida/Surajpur area, GautamBuddha Nagar district, where single birds have been recorded mainly in the winter months, October to March. The breeding season is between February and May, and a recently fledged, apparently weak, juvenile found on the ground in mid-May in a residential colony in Gurugram,[187] is ominously suggestive of an escaped trafficked bird or, optimistically, a nearby nesting site. The records do not show a clear pattern, but the repeated sightings about Surajpur over the Laggar's late winter nesting season point (hopefully!) to a breeding site in the neighbourhood.

Laggar Falcon

Peregrine Falcon

Falco peregrinus

Winter Visitor Uncommon

An iconic, cosmopolitan species that breeds in all continents except Antarctica. Rather variable in morphology and habitats over this vast range; 19 subspecies are recognized, of which three occur in the subcontinent: the pale, Siberian tundra breeding *F. p. calidus*, that winters widely across South Asia; the Central Asian Desert race *F. p. babylonicus* that winters in Pakistan and northwestern India; and the resident *F. p. peregrinator*.

Peregrines are quite widespread as winter visitors in the Delhi area between September and April, affecting wetlands by preference – which usually teem with waterfowl in this season – but some take up residence for a while in the city as well, using tall buildings, water tanks, pylons and the like as hunting stations. Peregrines spend the day loafing on high perches and power pylons, with an occasional fly-around in their inimitable style, and hunt ducks, waders, pigeons and even larger birds at dawn and in the evenings. The race that is most often encountered is *calidus*, the other two are far scarcer with only a few reports from the area in recent years. The distinctive subcontinental race *peregrinator*, or Shaheen, has been recorded locally in winter, between the last week of August and April, from the river floodplains (where one adult female has regularly wintered every year since at least 2018) and occasionally from the Sultanpur National Park area. These seem to arrive a little earlier than *F. p. calidus* and are probably birds dispersing from their breeding grounds in the Himalayan foothills. The race *babylonicus*, or Red-capped Shaheen, has also been identified in mid-winter from the Yamuna floodplain, Okhla, in early January 2008, the first record from Delhi; from the Dhanauri wetlands, the GautamBuddha Nagar district in January 2015, February 2018, January 2019 and January 2021; and from the Sultanpur area in November–December 2017.

A *Peregrine: Race* calidus, *Dhanauri wetlands*

B *Peregrine: Race* peregrinator, *Yamuna* khadar

C *Peregrine: Race* babylonicus, *Dhanauri wetlands*

PSITTACIFORMES: PARROTS
PSITTACULIDAE: OLD WORLD PARROTS

A highly diversified family with a remarkable variety of forms in Indonesia and Australasia. African and Asian forms are less varied (though the Indian Ocean islands had several extraordinary endemics that are now extinct). India has but two genera, Hanging Parrots *Loriculus* and Parakeets *Psittacula*.

India: 12 species, Delhi: 3

▲ *Parakeets feast on grain put out by gardeners*

Alexandrine Parakeet *Psittacula eupatria*

Breeding Resident · Fairly Common

Ranges from the Indian subcontinent to Thailand and Indochina. Five subspecies are recognized, four occurring within Indian limits, with race *nipalensis* representing the species across northern India.

Impressive parakeets of wooded gardens in the Delhi area. Till the middle of the twentieth century, they were considered only spring visitors and even in the 1960s Ganguli had considered them 'not too common residents in small numbers'. They now breed freely between December and April – though they begin to explore possible nesting sites as early as October – in tree cavities and, commonly, in holes and crevices in human-made structures and buildings, in which they lay a clutch of three to five eggs.

Rose-ringed Parakeet, male

Rose-ringed Parakeet *Psittacula krameri*

Breeding Resident · Common

Native to parts of sub-Saharan Africa and the Indian subcontinent to Myanmar, but now established in many locations across the world through deliberate introductions or escapes. Four subspecies are recognized: two in Africa and two in India, *borealis* in northern India and *manillensis* in southern India and Sri Lanka.

Familiar, noisy residents of urban habitats and widespread throughout the Delhi area, breeding from early spring in holes in trees, ruined monuments and inhabited buildings, and commonly adopting nest boxes. They feed on fruit, ripening sunflower seeds, flower petals and grain, large numbers collecting in orchards and grain storage depots, or at temples where they are fed by devotees. Flocks commute between communal roosts which are often located in city gardens and seasonal feeding areas and are a feature of Delhi's evening skyscape.

Plum-headed Parakeet *Psittacula cyanocephala*

Breeding Resident · Fairly Common

An endemic parakeet of the subcontinent, in northern Pakistan, the Nepal *terai*, India (except the Northeast) and Sri Lanka. Monotypic.

These colourful parakeets are localized breeding residents about Delhi, in wooded areas. Hutson, writing of the 1940s, had considered them occasional wanderers in the Delhi area, and Ganguli believed they began nesting in Delhi only in the early 1950s. Their numbers have increased since but tend to fluctuate year to year. They are less prevalent in urban habitats than Delhi's other two species, and favoured areas include the New Delhi Ridge where they collect in flocks to feed on grain put out by gardeners, and groves of large trees, especially with fruiting figs, as at Okhla and parts of the Southern Ridge.

A male Plum-headed Parakeet

PASSERIFORMES: PASSERINES
PITTIDAE: PITTAS

A family of stocky, short-tailed, colourful birds with a largely terrestrial lifestyle that are highly diversified in Southeast Asia, but a few species extend northwards to Japan, southwards to Australia and westwards as far as Africa. Pittas eat a variety of insect and other invertebrate life picked up from the leaf litter and build large globular nests of twigs, etc. with a side entrance, sometimes low down in a thicket but often placed quite high in a tree. Four to six eggs form a typical clutch. One species is a summer visitor to the Delhi area.

India: 6 species, Delhi: 1

Indian Pitta *Pitta brachyura*

Localized Summer Breeding Visitor Fairly Common

Endemic to the subcontinent, a breeding summer visitor to forests in its northern half, wintering in the peninsula and Sri Lanka. Monotypic.

Ganguli, writing in her *Guide to the Birds of the Delhi Area*, mentions only 'winter' records from the 1950s, and there are a few later reports of single birds from the Delhi area, mainly in June and August–October. These were considered passage records until the summer of 2015, since when several pairs were found to be breeding regularly in the summer months in the richer intact patches of forest of the Southern Ridge and Aravalli Range outcrops in Haryana, as at Mangar Bani, Faridabad district; Bhondsi, Gurugram district; Manethi, Rewari district, and doubtless elsewhere. They reach these sites around the second week of May, announcing their arrival with their arresting whistles, building nests by June and feeding young by July and August. These sites have been explored intensively by local birdwatchers only in recent years, and rather than representing a range extension, their nesting in the neighbourhood of Delhi was probably overlooked in the past. Indian Pittas breed along the Aravalli Ranges as far north as the Sariska National Park, and this is one of a handful of 'Aravalli' species that get a toehold in the Delhi area along the Southern Ridge biodiversity corridor.

Indian Pittas nest at Mangar Bani on the Southern Ridge

PASSERIFORMES: PASSERINES
CAMPEPHAGIDAE: MINIVETS AND CUCKOOSHRIKES

The minivets of Asia, subfamily Pericrocotinae, are slim, elegantly long-tailed, arboreal birds in bright reds, yellows, greys and black, with one widespread resident species in the Delhi area, one winter migrant from the Himalayas and a third thinly distributed species that had so far been considered a scarce and erratic wanderer into the Delhi area in winter, but recent summer records throw up the possibility that it just might nest here. The Cuckooshrikes, subfamily Campephaginae, birds of woodlands from Africa to Australasia, are rather thickset birds in mainly grey, white and black, not closely related to either cuckoos or shrikes. Both are primarily insectivorous, gleaning their invertebrate food from trees. Their nests are typically neat cups of vegetable material anchored to a small branch, camouflaged with lichens and spiderwebs. They are represented by three species in the Delhi area, one occasional, one breeding visitor and one vagrant.

India: 14 species, Delhi: 6

◄ *A male Small Minivet, a woodland resident*

White-bellied Minivet *Pericrocotus erythropygius*

Winter Visitor Scarce

Endemic to India, localized and generally uncommon birds of grass and acacia scrub habitats in western and central India. Mainly resident, but subject to erratic movements outside the breeding season. Monotypic.

From historical observations by Basil-Edwards in the 1920s, who found them in *babool* woodland in parties of up to six, and by Hutson in the 1940s, who considered them 'not at all common' in the Delhi countryside, these birds seem to have been uncommon visitors in winter about Delhi between November and March. After the 1940s, Ganguli remarked that they had become very scarce. Most reports since have also been in winter, from grass and acacia scrub habitats in south Delhi and Haryana: Mehrauli and Tughlaqabad in the 1960s when these areas were less built up and more recently from the Southern Ridge (Asola–Bhatti Wildlife Sanctuary south to Sohna, Manesar and Rewari), with a few records from the Sultanpur National Park as well. However, parties of up to five birds are not infrequently recorded in May and June from Southern Ridge habitats every year, especially from *babool*- and *kikar*-dominated woodland in the Mangar Valley, Faridabad and Rewari districts, which may be lingering winterers, but raise the possibility that these birds may nest in the Delhi area. Their typical habitat, very different from their co-generics, is certainly available aplenty on the Southern Ridge.

Small Minivet *Pericrocotus cinnamomeus*

Breeding Resident · Fairly Common

Common residents of light woodland habitats over much of the Indian subcontinent and across Southeast Asia to Java. Nine subspecies are recognized, of which five inhabit the subcontinent and vary in intensity of coloration; race *peregrinus* inhabits the Gangetic Plain of northern India.

Small Minivets are quite widespread in the Delhi area, frequenting patches of original woodland in the larger parks and gardens, groves in rural areas and on the Ridge. They breed between March and September, laying a clutch of usually three eggs in a tiny cup nest of soft plant material bound with cobwebs anchored to a fork of a small tree, often *ronjh* or a similar thorny one, where it is extremely well camouflaged.

Long-tailed Minivet *Pericrocotus ethologus*

Winter Visitor · Uncommon

Nest commonly along the Himalayas eastwards to China and Southeast Asia, the Himalayan nesters descending to foothills and northern Indian plains in winter. Six subspecies are recognized: two within Indian limits, of which the West Himalayan race *favillaceus* presumably visits the Delhi area in winter.

These eye-catching minivets are uncommon visitors to the Delhi area between November and March, when small parties may frequent wooded gardens, forest patches as at the Sultanpur National Park and woodland habitats of the Ridge. They are typical constituents of the mixed parties of insectivorous species that roam the jungles, a feature of the Himalayan hill forests but rather less in evidence in the plains.

Long-tailed Minivets are winter visitors to the area.

Large Cuckooshrike *Coracina macei*

Non-breeding Visitor · Scarce

Range over much of the Indian subcontinent though scarce or absent in the western parts, southern China and mainland Southeast Asia. Eight subspecies are recognized over this range, including the large *nipalensis* of the Himalayan foothills and northeastern India, and *macei* of central and southern India.

Delhi is located rather to the west of this species' breeding range in the northern plains and appears to be only an irregular visitor here. Records from the Delhi area are scattered mainly between July and March.[188] These records have not been racially assigned, though Ganguli suspected a pair seen in March 1969 to be of the northern race, while one male at Rewari on 1 March 2023 is clearly of the peninsular race *macei*. There are no reports during their nesting period April to June; the records are best interpreted as birds dispersing outside their nesting range in the non-breeding season.

Black-winged Cuckooshrike *Lalage melaschistos*

Vagrant

A cuckooshrike of the Himalayas to East and Southeast Asia, with four recognized subspecies, the nominate in the Indian subcontinent. An altitudinal migrant in the Himalayas, wintering in the foothills, some birds moving further south as far as central and eastern India.

The winter range of this species in peninsular India is mainly to the east and south of Delhi, and only accidental stragglers have been reported from the area. The first record was an immature recorded at Sultanpur in November 2005. There have been three reports since: a male at Dhaula Kuan, Delhi, and an immature at Mangar Bani, Faridabad district,[189] both in late December 2021, and one at Sunder Nursery, Nizamuddin, New Delhi, on 7 February 2022. This cuckooshrike has also been recorded once, in early February 2020, from the Haiderpur Wetland, Hastinapur Wildlife Sanctuary, which is located closer to the Himalayan foothills and just to the northeast of the limits of the Delhi area as defined here.

Black-headed Cuckooshrike *Lalage melanoptera*

Localized Summer Breeding Visitor Uncommon

A Black-headed Cuckooshrike at Mangar Bani

A subcontinental near-endemic, with two subspecies, the nominate in the West Himalayan foothills, *sykesi* over the rest of its range in India and Sri Lanka. Mainly a breeding summer visitor in northern and central India, resident in the peninsula.

This cuckooshrike was formerly known in the area from just three reports: a pair observed feeding two juveniles at Hastinapur in August 1981,[190] and two February records of presumed migrants from Sultanpur National Park, in 2001 and Okhla in 2012. With more intensive coverage and observation since 2015, they have been found to be breeding summer visitors to the Delhi area from late March to early October, recorded annually from the better-wooded habitats of the Southern Ridge, e.g. the Asola–Bhatti Wildlife Sanctuary, Mangar Bani in the Faridabad district, Bhondsi Nature Park and Manesar in the Gurugram district, Haryana. They nest here in April–May, building a small, shallow saucer of dry grass plastered with cobwebs in a forked branch of a tree, which holds a clutch of three eggs. There is just one winter record from the JNU campus, south Delhi, in December. Males attract attention as they sing freely on arrival, and this is another species that finds a foothold in the Delhi area in the finger of suitable habitat that extends along the Aravalli Ranges from the south.

PASSERIFORMES: PASSERINES

ORIOLIDAE: ORIOLES

These gorgeous, showy birds are a widespread Old World family, mainly frugivorous in diet although insects and other invertebrates may also be taken. Delhi hosts one of the loveliest of the family, the Indian Golden Oriole, whose liquid calls announce the advent of summer in the city's gardens and parks.

India: 6 species, Delhi: 2

Indian Golden Oriole *Oriolus kundoo*

Summer Breeding Visitor · Fairly Common

Ranges from India northwards to Central Asia; summer breeding visitor in the northern parts of its range, wintering in southern India and Sri Lanka. Split from the Eurasian Golden Oriole *O. oriolus* on account of differences in plumage, voice and absence of intergradation. Monotypic.

Delightful breeding visitors to the area from mid-March to October, these orioles are quite common in city parks and gardens, rural groves and woodland, and roadside avenues. They nest in such habitats from April to July, their finely built nests like small, deep pouches suspended by the rim from the fork of a small branch, in which their eggs, usually three in a clutch, are laid. Orioles' nests are often deliberately located near those of aggressive species such as drongos *Dicrurus* which confers a degree of protection.

An Indian Golden Oriole feeding on pipal figs

Black-hooded Oriole *Oriolus xanthornus*

Winter Visitor · Rare

A mainly resident oriole of India east to southern China and across much of Southeast Asia. Of its five subspecies, three occur within Indian limits: the nominate in northern India, *maderaspatanus* in the peninsula and *reubeni* in the Andaman Islands.

Delhi itself lies rather to the west of the normal Indian range of this widespread and common South Asian oriole, but a few individuals, mostly immature birds dispersing after the breeding season, have been reported from the area.[191] These orioles become progressively more regular towards the eastern districts of the area, and recorded through the year from Meerut, the Hastinapur Wildlife Sanctuary and the Ganga *khadar*, though even here most frequently as immatures.

PASSERIFORMES: PASSERINES

VANGIDAE: WOODSHRIKES

The four species of woodshrikes of southern Asia were in their own family, Tephrodornithidae, related to cuckooshrikes, but molecular analyses suggest that they are part of a larger clade that brings together the remarkable Malagasy vangas and a few other African and Asian genera. Woodshrikes are now placed in the family Vangidae, subfamily Tephrodornithinae.

India: 4 species, Delhi: 1

Common Woodshrike *Tephrodornis pondicerianus*

Breeding Resident · Fairly Common

Range across the Indian subcontinent to Southeast Asia. Three subspecies are recognized, with two in South Asia, *pallidus* in Pakistan and northwestern India (including the Delhi area), nominate in the rest of the country.

Woodshrikes require tree growth, neither too open nor too dense, and are typical birds of Ridge habitats, urban parks and gardens with patches of natural vegetation, and *Acacia/Prosopis* woodland elsewhere, where their rising series of plaintive whistles are a characteristic feature. They nest between March and June in the Delhi area, building a cup of twigs and bark wedged in a horizontal fork of a branch of a thorny tree. A typical clutch is of three eggs.

AEGITHINIDAE: IORAS

Ioras are a small family of four pint-sized attractive birds endemic to southern Asia. Breeding males of the two Indian species are largely black-and-yellow but with strong black-and-white contrasts that show to best effect during the males' aerial courtship displays.

India: 2 species, Delhi: 1

Marshall's Iora *Aegithina nigrolutea*

Localized Breeding Resident · Scarce

Endemic to the Indian subcontinent, a bird of dry open woodland biotopes of the Aravalli foothills and in northern Gujarat, with isolated populations in the southern peninsula as well. Monotypic.

Basil-Edwards found these ioras uncommon in babool woodland at the site where New Delhi was being built in the 1920s, and Ganguli recorded them from the New Delhi Ridge, from where they continued to be decreasingly reported until at least June 1986.[192] There were no reports from the Delhi area for over two decades thereafter and were feared locally extirpated, but they have been recorded again since 2012 and confirmed nesting on the Southern Ridge. Their nests are shallow cups of twigs and bark, plastered and anchored with cobwebs to a fork in a small tree and hold a clutch of two or three eggs. Marshall's Ioras may have withdrawn from central Delhi but are still welcome residents of the Delhi area.

PASSERIFORMES: PASSERINES
RHIPIDURIDAE: FANTAILS

Fantails are a family of Australasian origin that extends as far westwards as India, where we have three species. The Delhi area hosts one species of these active, attractive birds. They build a tiny neat nest of grass stems and cobwebs in an upright forked twig but make little effort to conceal it. As a result, fantails often suffer low nesting success.

India: 3 species, Delhi: 1

▲ *A White-browed Fantail on the Southern Ridge*

White-browed Fantail *Rhipidura aureola*

Breeding Resident · Uncommon

Mainly resident in the drier, more open woodland areas over much of the subcontinent, eastwards to mainland Southeast Asia. Three subspecies are recognized, of which nominate *aureola* is resident in Pakistan and northern India, and *compressirostris* in the peninsula and Sri Lanka.

These delightful fantails appear to have been commoner about Delhi in the past. Ganguli considered them local residents in small numbers in gardens, groves and woodland in the 1960s and mentioned old breeding records in the city itself. She believed they were decreasing in numbers even then, and the decline seems to have continued since in Delhi, paralleled over parts of the Gangetic Plain as well. They are, however, still regularly reported throughout the year from the Delhi area, usually single birds or pairs, but with most reports in winter, from acacia woodland, brushwood and groves and well-wooded gardens, etc., both in the city and rural areas. They are confirmed as breeding in small numbers in woodland pockets on the Southern Ridge, including in the Asola–Bhatti Wildlife Sanctuary and contiguous habitats extending southwards therefrom, in Delhi and Haryana. There is still plenty of habitat available for this species in the area, and it commonly occupies gardens in urban areas elsewhere, so reasons for its decline remain mysterious.

PASSERIFORMES: PASSERINES
DICRURIDAE: DRONGOS

Drongos are an African-Asian family with their greatest diversity in southern Asia, where some species have highly decorative and elaborate tail features. Insectivorous, drongos typically hunt from a perch by sallying forth for insects and commonly join mixed-species flocks of birds that traverse the woodlands. They have a varied repertoire of calls, from harsh scoldings to loud, melodious notes, sometimes mimicking local bird or even mammal species. Pugnacious and fearless, they will unhesitatingly mob raptors and larger birds if they threaten or even innocently approach their nests (hence, the colloquial 'King Crow' or *kotwal* in Hindi). Milder-natured species such as green pigeons or orioles often build in the proximity of a drongo's nest, as can be often observed in the Delhi area, to benefit from the protection it offers.

India: 9 species, Delhi: 5

◄ *Black Drongos, juvenile above, adult below*

Black Drongo *Dicrurus macrocercus*

Breeding Resident, also Passage Migrant · Common

Wide breeding range covering the Indian subcontinent, China and Southeast Asia. Northern populations in Pakistan, northwest India and much of China are migratory and vacate their nesting areas in winter. Seven subspecies are recognized, of which two inhabit the Indian mainland: *albirictus* in the north and nominate *macrocercus* in the peninsula.

Common about Delhi, in open country and cultivation, the Ridge habitats and larger city gardens, Black Drongos are subject to some local movements. Many that nest in city parks vacate these areas in winter and move to the countryside, returning to their urban breeding territories in March. They nest between April and early June in the Delhi area, laying three or four eggs in a shallow cup of twigs, grass etc., built usually towards the extremity of a branch of a large tree. In addition, Black Drongos that breed to the northwest of the Delhi area can be observed overflying the city on passage to and from their wintering grounds further south. Thus, between mid-March and early April, when many local pairs have already established territories and begun building nests, others can be seen moving in a northwestern direction, singles or loose parties following each other at about 30 m height in the mornings till about 10 a.m. From late September through October, they migrate southwards, moving down the Yamuna floodplain and over New Delhi city. Black Drongos often attend cattle for the insects they flush and can be very noisy and obtrusive in the evenings with their Shikra-like calls.

Ashy Drongo *Dicrurus leucophaeus*

Passage Migrant Uncommon

Nest from the Himalayas eastwards to China and over much of Southeast Asia and some of the peninsular Indian hills as well. Himalayan breeders spread in winter across most of the Indian subcontinent as far south as Sri Lanka. Fourteen subspecies are recognized, with two in India, *longicaudatus* breeding in the western and central Himalayas and *hopwoodi* from Bhutan eastwards to southern China.

Ashy Drongos are regularly recorded during passage through the area in late March–April and again in September–October, singly, in pairs or in small parties in wooded corners of city gardens and groves in the countryside. A few remain about Delhi through milder winters.

Ashy Drongo

White-bellied Drongo *Dicrurus caerulescens*

Localized Breeding Resident Uncommon

Endemic to India and Sri Lanka and are typical residents of monsoon forests from the Himalayan foothills southwards through the peninsula, including in the Aravalli Range. Three subspecies are recognized: nominate *caerulescens* in India with two others in Sri Lanka.

White-bellied Drongos were formerly considered only vagrants in the Delhi area in winter, with scattered records from parks and woodland between October and April. Since 2015, however, these drongos have been confirmed to nest regularly in the better-wooded habitat patches on the Southern Ridge, adding to the list of 'Aravalli' bird species that nest at these northern outposts of the range that still retain some of their original vegetation. Their breeding season is mainly March and April; they are seen at these sites through the winter as well, so at least some are resident here even if others disperse to account for the vagrancy reports elsewhere.

White-bellied Drongo

Hair-crested Drongo *Dicrurus hottentottus*

Vagrant

Resident of well-wooded Himalayan foothills and forested hill ranges of peninsular India, extending eastwards to China and Southeast Asia. Individuals and small parties are known to wander occasionally from the Himalayan foothills into the northern plains, but these seasonal movements depend on nectar supplies. Hair-crested Drongos appear to be inordinately fond of flower nectar, including of eucalyptus, and numbers gather to feed at these trees when a grove is flowering. Fourteen subspecies are recognized, represented by nominate *hottentottus* in India.

There are a few recent records of these drongos, all of single birds that have presumably wandered into the Delhi neighbourhood from the *terai*: at the Yamuna *khadar*, Wazirabad, north Delhi, on 9 April 2017; at the Aravalli Biodiversity Park, Gurugram, on 19 November 2019 and at nearby Rajokri, south Delhi, on 1 February 2020, both records possibly pertaining to the same individual; one at Mangar Bani, Faridabad district, that remained at that site at least from mid-September to mid-October 2021; on 4 June 2021 at the Bhondsi Nature Park, Gurugram district; at the Sultanpur National Park on 11 December 2021; and twice at Okhla on 7 August and again on 7–10 October 2022. There are a number of winter records and even some in midsummer, from the Haiderpur Wetland, including in parties of as many as 15 together; this site, however, is considerably to the northeast of Delhi and much closer to the normal range of this drongo in the UP *terai*.

Greater Racket-tailed Drongo *Dicrurus paradiseus*

Vagrant

Resident of semi-evergreen monsoon forests of India and Southeast Asia. Thirteen subspecies are recognized over its range, of which three represent this species in mainland India – *grandis* in the north, *rangoonensis* in the east and nominate *paradiseus* in the peninsula – with three other races in Sri Lanka, and the Andaman and Nicobar Islands.

These superb drongos are vagrants to the Delhi area. H. Alexander mentions a 'doubtful' 1949 record, but there have been at least six recent occurrences reported from the immediate environs of Delhi, all of the north Indian race *grandis*: at Usmanpur, Yamuna *khadar*, north Delhi, on 14 February 2006; at Sunder Nursery, New Delhi, on 28 June 2016; at Mangar Bani, Faridabad, one in the third week of May 2017, two there on 20 June 2021, and several observations at the same site from 10 July, when first recorded, through at least early November 2022 and one at Surajpur on 18 February 2023. Typically a forest species, this drongo is known to occasionally wander to isolated woodland patches across considerable distances of open country, which perhaps explains its appearances in the Delhi area. It is reported more frequently towards the eastern part of the area along the Ganga *khadar* in the Hastinapur Wildlife Sanctuary and adjacent Haiderpur Wetland, that are much closer to its usual residential range along the Himalayan foothills and *terai* from Kumaon eastwards, with several records in most months of the year, although there is no confirmation of breeding at that location.

PASSERIFORMES: PASSERINES
MONARCHIDAE: MONARCHS

The Monarchids are a family of strong-billed insectivorous birds, with fly-catching habits, of woodland habitats in Africa, southern Asia and Australasia, the males of several species adorned with crests and spectacularly long tail feathers. Highly diversified in the Australo-Pacific region, we have two of its loveliest and most graceful members on the Delhi list, the Black-naped Monarch and the Indian Paradise Flycatcher.

India: 4 species, Delhi: 2

▲ *Indian Paradise Flycatcher, a female at her nest on the Southern Ridge*

Black-naped Monarch *Hypothymis azurea*

Vagrant

Widespread in forest and wooded country along the Himalayan foothills and much of eastern and peninsular India, Sri Lanka and Southeast Asia to the Philippines. Twenty-three subspecies are recognized: *styani* in much of mainland India and Indochina (with three other races within subcontinental limits in the Andaman and Nicobar Islands and Sri Lanka). Many other subspecies are isolated island populations.

Black-naped Monarchs are vagrants to the area. Three birds in female/juvenile plumage were recorded at the Mangar valley, Faridabad district, on 15 July 2020, and at least one remained in the area till the end of that month, the first and only record for Delhi to date. Black-naped Monarchs have been reported from Sawai Madhopur and Jaipur, Rajasthan, and even nesting at Jasrota, Jammu, in June 2020, which together appear to represent an unusual flurry of reports in midsummer 2020 west of this monarch's usual range.

Indian Paradise Flycatcher *Terpsiphone paradisi*

Summer Breeding Visitor and Passage Migrant · Uncommon

Ranges from Central Asia to the Indian subcontinent with three subspecies: *leucogaster,* a summer visitor, nesting in Central Asia from Uzbekistan and Afghanistan to the Himalayas and adjacent plains of northern India, wintering further south; nominate *paradisi,* resident in peninsular India from the Aravalli and Vindhya ranges southwards; and *ceylonensis* in Sri Lanka.

Indian Paradise Flycatchers, apparently of the nominate peninsular race, are scarce summer breeding visitors to the Delhi area. They seem to have bred more commonly formerly, with Hutson mentioning nests in Delhi gardens in the 1940s. By the 1960s, however, Ganguli had remarked that Paradise Flycatchers rarely attempted to breed in Delhi. However, nesting, or attempts at nesting, have been recorded almost annually in recent years but localized in wooded Ridge habitats. A female observed on a nest at Dhaula Kuan, New Delhi Ridge, in June 2017 is the first from the New Delhi area after many decades; other breeding records include pairs nest building at Sanjay Van, on the South-Central Ridge, in June and July with fledged chicks in August 2017, and nests in woodland patches at the Bhondsi Nature Park and in the Mangar Valley, Southern Ridge, with pairs feeding chicks in July. Their nests are small, open cups of vegetable fibres, lichens and cobwebs, often placed in a bare fork of a branch. Generally, four eggs are laid, and since both sexes share incubation duties, a male Paradise Flycatcher can be ridiculously conspicuous in such exposed sites with his long tail hanging over the edge of the nest. Birds of the northern race *leucogaster* are more frequent, but still uncommon, passage migrants through the area from late March through May in spring, and between August and October in autumn, with as many as nine reported from a single site on the Southern Ridge on 31 March 2019.

An adult male Indian Paradise Flycatcher, white phase

PASSERIFORMES: PASSERINES
LANIIDAE: SHRIKES

A mainly Old World family, but with two species in North America, shrikes are boldly patterned, aggressive birds with sharp, hooked beaks, that hunt their food that ranges from insects to small mammals, reptiles and even birds, often impaling them on thorns and spines of bushes for consumption later. The family is represented in the Delhi area by three handsome resident shrikes, one of which appears to have lost ground for reasons that are not quite clear.

India: 13 species, Delhi: 6

Red-tailed Shrike *Lanius phoenicuroides*

Vagrant? / Reconfirmation desirable

Nest in Central Asia southwards to Afghanistan and winter in East Africa, with small numbers migrating through Pakistan and the Indian border states of Rajasthan and Gujarat in autumn.[193] Formerly considered conspecific with *L. isabellinus*. Rasmussen & Anderton state that 'birds reported as this in winter from extreme western India (mostly Kachchh) are closer to Isabelline Shrike; pure *phoenicuroides* winter extralimitally.'

References to a few birds wintering about Delhi,[194] reports of immatures from the Yamuna *khadar*, and an observation of an apparently adult male from the Najafgarh Jheel, Gurugram district, in mid-winter,[195] cannot be categorically identified as this species.

Isabelline Shrike *Lanius isabellinus*

Winter Visitor Uncommon

Nest in eastern Russia, Mongolia and western China with three subspecies: the warm buff nominate *isabellinus* (Daurian Shrike) in the northern parts of its range which winters mainly in Africa, and the paler, greyer, races *arenarius* and *tsaidamensis* (Chinese Shrike) of Xinjiang and Mongolia, which winter in Pakistan and western India.

These shrikes are regular but uncommon visitors to the Delhi area from September to April, affecting dry open country, tamarisk scrub and desiccating flats near marshes. Rasmussen & Anderton claim that race *isabellinus* is 'not definitely known for our region', but *isabellinus*-type individuals have been verifiably recorded in western India and also identified as such at Basai in December 2002.[196]

Isabelline Shrike, in winter

Brown Shrike *Lanius cristatus*

Winter Visitor Scarce

Eastern Palaearctic shrikes that nest in central and eastern Siberia, northern China and Japan, and winter in India mainly to the east and south of Rajasthan, Haryana and Gujarat, and extralimitally in much of China and Southeast Asia. Four subspecies are recognized; birds wintering in northern India are mainly of the nominate race, while the distinctive *lucionensis* also winters sparsely in southern India and Sri Lanka.

Hutson deemed them winter visitors in small numbers between September and March in the 1940s but may have confused them with Isabelline Shrikes which he saw just once but are actually much more frequent. Nevertheless, Brown Shrikes have been recorded infrequently between September and March, from the Delhi area: from Okhla and elsewhere in the floodplain, Jhajjar and Sonipat districts, and with several recent records from the well-watched Sultanpur National Park neighbourhood, Haryana. They can be expected to appear with greater frequency towards the east of the area, along the Ganga *khadar* in UP.

Bay-backed Shrike *Lanius vittatus*

Breeding Resident Fairly Common

A Bay-backed Shrike in the Aravallis

Widespread from eastern Iran and Afghanistan to India (except the northeastern parts). Two subspecies are recognized: nominate *vittatus* in India.

These brightly patterned shrikes are fairly common residents in the Delhi area, in open woodland in rural areas and village precincts, neglected corners of the larger city parks and groves and the less densely wooded areas of the Ridge. They breed in such habitats between March and July–August, building a substantial cup nest of twigs lined with hair and feathers in a fork of a small thorny tree, often *babool* or *ronjh*, to hold a clutch of four to six eggs. Their numbers about Delhi may have declined in recent years, as they seem to be missing from some areas where they were formerly frequently recorded as nesting and resident.

Long-tailed Shrike *Lanius schach*

Breeding Resident · Fairly Common

The Long-tailed Shrike, the most frequently encountered shrike of the area

A widely distributed Asian species, from the Indian subcontinent to China and across Southeast Asia. Nine subspecies are recognized, of which three inhabit India: *erythronotus* in the north, *caniceps* in the peninsula and the black-headed *tricolor* in the east and Northeast.

These shrikes, of the race *erythronotus*, are common residents of the area and have evidently increased since the 1960s when Ganguli had called them the 'least numerous of the three resident shrikes'. Since the 1990s certainly they are the most frequently encountered and conspicuous shrikes in bush and grass country and open woodland and are also partial to the vicinity of marshes and wetlands. They breed from April to July, building a bulky cup nest of thorny twigs, grass and moss, lined with finer material, and placed in a thorny bush or a small tree; fledged young are regularly seen in July and August. Their numbers seem to increase in winter and possibly include some migrants from the northwestern parts of its range, where this same race, *erythronotus*, is migratory, wintering in north India.

Great Grey Shrike *Lanius excubitor*

Breeding Resident · Now Scarce

Great Grey Shrike

A widely distributed species, breeding from northern and Central Europe to Central Asia, and southwards to the northern half of Africa, Arabia and much of the western and northern Indian subcontinent. Northern populations are migratory, the others mainly resident. Twelve subspecies are provisionally recognized, of which four occur in South Asia, but only one, the resident Indian race *lahtora*, concerns us in the Delhi area.

These handsome shrikes were fairly common residents of the dry open country about Delhi in the past[197] and remained so in such habitats and in open cultivated areas with a few thorny trees, till the 1970s at least. They bred regularly in such habitats, their nests were deep cups of twigs and grass built typically in an isolated *babool* tree standing in cultivation. They are now much scarcer, with no recent breeding records. This change in status is a mystery; there is still suitable habitat available, yet today they are the least common of our regularly occurring shrikes. A decrease in numbers of this species has been noted elsewhere in India too.[198]

PASSERIFORMES: PASSERINES

CORVIDAE: CROWS AND ALLIES

Treepies, in subfamily Crypsirininae, and typical crows, in Corvinae, are two (of five) clades in a varied, cosmopolitan family and include some of the most intelligent of birds: adaptable, capable of learning, with a brain-to-body mass ratio said to equal that of great apes. Delhi hosts three Oriental Region species as familiar residents and a race of the Raven has been recorded as a vagrant.

India: 23 species, Delhi: 4

▲ *Rufous Treepie*

Rufous Treepie *Dendrocitta vagabunda*

Breeding Resident · Common

One of a small group of arboreal magpies of the Oriental Region, Rufous Treepies are the most widely distributed of the genus in the Indian subcontinent and across Southeast Asia. Nine subspecies are recognized, five within Indian limits, where the nominate race *vagabunda* of northern India represents the species in the Delhi area.

A graceful, conspicuously patterned bird and a widespread and common resident of the area, inhabiting wooded country of all types, urban gardens, roadside trees, light woodlands, etc. where their loud calls – some melodious and pleasant, others harsh and scolding – are familiar sounds. They breed about Delhi in summer from April to June, laying a clutch of four or five eggs in their nests, which are relatively small, cups of thin twigs lined with softer material, often in an isolated tree in the countryside.

House Crow *Corvus splendens*

Breeding Resident Common

Essentially South Asian, but introduced and naturalized, deliberately or as stowaways on ships, to many parts of the world. Five subspecies are recognized, the birds about Delhi probably intermediate between the paler northwestern *zugmayeri* and nominate *splendens* of most of India.

Commensal with humankind, House Crows are very much at home in urban Delhi. They breed throughout the area between May and August, building untidy open cup structures of sticks which are often very visible in trees and even on power pylons, in which typically four or five eggs are laid. Outside the nesting season, they roost communally in groves of large trees. House Crows are cunning, adaptable creatures, omnivorous and opportunistic in their food habits; they will scavenge for scraps, kill and eat any living thing they can overpower and be a menace to eggs and young of other birds. The only creatures that regularly get the better of House Crows are Koels *Eudynamys scolopaceus*, brood parasites which foist their own eggs and chicks on them and are often seen being chased by the crows, screaming at the top of their voices.

Large-billed Crow *Corvus macrorhynchos*

Breeding Resident Scarce

As a species, ranges from the Indian subcontinent northeastwards to China and Japan, and Southeast Asia. Up to 13 subspecies are recognized, which some taxonomies split into four species, including three within Indian limits: Indian Jungle Crow *C. culminatus*, Eastern Jungle Crow *C. levaillantii* (in eastern India, extending into Southeast Asia) and the Large-billed Crow *C. macrorhynchos* (with two races in the Himalayas, several others in East Asia) on grounds of vocal, morphological and other differences. The Indian Jungle Crow is then endemic to India and Sri Lanka.

A scarce bird about Delhi, which is towards the western fringe of its range in the Gangetic Plain. It was listed as a vagrant in Delhi prior to 1950, and Ganguli mentions only very few sightings from the 1960s. They have increased since, albeit slowly, and are best considered scarce residents with a few nesting records in our area.

Common Raven *Corvus corax*

Vagrant

This corvid, the largest bodied of all passerine birds, is a very widely distributed species across the Northern Hemisphere in wilderness areas of all types except tropical rain forests. Eight subspecies have been named over this range, of which two extend to within Indian limits: *tibetanus* in the high Himalayas and *laurencei* in northwestern India. It seems to have bred sparingly in the drier landscapes in Haryana westwards of Delhi over a century ago.

This corvid is a vagrant in the area. Frome mentions 'a doubtful sighting' from Gurugram, Haryana, in December, and a pair near Hauz Khas, Delhi, in January, probably in the early 1940s. The only later records are of two on the New Delhi Ridge on 20 March 1996 and one at the Bhindawas Bird Sanctuary on 15 May 2011.[199]

PASSERIFORMES: PASSERINES
STENOSTIRIDAE: FAIRY FLYCATCHERS

This lately set up Afro-Asian family of small insectivorous birds represents a clade comprising a few African flycatchers and three Asian species: two canary-flycatchers in the genus *Culicicapa* and one aberrant fantail, *Chelidorhynx*. These are incessantly active, cheerful birds that dart to and from their perches to capture flying insects in mid-air and regularly join mixed-species flocks. They build tiny, well-camouflaged nests of vegetable fibres and cobwebs on the moss-covered branch of a forest tree. We are concerned with only one species, as a winter visitor, in Delhi.

India: 2 species, Delhi: 1

◄ *Grey-headed Canary-flycatchers enliven the woodlands in winter.*

Grey-headed Canary-flycatcher *Culicicapa ceylonensis*

Winter Visitor · Uncommon

Breeds in forested hills from the Indian subcontinent through Southeast Asia and spreads into the foothills and plains in winter. Five subspecies are recognized, of which two inhabit India: *calochrysea* that breeds in the Himalayas and winters in the northern plains and *ceylonensis* in southern India. These are uncommon winter visitors to the Delhi area between October and March, when single birds, or twos and threes together, frequent the better-wooded corners of parks, gardens and groves. They are active, lively flycatchers that often join mixed hunting parties of small insectivorous birds, when their loud and cheerful calls are unmistakable indicators of their presence.

PARIDAE: TITS

These are an Old World and North American family of small woodland birds with generalist lifestyles and broad uniformity in structure and habits, with the exception of four aberrant species, one of which figures on the Delhi checklist as a rare spring migrant. They are mainly insectivorous, though they will readily take seeds, buds and berries as well and are typical members of mixed-species flocks. They nest in holes, usually in trees in the Delhi area, but elsewhere also under eaves of houses, weep-holes in retaining walls and similar sites, in which normally four to six eggs may be laid on a pad of dry grass, moss, hair and similar material.

India: 15 species, Delhi: 3

Fire-capped Tit *Cephalopyrus flammiceps*

Passage Migrant · Scarce

Fire-capped Tit, Bhondsi

Breeds in broadleaf forests at mid-elevations in the western Himalayas as far east as Nepal, and in China; the Himalayan population winters in central India from eastern Rajasthan and southern UP south across Madhya Pradesh. Fire-capped Tits are aberrant members of the family, formerly classified with the Penduline Tits but their nesting biology is like the true Tits. Two subspecies are recognized: nominate *flammiceps* in the western and central Himalayas and *olivaceus* in the Eastern Himalayas to southern China.

These bright little tits are scarce passage migrants through the area in late February and March, reported from about Delhi at long intervals until recently and possibly overlooked in most years. Historically, a small party had been recorded on 15 March 1957, when the species was considered 'accidental' in the area, and after nearly three decades, a group of five at Sunder Nursery, New Delhi, on 29 March 1986.[200] Since 2019, however, they are being reported regularly every year between mid-February and March from the woodland on the Southern Ridge, and particularly at the Bhondsi Nature Park, Gurugram district, with about 20 birds on 25 March 2019; more than 20 in the third week of March 2020; four there in late February 2021; at least 10 in the first half of March 2022 and 14 there from the third week of February till the first week of March 2023. One, in non-breeding plumage, was recorded at the Bhondsi Nature Park in December 2020, the first mid-winter record from the area; this is not entirely unexpected, however, as there are a few mid-winter records from Ranthambore, Bharatpur and Ambala too;[201] and another at the Mangar Valley, Faridabad district, on 30 October 2021, represents the sole autumn record from about Delhi. These reports suggest that this species is a very localized, but regular, spring passage migrant in small numbers through the Delhi area, in late February and March, with odd individuals recorded in October and mid-winter as well, rather than a mere vagrant as hitherto believed.

Cinereous Tit *Parus cinereus*

Localized Breeding Resident · Fairly Common

Resident of southern Asia from the Indian subcontinent eastwards to Indonesia, nest in the Himalayas and peninsular hills, spreading in winter into the adjacent plains as well. Twelve subspecies are recognized over this range, five occurring within Indian limits. Race *stupae* of peninsular India extends west to the Aravalli Range and to the Delhi area.

These tits were formerly considered vagrants about Delhi in winter, with just a handful of reports from Southern Ridge habitats, urban gardens and woodland patches in rural areas. Since 2015, however, they have been found to be fairly common but localized breeding residents, nesting in holes in *babool* woodland, either natural or excavated by other species such as woodpeckers, in these South-Central and Southern Ridge landscapes. Since they are resident at the Sariska National Park in the northern Aravallis, 150 km to the south, it was only to be expected that they would be discovered nesting here – yet another species of Aravalli biotopes whose breeding range extends into the Delhi area along the Southern Ridge.

White-naped Tit *Machlolophus nuchalis*

Localized Breeding Resident · Scarce

Endemic to India with a restricted and fragmented range in dry deciduous and thorn forest in the western and southern parts of the country. In Rajasthan, they are known from the Jaipur and Sikar neighbourhoods, less than 200 km from the limits of the Delhi area. Monotypic.

These rather unusual tits have been regularly recorded from within the limits of the Delhi area since 2016, when they were first reported from dry mixed woodland on the hilly outliers of the Aravalli Ranges in the Rewari (at Manethi) and Mahendragarh districts of southern Haryana. The temporal spread of the observations all through the year is strongly suggestive of these tits being local residents and nesting in these hilly thorn scrub woodland habitats. They have evidently been overlooked in these ornithologically unexplored areas in the past.

PASSERIFORMES: PASSERINES

REMIZIDAE: PENDULINE TITS

These are tit-like birds of Eurasia, Africa and North America, that build soft, pear-shaped nests of plant fluff and cobwebs that hang from twigs ('penduline') of marsh-side vegetation, usually over water.

India: 1 species, Delhi: 1

White-crowned Penduline Tit *Remiz coronatus*

Winter Visitor · Rare

Winter visitors to the northwestern subcontinent from their breeding grounds in Central Asia, affecting varied habitats in the non-breeding season, including tamarisks, scrub on reservoir embankments, and reed beds (shunned in breeding season, but popular in winter). Monotypic.

White-crowned Penduline Tits are rare winter visitors so far east on the subcontinent, with a few reports of single birds or in parties from waterside vegetation and reed beds in the area, all between November and February. First recorded in the area by the river at Okhla, parties of four to twelve birds in scrub, in December 1981, having been possibly overlooked earlier, with later reports from the same site in February 2006, November 2011 and late December 2020. They were also recorded from the Sultanpur National Park environs in February and November 2019, through the 2021–22 winter from December to February and again in the following 2022–23 winter; from the Mandothi wetlands, Jhajjar district, in December 2020, as well as in the Ganga *khadar* at the Haiderpur Wetland, Hastinapur Wildlife Sanctuary, in December and February 2020.

White-crowned Penduline Tit, at the Budhera reed beds

PASSERIFORMES: PASSERINES

ALAUDIDAE: LARKS

A predominantly Old World family of open country ground-living birds, superficially similar but actually quite diverse, and their habitat, habits and song flights are all useful pointers in identification. Larks nest on the ground, a cup of fine grass in a shallow scrape or dried hoofprint, and in some species, surrounded with a low parapet of gravel. Construction is usually by the female, as is incubation of a clutch of usually three or four eggs.

Delhi's dry open landscapes suit them well, and the family is well represented in the Delhi area by eight breeding residents and five winter migrants from Central Asia. The local movements of some of our resident species are poorly known.

India: 22 species, Delhi: 13

The Oriental Skylark, a bird of grassland and cultivation

Rufous-tailed Lark *Ammomanes phoenicura*

Breeding Resident? Scarce

A subcontinental endemic, widespread resident across the drier parts of India, from Rajasthan, Haryana and Delhi southwards through the peninsula, in open habitats, grasslands of the Deccan Plateau, even fallow fields. Subject to local movements and apparently mainly a summer visitor towards the northwestern limits of its range. Monotypic.

This lark's range extends to Delhi along the Southern Ridge, north to Tughlaqabad. They are probably scarce breeding residents in these dry, stony habitats and with recent records from open ground and sandy flats near the Sultanpur National Park as well. Most reports fall between March and November, and there are observations of small parties of three or four flying in a northward direction in April–July, the significance of which is unclear. Their normal breeding season is March–April, and they almost certainly nest in the Delhi area though there has been no evidence, direct or circumstantial, of breeding – or even birds performing their distinctive flight displays – so far.

The Ashy-crowned Sparrow Lark, a common lark of open, bare ground

Ashy-crowned Sparrow Lark

Eremopterix griseus

Breeding Resident · Fairly Common

Another subcontinental endemic of dry, open habitats. Monotypic.

These quaint larks are fairly common residents of rural areas about Delhi, affecting sandy flats and open waste ground on the edges of cultivation and bare overgrazed land around villages. They are most in evidence in the summer between February and September when birds are displaying, and it is possible that some move out of the Delhi area in the coldest months, though they are recorded throughout the winter. The *Atlas*, however, had considered them summer visitors, breeding before the rains set in with perhaps a second brood also after the monsoon.

Singing Bushlark *Mirafra javanica*

Localized Summer Breeding Visitor · Scarce

A patchily distributed species in Africa, the Indian subcontinent, eastwards to Australia. Twenty subspecies are recognized; the South Asian race *cantillans* in India, described by Hugh Whistler as 'a curiously local bird, restricted in places even to particular fields ...'[202]

Singing Bushlarks appear to be localized breeding visitors to grassy flats in the Sultanpur National Park neighbourhood, Haryana, and possibly elsewhere in the Delhi area as well. They had not been reliably recorded from the Delhi area historically, until a singing bird was first identified from the Sultanpur National Park area in June 2004.[203] In recent years, however, and especially since 2016, these larks have been recorded annually from this neighbourhood and display a clear seasonal pattern between May and September, with males in song in July–August, coinciding with their nesting season. One record from Greater Noida, GautamBuddha Nagar, in early August 2018 is the only other known locality for this species in the Delhi neighbourhood.

Bengal Bushlark *Mirafra assamica*

Breeding Resident · Uncommon

Bushlarks of damp grassland in the *terai* and northern Gangetic floodplains, east to Assam and Bangladesh. Monotypic.

The first records of Bengal Bushlarks from the Delhi area were in the 1990s, when a number of birds were observed displaying and in song at Okhla and Madanpur *khadar*, Delhi, between March and October. They have since been reported more widely from damp habitats in the Yamuna floodplain and near various other wetlands in the area. Their structure and jizz are unmistakable, as are their characteristic song flights, in which the bird flies up about 50 m and then circles in undulating scallops, alternating quick flutters up and short glides down with open wings, as it repeats a brief note followed by a buzz, at each peak.

Indian Bushlark *Mirafra erythroptera*

Breeding Resident · Fairly Common

Indian Bushlarks prefer open scrub habitats.

Endemic to India and Pakistan, in dry scrub and often rocky habitats. Monotypic.

Indian Bushlarks are common residents of the Delhi area in stony scrublands of the Southern Ridge, sparse grass-and-bush country in the floodplain along river embankments and uncultivated scrubby patches in the countryside. They may abandon former habitats, such as the New Delhi Ridge, if the vegetation becomes too overgrown, dense and tall. They breed from March through the monsoon to October, when their display flights and song are a typical feature of their habitats.

Horned Lark *Eremophila alpestris*

Vagrant

Widespread larks of open landscapes of the Holarctic region, in a range of latitudes from the Arctic south to Central Asia. In the Himalayas, where they nest at high altitudes, they normally do not descend below the treeline in winter. Up to 42 subspecies are recognized over this vast range, two breeding within Indian limits.

One individual of this lark was recorded at the Madanpur *khadar* with a party of Tree Pipits *Anthus trivialis* on 30 September 2001.[204] This is the only record of the species south of the Himalayas.

Greater Short-toed Lark

Calandrella brachydactyla

Winter Visitor · Common

Greater Short-toed Lark

These larks breed from the Mediterranean region to Central Asia and winter southwards and eastwards to North Africa, the Middle East and South Asia. Seven subspecies are recognized over this range, of which the Central Asian race *longipennis* winters in India on semi-arid plains, mainly towards the west and northwest.

These larks are locally common in the Delhi area between September and April in fallow land, dry plains and overgrazed ground, in flighty flocks that wheel over the flats in synchrony, settling every now and then to feed. They are frequent in the floodplain, fallow fields and open dry ground near wetlands, as about Sultanpur, Haryana.

Mongolian Short-toed Lark *Calandrella dukhunensis*

Passage Migrant? / Reconfirmation desirable

The easternmost populations of the Greater Short-toed Lark, breeding from Mongolia to Tibet, that were formerly subsumed as race *dukhunensis* of that species, have recently been split as a species in their their own right, as Mongolian (or Sykes's) Short-toed Lark, which genetic studies suggest is actually closer to Hume's Short-toed.[205] Winter in the Gangetic Plain and the peninsula, mainly to the east and south of the wintering range of the Greater Short-toed *C. b. longipennis*. Monotypic.

This species has been identified from photographs as passage migrants about Delhi only lately, probably correctly. Criteria for separating Mongolian from Greater Short-toed Larks in the field have, however, only recently been defined and earlier records have not been subspecifically identified; the precise status of Mongolian vis-à-vis Greater about Delhi cannot be determined with the data available at present.

Hume's Short-toed Lark

Hume's Short-toed Lark

Calandrella acutirostris

Winter Visitor Uncommon

Breeds in Afghanistan, Baluchistan, parts of Central Asia, and Tibet, and migrates in winters to the northern subcontinent. Two subspecies are recognized, the nominate and *tibetana*, both occurring in northern India.

These larks had never been recorded historically from Delhi, certainly overlooked, but recent observations suggest that they are uncommon visitors from November (with unconfirmed reports as early as September) to mid-March. Identification problems confound records. Most records are from the well-studied Sultanpur National Park neighbourhood, from near Sonipat, and the Yamuna *khadar*, where they seem to prefer sparse grass cover rather than bare ground,[206] and do not usually mix with flocks of Greater Short-toed Larks.

Bimaculated Lark *Melanocorypha bimaculata*

Winter Visitor Scarce

Breeding summer visitors to Turkey, the Caspian basin and Central Asia, that migrate in winter to northeastern Africa, the Middle East, Pakistan and India as far east as the Yamuna River. Monotypic.

Delhi is at the eastern edge of the wintering range of these robust, thick-billed larks, and they tend to be rather scarce here. There was just one historical record till the 1970s from the Yamuna floodplain, but it has since been recorded irregularly – and not every year – in small parties in dry fields and bare grazing ground near villages, and short-grass flats in the neighbourhood of marshes, as in the environs of the Sultanpur National Park, and in the river floodplain. Small numbers often associate with flocks of Short-toed Larks and in flight may be readily picked out by their larger size and proportionately shorter tails.

Turkestan Short-toed Lark *Alaudala heinei*

Vagrant / Taxonomy dependent

Inhabits the bare, arid plains habitats from Eastern Europe and Turkey to Central Asia, the latter populations wintering to southwestern Asia. Three subspecies are recognized, of which two have been recorded from within Indian limits: Iranian breeding *persica* is known to wander into the Indus Plain in winter with a few records from western India as well, including from Haryana, and Central Asian breeding *heinei* has been identified wintering in Rajasthan.

The first report of this lark from the Delhi area (and the easternmost from India) was one associated with a party of Sand Larks *A. raytal* at the Yamuna *khadar* near Faridabad on 31 December 2016.[207] Uncertainty remains, however, about the specific identification of this bird, whether of the Turkestan Short-toed Lark, subspecies *persica* or *heinei*, or one of the Asian Short-toed Lark *A. cheleensis* group, or even a Sand Lark, most likely *A. r. adamsi*.[208]

Sand Lark *Alaudala raytal*

Breeding Resident · Fairly Common

Locally common along the major river systems of the northern subcontinent and adjacent cultivation on sandy soils in the floodplain. Three subspecies are recognized: the western race *adamsi* of the Indus Plain, Rajasthan and Punjab, and the eastern nominate of the Gangetic Plain intergrade in Haryana, and the birds about Delhi may show intermediate characters; *adamsi* having a thicker bill than *raytal*.

This is a fairly common lark of the Yamuna and Ganga sandbanks and associated *khadar* habitats of the area. Their silvery coloration can be very obliterative on dry river sand, but their tinkling song is a feature of their preferred habitat, background music to the river landscape as it were, all through their breeding season from March to May. Birds sing from the ground, strutting about with raised crest, or in song flight with rapid wing flapping alternated with pauses, before parachuting down in a series of steps.

A Sand Lark at a Yamuna sandbank

Eurasian Skylark *Alauda arvensis*

Winter Visitor   Scarce

Widespread across temperate Eurasia, wintering southwards to North Africa, the Middle East, northwest India and China. Thirteen subspecies are recognized; the Central Asia breeding race *dulcivox* winters to India.

Eurasian Skylarks are scarce and irregular winter visitors to the Delhi area. Historically, both H. Alexander, writing in 1948, and Ganguli, of the 1960s, mention a few records of this Skylark from the floodplain and the Gurugram district. In years when they do appear, they have been recorded in small parties in mid-winter, December–January, in open flats, fallow fields and stubbles, mainly on the Haryana side of the Delhi area. One, photographed at Basai, Gurugram, in mid-June[209] was out-of-season, possibly an over-summering injured bird or, more remotely, an escape.

Oriental Skylark *Alauda gulgula*

Breeding Resident   Common

Widespread over the southern half of the Asian continent, from Central Asia to China, the Indian subcontinent and Southeast Asia. Eight subspecies are recognized, four within Indian limits, of which the pale race *inconspicua* of Central Asia extends to northwestern India.

The resident race of Oriental Skylark about Delhi is probably *inconspicua*, a common bird of the open countryside, in the river floodplain, in cultivation and damp grassy ground near *jheels*. Their exuberant and sustained song, delivered from the air while hovering on vibrating wings, is a characteristic feature here all through spring and summer, especially in the nesting season between March and June, less often in other months.

Crested Lark

Crested Lark *Galerida cristata*

Summer Breeding Visitor   Common

A very widespread species in Europe, the northern half of Africa and Asia, extending to the subcontinent across the Indo-Gangetic Plain, with no less than 35 subspecies recognized over its range. The race *chendoola* represents the species in the northern subcontinent east to Bihar.

These are common larks about Delhi in rather dry, often sandy areas in urban outskirts, railway yards, village environs, overgrazed land, along rural paths and similar bare ground in the floodplains. They keep in pairs or small loose parties, breeding in such habitats between March and June, when their modest song delivered from the ground or in a short, low display flight, is a familiar feature.

PASSERIFORMES: PASSERINES
CISTICOLIDAE: PRINIAS AND CISTICOLAS

The former huge and diverse family of Old World Warblers, Sylviidae, included several distinct clades that have now been reconstituted, following extensive molecular phylogenetic studies, as separate families, amongst which Cisticolidae, Acrocephalidae, Locustellidae, Phylloscopidae and Cettiidae are represented in India. Many of these lineages are only distantly related to one another. The Sylviidae, as defined after this reorganization, now constitute a much smaller family which appears more closely related to some families of babblers.

Within the Cisticolidae family, four clades or subfamilies have been identified, and two of these, Cisticolinae and Priniinae, are represented in India. The family is primarily subtropical and tropical, most highly diversified in Africa, while prinias and tailorbirds are well represented in southern Asia as well. They live in a variety of habitats, from grasslands and reed beds to shrubby woodland and gardens, and build various types of nests, including a loose ball of grass with a side entrance; or an oval purse, finely woven from grass, rootlets and other vegetal fibres, sometimes domed with a side entrance near the top; or, most famously, tailorbirds and some prinias draw and stitch together the edges of either one large leaf or several smaller leaves, using strands of spider silk, vegetable floss, cotton and similar fibres threaded through holes punctured around the perimeter of each leaf, to create a pocket that holds the nest. The actual nest itself is a finely woven cup of grass lined with feathers and floss, in which typically three to five eggs are laid. The eggs, too, are very variable, either spotted on a base colour of pale pink or blue, unspotted, or, notably in Ashy *Prinia socialis* and Yellow-bellied Prinias *P. flaviventris*, deep, uniform mahogany red.

India: 16 species, Delhi: 9

Common Tailorbird *Orthotomus sutorius*

Breeding Resident · Common

Distributed throughout the Oriental region, with nine subspecies over this range, three of which occur within Indian limits. *O.s. guzuratus* occurs in most of India, with two other races in the Northeast.

Well-known in spite of their tiny size, tailorbirds are common residents of urban and suburban Delhi, in city gardens, woodland patches and groves, in fact wherever there is some shrubbery to provide them a modicum of cover. They are noisy, confiding birds and their far-carrying notes are familiar sounds of the city and countryside. They breed from May to September, when males develop very long central tail feathers, and build their famed nests that give the birds their name, to hold their clutch of three speckled eggs.

A male Common Tailorbird in breeding season with its long tail

Rufous-fronted Prinia *Prinia buchanani*

Breeding Resident Fairly Common

Endemic to India and Pakistan, a characteristic bird of dry scrub habitats in the northern and central subcontinent. Monotypic.

Fairly common but local in the area, inhabiting low thorn scrub interspersed with bare ground as is typical of the eroded ravines of the Southern Ridge. They are restricted by this habitat choice and seem to vacate habitats that get too overgrown, such as the New Delhi and South-Central Ridge areas where they occurred previously but which are now overrun by a dense growth of invasive *vilayati kikar*. They normally keep in small parties, and their spirited songs enliven their native haunts during their nesting season from April to July; their nests are of the usual prinia pouch style, placed in a bush or grass clump.

Grey-breasted Prinia *Prinia hodgsonii*

Breeding Resident Fairly Common

Range widely from India through mainland Southeast Asia in open woodland and scrub. Six subspecies are recognized, of which three occur within Indian limits. The birds about Delhi belong to subspecies *hodgsonii* of northern and central India, which has distinct breeding and non-breeding plumages.

These prinias are rather local residents in the Delhi area, in gardens and groves, especially *babool/kikar* woodland with a well-developed scrub understorey on the Ridge. They breed between April and September and sing incessantly during these months. Their nests resemble tailorbirds' nests, but usually employ a single broad leaf whose edges are drawn together to form a cone that holds the nest. Like Rufous-fronted, they are usually in parties that work the canopy for insects; they are also very fond of flower nectar.

Delicate Prinia

Delicate Prinia *Prinia lepida*

Breeding Resident Fairly Common

Recently split from the Graceful Prinia *P. gracilis* based on differences in morphometrics, plumage, vocalizations and DNA divergence;[210] with five subspecies that range from Turkey to northern India, two of which are resident in South Asia: nominate *lepida* in the Indo-Gangetic Plain and *stevensi* in Northeast India.

Delicate Prinias are fairly common in the sparse grass- and tamarisk-covered sandbanks and *khadar* lands of the Yamuna and Ganga floodplains, sometimes expanding into coarse grassy margins of neighbouring cultivation and embankments along canals. They breed between April, when birds begin singing their weak reeling songs, and August, and build their oval pouch-like nests of dry grass anchored in a tamarisk bush or in tall grass tussocks.

Jungle Prinia *Prinia sylvatica*

Breeding Resident Fairly Common

Jungle Prinia

Endemic to India and Sri Lanka, with five subspecies. Delhi is within the range of the race *gangetica* of the northern Indian subcontinent, with a marked difference between its summer and winter plumages.

There were historical records of these large prinias from the Delhi neighbourhood till the 1960s,[211] but they then seem to have dropped out of sight for over four decades, possibly because their favoured habitats were not adequately studied in the right season, when it is almost impossible to overlook their presence. Over the last decade or so, they have been found to be common in grass-and-scrub habitats and broken country of the Southern Ridge in Haryana. Here, their loud songs and courtship flights are a conspicuous feature of their nesting season, June to September. Their nests are loosely woven balls of grass placed low down in bushes. This is a resident species like all prinias, but in winter when they are silent, they can be easily overlooked.

Yellow-bellied Prinia *Prinia flaviventris*

Breeding Resident Fairly Common

The Yellow-bellied Prinia prefers reed beds in wetland areas.

Distributed from the Indo-Gangetic Plain widely across Southeast Asia. Seven subspecies are recognized over this range, with two in the subcontinent: *sindiana* in the Indus River drainage and nominate *flaviventris* in northern India.

Yellow-bellied Prinias were unknown from the Delhi area before the 1970s when they were first recorded from reed beds by the Yamuna at Okhla.[212] These colourful prinias are distinctive birds, unlikely to have been missed earlier at these well-studied sites – they seem to have expanded their range naturally into the Delhi area in the 1970s, intriguingly about the same time as did two other species that occupy the same reed bed habitat, Striated Grassbirds *Megalurus palustris* and White-tailed Stonechats *Saxicola leucurus*. They were also reported from Hastinapur, Meerut district, around the same time.[213] The subspecies involved is uncertain but presumed to be *sindiana*. They are now widespread in wetlands and reed beds across the Delhi region and breed freely between March and August when their distinctive tinkling song – very different from the 'jangling bunch of keys' performances of other small prinias – is a very pleasant feature of its habitat. They build the usual prinia-style pouch nest, which is usually anchored in grass clumps. Four red eggs in a clutch seems to be the norm.

The Ashy Prinia, a bird of Delhi's gardens and parks

Ashy Prinia *Prinia socialis*

Breeding Resident · Common

Endemic to the subcontinent with four subspecies. Represented about Delhi by subspecies *stewarti* of northern India, that shows a marked difference between summer and winter plumages.

Ashy Prinias are common and familiar birds of urban gardens, hedges and shrubbery near villages, grass-and-scrub environs of wetlands and tall cultivation such as sugarcane. They breed through the monsoon from May to September in the Delhi area. They build at least two types of nests, either like a tailorbird's, or the standard prinia-style pouch with a side entrance near the top, in a bush or potted plant and commonly in gardens in urban habitats. These hold a clutch of usually four glossy mahogany-red eggs; more than one brood may be raised in a season.

Plain Prinia *Prinia inornata*

Breeding Resident · Common

A widely distributed species from India to China and in mainland Southeast Asia. Ten subspecies are recognized, four in India, and represented about Delhi by subspecies *terricolor*, the paler race of the northwestern subcontinent with distinct summer and winter plumages.

These are common birds of the Delhi neighbourhood and widespread in various open habitats, in cultivation, along canal embankments and seepage marshes, and coarse grass areas near wetlands. They breed between March and September, when their jingling songs are a familiar sound of the countryside. More than one brood may be raised in succession in a season. Their nests, of the typical prinia type, are built in a thorn bush or a grass clump, often at the edge of cultivated fields or embankments.

Zitting Cisticola *Cisticola juncidis*

Breeding Resident · Common

Widely distributed in southern Europe, much of Africa and Asia. Seventeen subspecies are recognized over this range, with race *cursitans* throughout the subcontinent.

Zitting Cisticolas are common residents of the countryside about Delhi in open grassland areas and fields of wheat and paddy, where they breed between March and October, laying four or five eggs in a deep cup woven of fine grass, cobwebs, etc., which is anchored and well hidden amongst grass stems, possibly rearing more than one brood every season. Their aerial display flights and 'zitting' notes are a feature of their habitat in this season.

Zitting Cisticola

PASSERIFORMES: PASSERINES
ACROCEPHALIDAE: REED WARBLERS

Reed Warblers, as a family, range very widely across the Old World and Australasia, with three genera represented in India, *Iduna*, *Arundinax* and *Acrocephalus*. They live in a variety of habitats, from marshes to dry woodland, even gardens. Primarily insectivorous, although the Clamorous Reed Warbler *A. stentoreus* has been observed taking even small fish in a marsh.[214] The Delhi area hosts one locally breeding species and several migrant visitors from Central Asia.

India: 13 species, Delhi: 6

Booted Warbler *Iduna caligata*

Passage Migrant · Fairly Common

The more northerly breeders of our two sibling *Idunas*, nesting mainly in Russia and Kazakhstan, migrating in winter to peninsular India. Monotypic.

These warblers traverse the Delhi area as passage migrants in September–November and on northward passage from early April to mid-May when they are more in evidence and singing freely. They affect parkland, gardens and other lightly wooded habitats, and often seem to migrate in small loose parties. Both this and Sykes's Warblers *I. rama* may be found in the same areas on passage, though Booted often forages in low bushes and even on the ground, while Sykes' usually feeds in trees.

Booted Warbler

Sykes's Warbler *Iduna rama*

Passage Migrant · Fairly Common

Previously regarded as a subspecies of the Booted Warbler *I. caligata*, now separated on the basis of genetic evidence and the fact that they nest sympatrically as distinct species in Kazakhstan. Sykes's Warblers breed as far south to Afghanistan and Pakistan, and winter in peninsular India. Monotypic.

As passage migrants, they tend to precede the Booted Warblers in autumn, arriving by August or even late July and have largely passed through by October, while the Booted arrive by late August or September and their passage continues till November. In spring, Sykes's main movement takes place in April and the first week of May, while Booted Warblers linger on till late May.

A Moustached Warbler in a characteristic posture with cocked tail

Moustached Warbler

Acrocephalus melanopogon

Winter Visitor Uncommon

Reed Warblers that nest in southern Europe, the Black Sea region and Central Asia, wintering in West Asia, the Caspian basin and the northwestern Indian subcontinent. Three subspecies are recognized; the eastern race *mimica* winters to the Delhi area.

Moustached Warblers are regular visitors to the Delhi area from September to May but tend to be localized and site specific. There are historical records from the 1940s and earlier but appear to have been overlooked thereafter until winter 2001, since when they have been found to be regular winter visitors at several marshes, especially with patches of *Typha* reeds, across the area. Fledged birds have also been netted in early September at the Basai marshes in suitable breeding habitat, and it is just possible that they may nest occasionally.[215]

Paddyfield Warbler

Paddyfield Warbler

Acrocephalus agricola

Mainly Passage Migrant Uncommon

Nest in central Eurasia, with the bulk of the population wintering in India. Three subspecies formerly recognized, but now generally considered monotypic.

These reed warblers are mainly passage migrants through the Delhi area, regularly reported between September and November in autumn and February to early May in spring, with some mid-winter reports as well. They prefer reed beds and wet waterside vegetation, though they may appear in drier habitats occasionally during migration, but typically keep low down, close to the ground, unlike Blyth's Reed Warblers which are often in trees and tall hedges. Their main passage in spring takes place earlier than that of Blyth's.

Blyth's Reed Warbler

Acrocephalus dumetorum

Passage Migrant   Common

Blyth's Reed Warbler, a very common passage migrant in the area

Nest in Eastern Europe eastwards to Central Asia, with the bulk of the population wintering in the Indian peninsula and Sri Lanka. Monotypic.

Blyth's Reed Warblers are very common passage migrants through the Delhi area, where they are widespread in parks, gardens, shrubbery and woodland of all sorts. The southward influx starts mid-August and continues till October; they are largely absent in winter, while spring passage may begin as early as February but peaks between April and mid-May, even early June, for they are amongst the last of our migrant warblers to depart. At this time they seem to be in every tree and are singing freely, a sustained, flowing series of mixed guttural and sweet notes.

Clamorous Reed Warbler *Acrocephalus stentoreus*

Summer Breeding Visitor and Passage Migrant   Fairly Common

Clamorous Reed Warbler

Widespread as a breeding species from West Asia and the Middle East to southern China. Nine subspecies are recognized, two within Indian limits, race *brunnescens* occurring variously as a breeding or non-breeding visitor in most of the country, with *amyae* resident in the Northeast.

These large reed warblers can be rather common and noisy during their nesting season between March and August wherever extensive reed bed habitat is available in the Delhi area. They build substantial cup nests of dry reeds and vegetal shreds, anchored amongst stems of reeds *Typha* or *Phragmites* standing in water in reed beds, which hold their clutches of usually four eggs. Peak numbers, however, occur in March–April and September–October when they are also more widespread in canal seepage marshes and similar sites, which suggests that birds that breed further north also pass through the Delhi area. Fewer birds are seen in winter when they are silent, but they are not entirely absent even during the coldest months. Indeed, Ganguli, writing of the 1960s, considered them to be mainly passage migrants with small numbers remaining to breed.

PASSERIFORMES: PASSERINES

LOCUSTELLIDAE: GRASSBIRDS

Widespread across the Old World, the Locustellidae show two distinct clades or subfamilies, Locustellinae and Megalurinae. Birds of grass-and-scrub habitats and reed beds, they include some of the largest 'former warblers' with loud, distinctive songs and conspicuous aerial song flights in the nesting season. Outside the breeding season, however, they can be elusive. One species, the Striated Grassbird *Megalurus palustris*, appears to have colonized the Delhi area only in the 1970s, at about the same time when a couple of other reed bed-and-marsh-haunting species were first reported here. They had not been recorded earlier, even though they are not easy to miss in their nesting season and have now established themselves firmly as part of our resident avifauna.

India: 13 species, Delhi: 3

▲ *Striated Grassbird*

Striated Grassbird *Megalurus palustris*

Breeding Resident · Uncommon

Residents of tall, swampy grassland and reed beds dotted with scattered bushes, distributed from northern India to Indochina, Java and the Philippines. Three subspecies are recognized, with race *toklao* in India and mainland Southeast Asia.

The largest of the family, Striated Grassbirds were unknown about Delhi before the mid-1970s,[216] and presumably expanded their range into the Delhi area in that decade. These are conspicuous birds with arresting calls and display flights and unlikely to have been overlooked earlier. They are now resident and quite widespread in suitable habitat in the Yamuna floodplain at Wazirabad and Okhla, canal seepage marshes, the Ganga *khadar* at Hastinapur and some of the larger wetlands, where their display flights are a conspicuous feature from March to October.

Grasshopper Warbler *Locustella naevia*

Passage Migrant · Scarce

Nest widely across Eurasia, wintering in Africa and India. Four subspecies are recognized, *L.n. straminea*, of Eastern Europe and Central Asia wintering locally in the Indian subcontinent. They are easily missed in winter when silent.

Ganguli, in her *Guide to the Birds of the Delhi Area*, mentions just four historical records of this species, in August and April. More intensive coverage in recent years suggests that they are regular, but apparently scarce, passage migrants in the Delhi area from the second week of August through September–October and in April, with a few winter reports in mid-November and February. They have been noted feeding in hyacinth-covered marshlands bordered by reeds and rank grass in the Yamuna floodplains, the Sultanpur National Park and the Bhindawas Sanctuary neighbourhoods, and even in dry, standing millet fields at Sohna, Haryana.

Bristled Grassbird *Schoenicola striatus*

Summer Breeding Visitor · Uncommon

A little-known species, reflecting the neglect of grassland ecology studies in South Asia. Endemic to the subcontinent, the Bristled Grassbird's habitat of grassland and reeds studded with scattered emergent bushes is monsoon dependent, increasingly threatened and fragmented, and the birds are consequently localized and apparently sporadic over their range. Monotypic.

Formerly considered rare in Delhi with just one record in July 1962, they were rediscovered at Okhla in 1996 with at least three males singing and displaying between July and September. Their calls and habits better known, they have since been found to be breeding visitors between mid-June and October to several wetlands of the area, including the *khadar* of both the Ganga and Yamuna Rivers, the Surajpur and Dhanauri wetlands, the Sultanpur and Bhindawas sanctuaries and their environs in Haryana, and doubtless elsewhere. There are no winter reports, and the birds presumably vacate the Delhi area after breeding, but their movements remain unknown. Hutson, writing of the 1940s, described 'a strange courting flight' for the Striated Babbler *Argya earlei* 'when one bird will fly round and round some 50 feet up, making a loud sparrow-like chirp',[217] a description that does not describe the behaviour of any babbler but is strongly reminiscent of this bird, which shares its habitat and, to an extent, its appearance.

The Bristled Grassbird, a summer visitor to the area, is rather enigmatic in its movements.

PASSERIFORMES: PASSERINES

HIRUNDINIDAE: SWALLOWS AND MARTINS

A cosmopolitan family of small, long-winged birds adapted for an aerial lifestyle; exclusively insectivorous, they catch flying insects on the wing with their short, broad beaks. Many are highly social, especially during the non-breeding season when they may gather in very large flocks to hunt over marshes and other insect-rich habitats, and at roosts. Breeding varies from solitary to colonial, and their feather-lined nests are either in tunnels excavated in sand bluffs, or half-cups or retort-shaped structures constructed of wet mud and attached under ledges of rock cliffs or artificial buildings, to accommodate their clutches of usually three or four eggs.

At least five species breed in the Delhi area while one, the very widely distributed Barn Swallow *Hirundo rustica*, is a common winter visitor. The status and movements of another, Pale Martin *Riparia diluta*, are yet to be fully worked out.

India: 14 species, Delhi: 7

▲ *A pair of Streak-throated Swallows collecting mud for their nest*

Grey-throated Martin *Riparia chinensis*

Breeding Resident · Common

Widely distributed across southern Asia, with two subspecies, the nominate *chinensis* being a common resident of the alluvial river systems of northern India and mainland Southeast Asia.

This is a common swallow of the area, by the Ganga and Yamuna Rivers and their tributaries, canal systems and the larger lakes. They breed colonially in the winter months between November and April when water levels are low, in burrows excavated by the birds themselves in alluvium banks left by the receding rivers after the rains. Earthen embankments, bluffs and cuttings around wetlands and canals in rural areas, may host large colonies, as at the Bhindawas Sanctuary. Erosion and flooding can affect nesting habitat suitability from year to year. Large post-breeding flocks of several hundred congregate with Streak-throated Swallows *Petrochelidon fluvicola* at Okhla and elsewhere from May to October, hunting over surrounding cultivated habitats as well in this season and roosting with other hirundines in reeds and marshes.

Sand Martin *Riparia riparia*

Vagrant / Reconfirmation desirable

Almost cosmopolitan in distribution, nesting in temperate latitudes across North America, Europe and Asia, wintering in South America, Africa, Arabia and southern Asia. Five subspecies are recognized, of which the darker and more strongly marked Asian subspecies *ijimae* winters in eastern India and Southeast Asia, straggling westwards into peninsular India. According to Rasmussen & Anderton, the status of Sand Martins in South Asia is unclear 'as range still confounded with Pale', but a specimen collected southwest of Delhi in March 1962 was assigned to this species.[218]

The identification of Sand and Pale Martins *R. diluta* is a challenge, and none of the photographs from the area are unequivocally *R. riparia*. One that offered direct comparison in the field with two Pale Martins on 7 October 2021 at Jhanjraula, Gurugram district, 'had a very distinctive dark brown breast band and darker brown upperparts as compared to Grey-throated, fitting the features of Sand Martin'.[219] Records of Sand Martins prior to the split (e.g. in Ganguli, *Guide to the Birds of the Delhi Area*, 1975) and recent claims of this species from about Delhi in winter, between September and February–March, are probably all Pale Martins.

Pale Martin *Riparia diluta*

Winter Visitor, also opportunistic breeder? Scarce

Split in the 1990s from Sand Martins of the Northern Hemisphere on grounds of sympatry and ecological differences; they have a more southerly breeding range in Asia that extends from Central Asia, Tibet and China to the northwestern subcontinent. Six subspecies are recognized, of which two may occur in the region: the Central Asian breeding nominate *diluta* and *indica* that nests in the northwestern parts of the subcontinent. Both disperse over the Indo-Gangetic Plain after their nesting season.

These martins are scarce and irregular visitors to the Delhi area with scattered reports over the year. Their status and movements have yet to be fully worked out. Two subspecies are likely to occur: the first, *indica* breeds along loess riverbanks in drier mountainous regions in Pakistan from February to April,[220] and winters in the plains, which suggests that midsummer records from the area may represent post-breeding dispersal by birds of this race. This race may breed in the Delhi area as well, at least opportunistically, as birds have been observed collecting nesting material and mating in appropriate habitat at the Yamuna sandbanks north of Wazirabad in late January 2020.[221] Records in the winter months and in spring (March/April), however, probably include wintering Central Asian *diluta* that breeds later, from May to August. Both subspecies, *indica* and *diluta*, are likely to be sympatric in winter about Delhi, along with the resident Grey-throated Martins.

Pale Martins, photographed near Gurugram

The Dusky Crag Martin in its typical habitat

Dusky Crag Martin *Ptyonoprogne concolor*

Breeding Resident Fairly Common

A mainly Indian species, though extending into Southeast Asia as well. Two subspecies are recognized; nominate race *concolor* in India.

These are fairly common resident swallows in the area, resident and nesting about ruins, in niches and corners of old forts and monuments, even the Central Secretariat Blocks in New Delhi, and modern multistorey structures elsewhere. A few birds inhabit natural features such as rock cliffs, or abandoned stone mines, in Southern Ridge landscapes. Ganguli called them 'resident in very small numbers', much too parsimonious a description in the light of its status today; although it is worth remarking that these martins appear to have disappeared from some areas of Gurugram since the last three or four years, where they used to breed fairly commonly in high-rise buildings until then. They nest between June and August, building a neat half-cup mud nest under an overhanging natural or human-made feature.

Barn Swallow

Barn Swallow *Hirundo rustica*

Winter Visitor Common

Nesting summer visitors throughout the Northern Hemisphere as far south as the Himalayas and migrate in winter to South America, sub-Saharan Africa and southern Asia. Seven subspecies are recognized, of which three, nominate *rustica* (that breeds across Europe and western Asia, south to the Himalayas), *gutturalis* (that breeds from the eastern Himalayas to China and Japan) and the rufous-bellied *tytleri* (that breeds in central Siberia), are recorded in winter in India.

Barn Swallows are common winter visitors about Delhi (presumed subspecies *rustica*, also perhaps some *gutturalis*), arriving as early as mid-June and leaving by March. They can be numerous on passage, their numbers increasing rapidly from July onwards. Barn Swallows are attracted by the flying insects that swarm over wetlands in the evenings, and very large numbers may collect to feed over the reed beds before settling in to roost along with other swallow species.

Wire-tailed Swallow

Hirundo smithii

Breeding Resident

Common

A pair of Wire-tailed Swallows, with the male on the right

Widely distributed over sub-Saharan Africa and southern Asia and typically associated with water. Two subspecies are recognized, *H. s. filifera* mainly resident in a major part of India, but a summer breeding visitor in the northwest.

These beautiful swallows are common residents about Delhi, where they may be seen in agile and elegant flight, their colours flashing in the sun, over waterbodies of all kinds, rivers, lakes, village ponds or canals, even over wet cultivation. They keep in pairs or family parties through the summer but may gather in groups of 10–15 from late September and October, their numbers falling thereafter over the winter months. These larger parties probably involve birds on passage as these swallows are mainly summer visitors in the northwestern subcontinent. They roost in reeds or twiggy vegetation overhanging water and nest between February and June, building a half-cup from pellets of wet mud and lined with feathers, which is attached bracket-wise under small bridges, canal headworks and similar human-made structures, also usually over water. Parties on migration in autumn, however, may wander some distance from waterbodies, perching on telegraph wires and hunting over open country and cultivation.

Red-rumped Swallow *Cecropis daurica*

Breeding Resident and Passage Migrant Uncommon

Wide breeding distribution over southern Europe, Africa and Asia. Nine subspecies are recognized over this range, three within Indian limits: *rufula* in the northwest, *nipalensis* nesting in the Himalayas and wintering southwards, and *erythropygia* resident in the peninsula.

These swallows nest locally about Delhi in March–May, with Tughlaqabad being the only known nesting site presently, though they undoubtedly nest in archaeological ruins elsewhere and in abandoned stone mines on the Southern Ridge. Their nests are beautiful retort-shaped structures with long entrance tunnels extending along the roof of a ledge, built of mud collected by the birds from damp areas – the retort is lined within with feathers and grass and holds a clutch of three or four eggs. Numbers, sometimes substantial, of the migratory Himalayan breeding race *nipalensis*, with a paler rump patch and finely streaked underparts, appear on passage from late September through November and again in April–May.

Streak-throated Swallow *Petrochelidon fluvicola*

Breeding Resident Fairly Common

Endemic to the subcontinent, mainly resident with some local movements. Monotypic.

These cliff-nesting swallows are locally common in the Delhi area around waterbodies of all sorts and particularly attached to masonry bridges over the larger canals. They breed between March and May when water levels are low, in large colonies of closely packed, pot-shaped, mud nests with tubular entrances, each holding three eggs, which are often clustered under bridges of the several irrigation canal systems that cross the area. They disperse after breeding and then swarm with Grey-throated Martins *Riparia chinensis* over the river, marshes or *jheels* and at large communal roosts in reed beds.

Common House Martin *Delichon urbicum*

Vagrant

Breeds across Eurasia and winters in Africa and South Asia. Two subspecies are recognized, of which race *meridionale*, that nests from the Mediterranean to Central Asia and Kashmir, winters locally in India.

The species is included in the Delhi list on the basis of two observations recorded on eBird checklists, both in April, and presumably vagrants during spring passage – two over the Yamuna in 1980 and one with swallows at Okhla in 1999 – both by competent observers familiar with the species. There have been no further records since.

PASSERIFORMES: PASSERINES

PYCNONOTIDAE: BULBULS

A large Afro-Asian family of small, lively birds. Some are forest species while many others, including Delhi's three resident bulbuls, are familiar garden birds and, with their crests, cheerful calls and sprightly demeanour, general favourites.

India: 23 species, Delhi: 4

Red-vented Bulbul *Pycnonotus cafer*

Breeding Resident Common

A common resident bulbul of the subcontinent and Myanmar, and widely introduced elsewhere, deliberately or accidentally. Nine subspecies have been recognized, six within Indian territory, of which race *humayuni* of north-central India occurs in the Delhi area.

Red-vented Bulbuls count among the commonest and most conspicuous components of our avifauna. Adaptable, lively birds, sociable but noisy and aggressive in demeanour, they breed freely in urban and suburban gardens and woodland, building open cup-shaped nests lined with plant fibres, etc. in bushes and hedges, in which they lay their clutches of usually three eggs. More than one brood may be raised. Ganguli, writing of the 1960s in her *Guide to the Birds of the Delhi Area*, suggested that many birds move out of the Delhi area in the cold months from October to February, but certainly no such movements are noticeable now.

Red-whiskered Bulbul *Pycnonotus jocosus*

Breeding Resident Common

Widespread in India except in the drier northwestern parts, ranging to southern China and mainland Southeast Asia; widely introduced elsewhere in the world. Nine subspecies are recognized over its range, of which six occur in India.

These sprightly bulbuls are familiar and cheerful inhabitants of Delhi's parks, gardens and woodland areas, where they breed between March and July, building open cup-shaped nests in bushes and hedges to accommodate their clutches of usually three eggs. This, it seems, was not always the case. Delhi is on the fringe of these bulbuls' normal range in the Gangetic Plain, and they appear to have spread in the Delhi area with the planned 'greening' of New Delhi in the first half of the twentieth century. Frome thought it to be a new arrival about 1941, and Ganguli noted that it expanded into North Delhi only in 1961.[222] The subspecies in Delhi is *pyrrhotis* of the Gangetic Plain, not *abuensis* of the southern Aravalli Ranges, so it has evidently expanded into the Delhi area from the eastern direction rather than the southern. It was unknown at the Sariska National Park, 150 km to the south,[223] with the first records there – the race involved is not mentioned – only in 2011.

White-eared Bulbul *Pycnonotus leucotis*

Breeding Resident Fairly Common

Bulbuls of relatively drier habitats from Iraq and the Persian Gulf to western India. Three subspecies, race *leucotis* in the northwestern subcontinent.

These are the least common of Delhi's three resident bulbuls and will not normally be found in Delhi's greener parks and avenues which are not to their liking, or in the cultivated countryside. On the other hand, they are lively and attractive residents of dry thorn scrub country of the Southern Ridge and other sparsely vegetated parts of the Aravallis, breeding between March and July in such habitats. When such areas become too overgrown and dense, these bulbuls tend to lose ground to the more aggressive Red-vented Bulbul *P. cafer*, as has happened on the New Delhi Ridge and several other areas.

The White-eared Bulbul, a bird of comparatively drier areas

Himalayan Bulbul *Pycnonotus leucogenys*

Vagrant, or escape?

A jaunty bulbul of bush-covered hillsides, hill forests and terraced cultivation at medium and low elevations in the Himalayan range.

One was photographed at the Yamuna Biodiversity Park, North Delhi, in April 2017, a genuine vagrant or a possible escape from captivity. There are winter records from just northeast of the area defined for this Checklist, from the Haiderpur Wetland, Hastinapur Wildlife Sanctuary, where a party of as many as 20 birds was reported in January 2021. Himalayan Bulbuls are largely resident in their range, but as these observations suggest, they do appear to wander further into the northern plains in peak winter.

PASSERIFORMES: PASSERINES

PHYLLOSCOPIDAE: LEAF WARBLERS

None of this species-rich and widespread Old World family nest in India except along the Himalayas, but a flood of these little birds spreads over the Indian peninsula every winter, both from the Himalayas and beyond.

Several species winter in, or migrate through, the Delhi area. Attention needs to be drawn to some unusual winter records of a few of these warblers from about Delhi in recent years. Whistler's *Phylloscopus whistleri* and Grey-hooded Warblers *P. xanthoschistos*, Himalayan breeders which were usually understood to be only altitudinal migrants along the Himalayan range, with just an occasional vagrant wandering further south into the plains in winter, are being recorded with increasing frequency from the area. This is probably the result of growing and more intensive ornithological coverage of Delhi and its neighbourhood in recent years, the presence of these species in winter having been overlooked earlier; or, less likely, a genuine change in the wintering pattern of these birds in the northern plains.

India: 36 species, Delhi: 16

▲ *Hume's Leaf Warbler, a common winter visitor*

Hume's Leaf Warbler *Phylloscopus humei*

Winter Visitor · Common

Summer visitors to Central Asia and the Himalayas, with an eastern population isolated in China as well; migratory, wintering in the subcontinent and Southeast Asia. Two subspecies, of which nominate *humei* is amongst the commonest of the family in northern and central India in winter, while *mandellii* winters in the Northeast.

A common warbler in the Delhi area in winter, widespread in any wooded area, in city parks and gardens, avenues, woodland patches and groves in the countryside from mid-September to April. Their loud, distinctly two- or three-syllabled chirps are a familiar roadside sound through the winter months.

Brooks's Leaf Warbler *Phylloscopus subviridis*

Winter Visitor Scarce

Nest in the conifer forests of the mountain ranges of northeastern Afghanistan and adjacent Pakistan, and winter in the Indus Plain in Pakistan and in northwestern India. Monotypic.

Regular but scarce winter visitors about Delhi between October and March, recorded mainly from areas of acacia woodland, especially in the environs of the Sultanpur National Park and on the Delhi Ridge. Before emigration in the second half of March, birds are in song on the New Delhi Ridge and then their accelerating trills of high, weak notes ending in a buzz are an unmistakable sign of their presence.

Lemon-rumped Warbler *Phylloscopus chloronotus*

Vagrant

Altitudinal migrants, nesting at higher elevations along the Himalayan range and reaching the foothills in winter; known to have wandered into the northern plains on rare occasions, exceptionally as far as the Bharatpur National Park, 160 km south of Delhi. Two subspecies are recognized, *simlaensis* breeding in the western Himalayas and nominate *chloronotus* east of central Nepal.

Vagrant individuals of these bright little warblers have been reported in recent years from south Delhi (presumed race *simlaensis*): first reported at the AIIMS campus in February 2015 and confirmed from the South-Central Ridge at Sanjay Van in January 2017, studied closely by many observers, at the Aravalli Biodiversity Park, Vasant Vihar, in the winters of 2019 and 2020, and the Southern Ridge (reported from Mangar Bani, Faridabad district), in October 2019 and 2021.

Tytler's Leaf Warbler *Phylloscopus tytleri*

Passage Migrant Rare

Summer breeding visitors to the northwestern Himalayas, migrating in winter to the Western Ghats in southern India. Scattered passage records in August–October and late March-April across the western half of the Indian mainland. Monotypic.

A historical record from the New Delhi Ridge on 23 August 1973 was considered 'probable' by the observer A.J. Gaston himself. A couple of later reports from Delhi cannot be verified. One, photographed at Bhondsi, Gurugram, in late March 2023 is, therefore, the only definite record to date from the area.

Sulphur-bellied Warbler *Phylloscopus griseolus*

Winter Visitor Uncommon

Breeds in the high plateaux of Central Asia southwards to Ladakh and winters in broken, rocky landscapes in central India. Monotypic.

Small numbers of these stocky, ground-loving warblers spend the winter in the Delhi area in suitable habitat on the Ridge, about ancient monuments, abandoned quarries and stony ground. Much commoner during passages in September–October and March–April, they also appear in atypical habitats in gardens, along river embankments, *babool* woodland, etc. They have the unique habit of creeping about rocks, stone walls, archaeological ruins and even tree trunks. They have a distinctive, hard call note.

Tickell's Leaf Warbler *Phylloscopus affinis*

Passage Migrant Scarce

Breeds in the high Himalayas west to east, but their movements to their winter grounds in eastern and southwestern India are mainly over the eastern parts of the peninsula and usually to the east of the Delhi area. In spring, they retrace their route northward, often in small parties, again keeping more to the eastern parts, then spreading westwards along the Himalayan chain as they return to their breeding grounds. Two subspecies are recognized, *perflavus* in the western Himalayas and the nominate in the central and eastern parts of the range.

There are scattered reports of these brightly coloured warblers from as far west as Delhi (presumed race *subflavus*), in both spring and autumn – old sight records in March 1960 and May 1968, and from the Delhi Ridge in October 1971–72;[224] and a number of recent reports from Delhi gardens, Okhla, the Sultanpur National Park environs and similar wooded sites, in September–November and February–March. Some reports, however, are vitiated by confusion with Sulphur-bellied Warblers, but a few mid-winter records, in December and January, are remarkable, far from the species' usual winter range in the peninsula.

Smoky Warbler *Phylloscopus fuligiventer*

Winter Visitor Rare

Breeds in the central and eastern Himalayas and adjoining Tibet, and winters in the foothills from Uttarakhand eastwards. Three subspecies are recognized: birds that nest in the Himalayas and winter to the foothills are of the nominate race *P. f. fuligiventer*, with the darkest underparts; the two other races winter mainly in northeastern India. These warblers tend to wander westwards and southwards of their usual winter range on occasion, even as far as the Bharatpur National Park. Terrestrial in habit, they typically frequent waterside habitats in winter.[225]

Smoky Warblers are rare, perhaps sporadic, winter visitors to the area, recorded on a few occasions from bushy vegetation, reeds and grass near wetlands and marshes and at the Surajpur Lake on 20 November 2022, and Dhanauri on 5 February 2023, both sites in the GautamBuddha Nagar district.[226]

Mountain Chiffchaff *Phylloscopus sindianus*

Vagrant

Breeds in the northwestern Himalayas, including Ladakh, and north to Tajikistan, and winters mainly in the Indus floodplain, with scattered records further eastwards (these are backed by specimens, others are possibly overlooked in the field) in Haryana and even northern Madhya Pradesh. Two subspecies are recognized: the nominate as above and *lorenzii* in the Caucasus ranges.

Vagrants have been reported from waterside vegetation at Okhla in December 2000 and at the Panipat Oil Refinery in January 2003,[227] the identification based primarily on the birds' distinctive vocalizations.

Common Chiffchaff *Phylloscopus collybita*

Winter Visitor Common

Widespread as a breeding bird from Europe across Eurasia to Central Asia, wintering further south. Six subspecies are recognized; Chiffchaffs of the eastern subspecies *tristis*, which nest in Russia and western Siberia, are distinct both in plumage and vocalisations from western races of the species, though there is some evidence of intergradation.

These 'Siberian' Chiffchaffs are common in winter about Delhi from mid-September to April, affecting *Acacia/Prosopis* and open woodland generally, though they are also fond of wet ground where they may feed in small parties on damp bare mud or irrigated lawns with bordering low twiggy vegetation. They sing enthusiastically in April before emigration.

Green-crowned Warbler *Phylloscopus burkii*

Vagrant

Breeds in broadleaf forests at mid-elevations in the Himalayas, from Himachal Pradesh to Bhutan, largely segregated altitudinally from sympatric Whistler's Warbler *P. whistleri* with which they were formerly considered conspecific. Many birds winter in Madhya Pradesh, Bengal and the northern Eastern Ghats and are reported occasionally on passage in the Gangetic Plain. Monotypic.

A vagrant individual was photographed, and its vocalizations recorded, at Rohtak, Haryana, on 31 May 2021.[228] The date is very late for a migrant, and this may possibly have been a waif blown off course. This is the first record of the species from the northern plains west of the Yamuna.

Whistler's Warbler *Phylloscopus whistleri*

Winter Visitor Rare

Breeds in the cool temperate forests of the Himalayas from northern Pakistan to Arunachal Pradesh, at higher elevations than its sibling species, Green-crowned Warbler *P. burkii*, and winters from 2,400 m down to the foothills and adjacent plains, rarely even further south, having reached Bharatpur, Rajasthan. Two subspecies are recognized: nominate *whistleri* in the Himalayas, *nemoralis* in Northeast India.

Y.M. Rai, who had studied the birdlife of Meerut district, UP, in the 1970s and 1980s, recorded a small party of these warblers in the Hastinapur forest as early as in March 1979, which he identified as *Seicercus burkii*; Green-crowned and Whistler's Warblers were considered races of a single species at that time, and it seems likely that this observation pertains to the latter. They have been recorded again from the Hastinapur Wildlife Sanctuary in recent years. Whistler's Warblers had not been reported from the immediate vicinity of Delhi, however, prior to the winter of 2009–10, when two remained at the Sultanpur National Park from mid-November to early February, studied closely by several observers. There have been several records since then, mainly of single birds from woodland patches in and around Delhi, and especially from the Southern Ridge.[229] These records suggest that Whistler's Warblers would be better categorized as rare winter stragglers about Delhi, being reported by the more intensive ornithological coverage of the Delhi neighbourhood in recent years, but, surprisingly, overlooked in the past.

Green Warbler *Phylloscopus nitidus*

Passage Migrant · Scarce, also rarely in winter

Summer visitors to the mountains of northern Afghanistan, Iran and the Caucasus, and migrate in winter to southern India and Sri Lanka. Common on spring passage across northern India, but their autumn movement is mainly on the western parts of the subcontinent, over Pakistan and Punjab. Isolated in their breeding range in the Central Asian mountain forests, they have diverged from their ancestral Greenish Warbler *P. trochiloides* complex, of which they were formerly considered a race, differing in their brighter coloration and voice. Monotypic.

Green Warblers are scarce passage migrants through the area, with a few records in spring, in March and April, from woodland habitats, and in autumn between October and November. Rare December–January records in recent years, from a couple of woodland sites on the Ridge and from the Sultanpur National Park, are unusual for this species which winters mainly further south in the peninsula. Passage birds in spring are likely to be in fresh plumage and more easily identified, and their distinctly three-syllabled call is another useful diagnostic characteristic.

Greenish Warbler *Phylloscopus trochiloides*

Passage Migrant · Fairly Common

Breeds over a vast range in Eurasia south to the Himalayas and winters in South and Southeast Asia. Four subspecies are recognized, three occurring within Indian limits, of which *ludlowi* and nominate *trochiloides* nest in the Himalayas, and *viridianus* further to the north, in east Europe and Russia. The entire population of the *ludlowi* and *viridianus* subspecies winters in India, mainly from the Nepal foothills southwards in the southern and eastern parts of the peninsula, and west to central UP. Race *trochiloides* tends to winter more towards northeastern India and further eastwards.

Greenish Warbler

Greenish Warblers are primarily passage migrants in the northwestern parts of the country, including about Delhi, with few validated records in winter. Both Ganguli and A.J. Gaston in his field studies on the New Delhi Ridge in 1971–73, had recorded this warbler as occurring only on autumn passage. They are, however, being reported commonly in recent decades on both passages, mid-August to October in autumn (most in the first half of September) and mid-March through April in spring, when birds are singing freely.

Large-billed Leaf Warbler

Phylloscopus magnirostris

Passage Migrant · Scarce

Relatively large warblers with unmistakable calls and song, breeding visitors to the Himalayas, west to east, and further into Myanmar and southern China, and nesting usually near streams in forested ravines. They winter in India in the Western Ghats and Sri Lanka. Monotypic.

An autumn passage migrant, with scattered records from woodland areas between mid-September and the third week of October.[230] Not reported from the area on northward passage, which takes place further to the east, then spreads westwards along the Himalayan range.

Large-billed Leaf Warbler

Western Crowned Warbler *Phylloscopus occipitalis*

Passage Migrant · Uncommon

West Himalayan breeding leaf warblers that winter in peninsular India. Monotypic.

These warblers are regular but uncommon passage migrants through the area, recorded from urban gardens, groves and woodland patches, between August and October (the movement peaking in the first half of September) in autumn, and between end February and mid-April in spring.

Grey-hooded Warbler *Phylloscopus xanthoschistos*

Vagrant

Familiar birds of the Himalayas and the hills of Northeast India, wintering to the foothills as is the norm, with a few recent records of single birds or small parties extending some distance into the plains as far to the south as Ludhiana, Punjab and Bijnor, UP, perhaps during cold waves. The recent records of these bright and cheerful little warblers in winter from Delhi are the southernmost to date in the north Indian plains. Four subspecies are recognized within their Himalayan range, the birds reaching Delhi probably of the west Himalayan race *albosuperciliaris*.

A vagrant was first recorded from the Sultanpur National Park on 13 November 2015, followed by another individual in December 2019 that remained at the site at least till 10 February 2020; a 'party of five or six' in the woodland at Mangar Bani, Faridabad, in late November 2020[231] with at least two remaining at the site through December; one at the Bhondsi Nature Park, about 11 km from this site, in the second week of December 2020 and again on 6 February 2021; and at Mangar Bani on 21 February 2021 (these December 2020–February 2021 records all plausibly represent dispersing individuals of the party that was originally reported from Mangar Bani in late November 2020). The records from the Delhi neighbourhood are unusual as these conspicuous, easily identifiable birds are unlikely to have been overlooked earlier. There are recent reports of single birds in January and February from the Hastinapur Wildlife Sanctuary as well. Grey-hooded Warblers are categorized here as vagrants to the Delhi area, but should more records accumulate in coming winters, may need to be redesignated as rare/erratic winter visitors.

PASSERIFORMES: PASSERINES

CETTIIDAE: BUSH WARBLERS

The Cettiidae or Bush Warblers, yet another clade extracted from the former Old World warblers radiation, bring together a varied group of small, mainly African and Asian, insectivorous birds that are diverse in appearance but apparently closely related from an evolutionary perspective. Delhi does not offer suitable habitat for most, but one species has appeared as a vagrant.

India: 17 species, Delhi: 1

Cetti's Warbler *Cettia cetti*

Winter Visitor Rare

Breeds from Europe and the Mediterranean region to Central Asia, largely resident in the west of its range, migratory in the east. Three subspecies are recognized, of which the eastern race, *albiventris*, which nests in Central Asia, spreads as far southwards as the northwestern subcontinent in winter, affecting swampy areas with reeds and bushes in the Indus Plain in Pakistan and in smaller numbers in Punjab, Haryana and Rajasthan in India.

This bush warbler appears to be a rare, but possibly regular, winter visitor to the Delhi area, which has been overlooked in past years because of its skulking habits. There is a sight record from Hisar, Haryana, in March 2003, a little beyond our limits. One that responded to playback in reed beds at Okhla on 7 December 2015 was the first indication of its presence in the area covered by this Checklist; since then, it has been recorded in December 2018–February 2019 from the Dighal wetlands and the Bhindawas Bird Sanctuary; its distinctive calls were heard at the latter site again in December 2019 and January 2020, and it was photographed in late December 2020 and February 2021 at the Budhera Water Treatment Plant, Gurugram district.

SYLVIIDAE: SYLVID 'WARBLERS'

The Sylviidae, as now reconstituted on the basis of DNA studies, appear to be more closely related to the Yuhinas and White-eyes Zosteropidae and some families of babblers, including Timaliidae, Pellorneidae and Leiothrichidae, rather than the former Old World warblers. Sylvid species in the genera *Sylvia* and *Curruca* are small, insectivorous birds of bush country, scrub and woodland, with their greatest diversity around the Mediterranean, even as a couple of species breed as far to the east as the northwestern subcontinent, in Pakistan, Kashmir and Ladakh. Two species are migrant visitors to Delhi, and two are vagrants on the eastern fringes of their wintering range and migration paths.

India: 6 species, Delhi: 4

Asian Desert Warbler *Curruca nana*

Vagrant

Breeds in the desert steppes of Central Asia, Mongolia and western China, wintering in similar desertic habitats of West Asia and Arabia, south and east to the Red Sea and the northwestern subcontinent. In India, the main wintering areas of these little warblers are in Rajasthan and Kachchh, wandering occasionally to Punjab and Haryana.

A vagrant individual, the only record for the area, was photographed at Surajpur, Greater Noida, on 10 October 2018.

Lesser Whitethroat *Curruca curruca*

Winter Visitor Common

The taxonomy of the Lesser Whitethroat group is notoriously complex and unsettled, given the intergradations between various forms across their wide range. The *Checklist* regards them as one species, Lesser Whitethroat, which breeds from Europe to Siberia and southwards to Afghanistan and the northwestern Himalayas, wintering in northeastern Africa, the Middle East and South Asia. Affect a range of habitats, including gardens, plantations and woodland, even oases in Central Asia. Six subspecies are usually recognized, often split into three species or subspecies groups: Lesser, Desert *C. minula* and Hume's *C. althaea* Whitethroats, all three of which are migrants to India.

Lesser Whitethroats of the races *blythi*, that nests in Siberia, and/or *halimodendri*, that nests on the Central Asian steppes, are common winter visitors from late August to May to the Delhi area, widespread wherever there are trees and bushes, in parks and woodland. Desert Whitethroats breed in arid Central Asia, wintering in drier habitats from Arabia to Rajasthan and Kachchh. They are only vagrants as far to the east as Delhi, first identified as such at Sultanpur in December 2001 and March 2002 (perhaps the same bird), at Sonipat in December 2004 and at Surajpur, UP, in September 2017. Hume's Whitethroats breed in the mountains of Iran, Kashmir and Ladakh, and winter in peninsular India; Ganguli considered them to be mainly spring passage migrants in the Delhi area, but a few are recorded through the winter as well, between October and April.

Eastern Orphean Warbler *Curruca crassirostris*

Winter Visitor Scarce

Breeds in open woodland and orchards with some bushy cover, from southeastern Europe and Turkey to southern Central Asia, Afghanistan and Baluchistan, and spends the non-breeding season in northeastern Africa, Arabia and South Asia. Three subspecies are recognized, the easternost race *jerdoni* wintering in the west-central subcontinent.

These large sylvids are scarce winter visitors between November and early April, with a few individuals reported, irregularly and not every year, with the majority of records from the babool/*kikar* woodland of the Ridge, the Bhindawas Wildlife Sanctuary and the Sultanpur National Park, Haryana.

Eastern Orphean Warbler

Common Whitethroat *Curruca communis*

Vagrant

Widespread, with a breeding range that extends from Europe and northwest Africa to southwestern Siberia and parts of Central Asia, the entire population wintering in sub-Saharan Africa. Four subspecies are recognized, birds of the pale eastern (Central Asian) race *icterops* cut across the western parts of the subcontinent during autumn passage to their wintering grounds in Africa.

Common Whitethroats are vagrants as far east as Delhi, with just a handful of records: mist-netted on the New Delhi Ridge in September 1971 and 1972, single birds reported from the Bhindawas Wildlife Sanctuary in September 2004, the Sultanpur National Park in February 2002 and again in October 2011 and 2018.[232]

PASSERIFORMES: PASSERINES

PARADOXORNITHIDAE: FULVETTAS, PARROTBILLS, ETC.

As with the former Old World warblers, phylogenetic studies have resulted in the former family Timaliidae, which had included a large, diverse set of miscellaneous species loosely called babblers, being reorganized into five lineages or families to better reflect the results of genetic research, viz. Paradoxornithidae (parrotbills and others from the former babbler complex, including Yellow-eyed Babbler *Chrysomma sinense*), Zosteropidae (yuhinas and white-eyes), a revised Timaliidae (tree and scimitar babblers), Pellorneidae (wren babblers), Leiothrichidae (laughingthrushes and allies). Many of these are forest birds, and the hill ranges of India host a particularly rich collection of taxa. Three of these families are represented in the Delhi area as well. Species in the family Paradoxornithidae typically forage through vegetation in small parties and strip leaf bases and bamboo in search of grubs.

India: 17 species, Delhi: 1

Yellow-eyed Babbler

Yellow-eyed Babbler *Chrysomma sinense*

Breeding Resident · Fairly Common

Widespread residents from India to South China and Southeast Asia. Four subspecies are recognized, two occurring within Indian limits, of which race *hypoleucum* occurs in the Delhi area.

These attractive birds are fairly common residents in the mixed grass-and-bush country of Aravalli-type habitats on the Ridge, in open woodland, clumps of tall grasses and scrub on margins of wetlands and along canal systems, and tall grass borders and cuttings in cultivated country. Their nests are deep cups of grass, lined with cobwebs, vegetable fibres and seed down, suspended in a clump of upright grass stems, in which three to five eggs are laid. They breed from April to August, when their characteristic songs are a feature of their habitat.

PASSERIFORMES: PASSERINES

LEIOTHRICHIDAE: BABBLERS AND LAUGHINGTHRUSHES

Leiothrichidae are another newly separated family following the division of the erstwhile Timaliidae and include several colourful laughingthrushes and assorted babblers that enliven the hill forests of the Oriental region.

Delhi's drier habitats host only four species of scrub country babblers of India in the genus *Argya*, noisy, gregarious birds which form a familiar part of the Delhi area's birdlife. Their nests are compact, sturdy cups made of slender twigs and grasses, lined with softer material, usually in thorny bushes and small trees, which hold their clutches of, typically, four blue eggs. They exhibit rich social complexities in their remarkable nesting habits – they live in cooperative breeding groups through the year, and each group may have one or more breeding pairs while other non-breeding members or 'helpers', possibly of previous broods, assist the parents in rearing their chicks (or cuckoo chicks, sometimes!).

India: 57 species, Delhi: 4

▲ *A Jungle Babbler, always in parties, hence 'Seven Sisters'*

Jungle Babbler *Argya striata*

Breeding Resident · Common

A subcontinental endemic, resident in most lightly wooded habitats across India. Five subspecies are recognized, represented in the Delhi area by subspecies *sindiana*, the paler race of northwestern India.

Jungle Babblers are common and familiar birds of gardens, woodland and groves in Delhi, preferring thicker vegetation with denser understorey than Common and Large Grey Babblers, though may coexist with them where habitats overlap. Gregarious through the year, they live in loquacious parties of eight to ten; they are brave birds and value teamwork and will quickly gather to sound the alarm and mob any threat, hawk, cat, snake or mongoose, that ventures into their beat. They breed from March to September with a cooperative breeding economy like the other babblers on the Delhi list. They are the usual hosts of the Pied *Clamator jacobinus* and Common Hawk Cuckoos *Hierococcyx varius*.

Common Babbler *Argya caudata*

Breeding Resident · Common

Endemic to the Indian subcontinent, widely distributed and resident across the drier parts of South Asia, typically in arid plains or low rocky hills with scattered thorn scrub and dry cultivations. Two subspecies are recognized, race *caudata* inhabiting most of the subcontinent.

Common Babblers are characteristic of the dry thorn scrub landscapes of Delhi and grass-and-bush country in the vicinity of cultivation, but will abandon areas if they get too overgrown, such as the New Delhi Ridge. They live in parties which do not break up even during the breeding season from March to June, and all members cooperate in rearing the chicks of breeding pairs. A.J. Gaston,[235] in his study of intra-group behaviour in babblers on the New Delhi Ridge, found a number of babbler groups on his study site, each of three to 20 birds but just one pair of breeders, the others acting as helpers at the nest. They are also parasitized by cuckoos, mainly Pied *C. jacobinus* and Common Hawk Cuckoos *H. varius*.

Striated Babbler *Argya earlei*

Breeding Resident · Fairly Common

Range from the northern subcontinent to Myanmar but tend to be constrained by their choice of habitat, requiring large areas of tall grass, rushes and reed beds, in marshes and by the great rivers of the plains. Two subspecies are recognized: *sonivia* in the Indus floodplains and nominate *earlei* aross the Gangetic floodplains from Haryana to Assam, and Myanmar.

Striated Babblers are locally common in marshes and reed beds along the Yamuna *khadar*. Though normally sedentary, canals, seepage marshes and temporary marshy habitats created by floods have enabled them to spread beyond the immediate floodplain to various wetlands across the Delhi area. They live in parties of 10–20 even during their nesting season between April and August, and like Delhi's other babblers, they too exhibit a cooperative breeding economy.[235] They are commonly parasitized by the Pied Cuckoo *C. jacobinus*.

Large Grey Babbler *Argya malcolmi*

Breeding Resident · Fairly Common

Endemic to India, from the Pakistan border to Bihar and southwards through the Deccan Plateau to Tamil Nadu. Monotypic.

These large, conspicuous babblers are common but locally distributed residents in the area. Basil-Edwards, in the 1920s, thought them to be the commonest babbler in the *babool* jungles of New Delhi, a position that has now been usurped by the Jungle Babbler *A. striata*. Large Grey Babblers inhabit cultivated country with largish trees, village environs, open grassy *babool/dhak* woodland and the Aravalli landscapes of the Southern Ridge, their favoured habitat being intermediate between the dry scrub beloved of the Common Babbler *A. caudata* and the gardens and lusher growth preferred by the Jungle Babbler. Gregarious like their congeners, they keep in noisy flocks through the year, often of a dozen or more birds. They breed between April and August, the whole group helping to feed the young of nesting pairs. Hutson, in his book *Birds about Delhi*, has left a fascinating account of such cooperative behaviour in a flock of Large Grey Babblers in Delhi.

PASSERIFORMES: PASSERINES
TICHODROMIDAE: WALLCREEPER

The unique, cliff-climbing Wallcreeper of the high mountains of Europe and Asia is the only species in its own family, Tichodromadidae. Inconspicuous as it climbs over the cliffs and rock faces of its montane habitat searching for insects, its striking bright red wing patches and white-spotted flight feathers flash like a butterfly in flight.

India: 1 species, Delhi 1

Wallcreeper *Tichodroma muraria*

Vagrant

Breeds at high elevations in the mountain ranges of Europe and Central Asia. Populations nesting in the higher Himalayas descend to the foothills in winter, even entering the adjacent plains during cold waves, where they may be found in substitute habitats such as walls of ancient monuments and forts, or stone-lined embankments. Two subspecies are recognized, the nominate in Europe, *nepalensis* in Asia.

Vagrants have reached Delhi at long intervals – historically, they have been recorded by Frome on the New Delhi Secretariat building between December and March in the 1940s; and by Ganguli at the Qutb Minar in January 1951 and Tughlaqabad in January 1962. The only later reports are from Humayun's Tomb, New Delhi, in October 1971 and Tughlaqabad in early 1981.[237]

SITTIDAE: NUTHATCHES

Small birds with short tails and strong feet, nuthatches are ideally suited to an acrobatic tree-climbing existence, moving up, down and sideways on branches and tree trunks with alacrity in their search for insects hiding in crevices in the bark. Widely distributed across the Holarctic and Oriental regions.

India: 9 species, Delhi: 1

Indian Nuthatch *Sitta castanea*

Former? Resident No recent records

Endemic to India, patchily distributed over the northern plains and the peninsula, in deciduous woodland, gardens and groves. Formerly treated as conspecific with Chestnut-bellied Nuthatch *S. cinnamoventris* of the Himalayas. Monotypic.

It seems, unfortunately, that the Delhi area may have lost this striking species, with the last recorded observations being a party of three in the Lodhi Gardens in December 1978 and one at Okhla in March 1979.[238] Ganguli considered them scarce residents in the 1960s and mentioned records from gardens and *babool* parkland. This nuthatch's disappearance from about Delhi is inexplicable; there are, however, observations scattered through the year from the Haiderpur Wetland, UP, just outside the area covered by this Checklist, thus raising the possibility that they may well be relocated after careful search in its eastern parts, e.g., from the Hastinapur Wildlife Sanctuary where suitable habitat is still available.

PASSERIFORMES: PASSERINES

CERTHIIDAE: TREECREEPERS

Tree-climbing birds with cryptic plumage and long, slender bills, in two subfamilies: the widely distributed treecreepers *Certhiinae* of the Holarctic and Oriental regions, with stiffened tail feathers to help prop them against tree trunks which they climb in a series of hops, and glean small insects and similar food items from the bark; and the very distinctive spotted creepers of India and Africa, *Salpornithinae*, which, while somewhat similar to the treecreepers in colour and pattern, are stockier, with shorter, rounded tail feathers, which are not used for support in climbing; they appear closer to the nuthatches in their movements.

India: 6 species, Delhi: 1

Indian Spotted Creeper *Salpornis spilonotus*

Breeding Resident · Rare, Localized

Endemic to India, but rather sparsely distributed and localized across the western and central parts of the country in deciduous forest and groves, believed to prefer mature trees with deep-fissured bark. Two subspecies are recognized: *rajputanae* in Rajasthan and the Aravallis, and the nominate in the rest of their range.

This rare and much-sought-after species had been recorded nesting at Hattin, Gurugram district, in the month of April, in the late nineteenth century, but since then records from the area have been sporadic. Ganguli presumed them to be scarce residents about Delhi and mentioned records from *babool* woodland and *shisham* groves in both summer and winter months, in the 1940s (quoting Frome and Hutson), in the 1950s, and in 1965 when she saw one herself in February at Okhla. The *Atlas* records an observation at Bhindawas in January 1987. After a gap of 31 years, Spotted Creepers have been found again within the boundaries of the Delhi area, in *babool* woodland in the Aravalli hills of southern Haryana, with several reports through the year since 2018 from Manethi, Rewari district, and Madhogarh Fort, Mahendragarh.

Indian Spotted Creeper, a much-sought-after, uncommon species

PASSERIFORMES: PASSERINES
STURNIDAE: STARLINGS AND MYNAS

Starlings and mynas are a varied lot, gregarious, commensal and boisterous birds, some species renowned for their powers of mimicry, while others can raise a real cacophony at their communal roosts. They inhabit a variety of habitats, from grasslands and scrub, farmland, urban areas to temperate and tropical forests. Many species, especially those living in human-modified habitats, have generalist lifestyles. Omnivorous and opportunistic in their food preferences, from flower nectar and fruit to insects and small vertebrate prey, even kitchen scraps and any discarded edible items. Many nest in holes in trees, buildings or earth banks, which are frequently contested by other hole-nesting birds such as barbets or parakeets against whom the garrulous mynas do not fare too badly, though some, such as Pied Mynas, build their own large, untidy globular nests from grasses and twigs. Some starling species, especially island endemics, can be rare, and several are extinct, while others are amongst the most adaptable of birds, having spread – with human aid or independently – and aggressively colonized other parts of the world. Common Starlings and Common Mynas would top the list.

Brahminy Starling, an attractive garden myna

India: 22 species, Delhi: 8

Common Starling *Sturnus vulgaris*

Winter Visitor · Fairly Common

Native to temperate Europe and Asia, south as far as Iran, Afghanistan, Kashmir and the Indus Plain in Sindh. Adaptable birds, Common Starlings have been introduced and flourish in North America, Australia and New Zealand. Largely resident, but the northern and eastern populations are migratory, the latter wintering in the Indian subcontinent. Twelve subspecies are recognized, with at least four wintering in northern India, the Siberian breeding *poltaratskyi* and Central Asian breeding *porphyronotus* being the most numerous, with *nobilior* and *humii*, that nest in Afghanistan and Kashmir, also spreading in to the northwestern parts of the subcontinent in winter.

At least two races, *poltaratskyi* and *porphyronotus*, are common winter visitors about Delhi between late September and early April. The races are practically impossible to separate in the field in fresh winter dress, but on close view and in good light, the patterns of green and purple gloss of their plumage are discernible by late winter before they emigrate, when racial identification may be attempted. They are social birds, keep in single species flocks during the day when they frequent irrigated cultivation, damp grazing grounds, landfills and rubbish dumps, but much larger numbers gather to roost with other mynas, wagtails, etc. in reed beds, such as at Okhla.

Rosy Starling *Pastor roseus*

Passage Migrant · Common

Breeds widely from southeastern Europe, the Caspian basin and Central Asia as far south as Afghanistan, and winters abundantly in peninsular India. Monotypic.

These handsome starlings are very common as passage migrants through Delhi. Autumn passage begins early in the area, from end July, peaks in August and continues till the end of October. Autumn numbers can be large but cannot match the spectacular passage in spring. This starts in early March, peaks in the first half of April and continues till the end of that month. Parties pass over Delhi in a continuing stream each morning, and thousands gather to roost at night in clumps of large trees in New Delhi gardens and temple groves. They do not spend the winter in the Delhi area, with only a few odd records from November to February. Spring flocks feed on figs and berries of *peepal*, *kareel* and *peelu* that are fruiting abundantly during their passage period, and *semal*, *dhak* and *dhauldhak* nectar. Hutson, in *Birds about Delhi*, provides an evocative account of Rosy Starling migration in Delhi in the 1940s.

Rosy Starling

Indian Pied Starling *Gracupica contra*

Breeding Resident · Fairly Common

Common bird of well-watered country from northern India to Southeast Asia whose range has gradually expanded westwards in Rajasthan and Punjab; now more common about Delhi as well compared to past decades. Three subspecies are recognized: the nominate across a large part of the northern and central subcontinent with two other races in the northeast, *sordida* restricted to northern Assam and *superciliaris* in Manipur and further east in Myanmar. This species was formerly treated as conspecific with the Siamese Pied Starling *G. floweri* and the Javan Pied Starling *G. jalla* as the Asian Pied Starling (or Myna), recently split on grounds of genetic divergence and morphology.

Ganguli had remarked that these starlings were seldom seen within city limits in the 1960s, but they are now frequently recorded in New Delhi, feeding in irrigated lawns and gardens and nesting freely in city parks. They remain particularly fond of marshy, reedy ground in the vicinity of wetlands and damp village pastures where flocks attend on grazing cattle. They breed between April and July, laying four to six eggs in a messy globular nest of straw, rags and miscellaneous rubbish, built in small trees or even on telephone poles in suburban areas.

Pied Starling

Brahminy Starling *Sturnia pagodarum*

Breeding Resident · Fairly Common

Practically endemic to the Indian subcontinent, a resident species of open wooded country and suburban gardens from northeastern Afghanistan to Bengal and southwards through the peninsula but are only a summer visitor in the northwest. Monotypic.

Strikingly handsome birds, Brahminy Starlings are widespread in gardens, groves and woodland about Delhi. Though quite common residents and present all year in the area, at least part of the population seems to disperse to the neighbouring rural countryside for the winter months. Numbers build up again from mid-March when they are still in small flocks before they pair off in April and begin nest building in a hole in a tree trunk or a building. Typically, four eggs are laid, and two or even three broods may be raised in succession. From late August till October, after nesting is over and chicks have fledged, they gather again in parties, roosting in trees and thickets by themselves, not with other mynas in reed beds. They feed on insects, fruit and berries and are fond of *semal* and *dhak* flower nectar.

Chestnut-tailed Starling *Sturnia malabarica*

Passage Migrant · Scarce

Breeds commonly in the Himalayan foothills in northern India and disperses over the Gangetic Plain in winter. Two subspecies are recognized: *malabarica* in most of their Indian range, with *nemoricola* in northeastern India and further east.

Delhi is rather to the west of the usual range of these small arboreal starlings in southern Asia, but a few do appear in the area during passage movements, with scattered records between September and early December, and in March–April, typically of single birds or small groups with parties of Common or Brahminy Mynas. Most spring records are of birds feeding on nectar from *semal* flowers, which are in bloom at that time.

Common Myna *Acridotheres tristis*

Breeding Resident · Common

Range from Uzbekistan in Central Asia across the subcontinent to Southeast Asia, widely introduced elsewhere. Two subspecies are recognized, the nominate in India and another in Sri Lanka.

These familiar mynas are at home everywhere in the area, in pairs and small flocks, breeding in holes in buildings and trees from March to July. They are omnivorous in the fullest sense of the term, scavenging and filching scraps from marketplaces and refuse dumps, attending domestic cattle for flushed insects and feeding on flower nectar from *semal*, *dhak* and *dhauldhak* trees in season. Their nests are untidy collections of twigs, grass and rubbish, tucked in holes or crevices of trees, buildings, etc., sites which are often disputed by these pugnacious mynas, or used in succession, with pairs of parakeets *Psittacula*, Brown-headed Barbets *Psilopogon zeylanicus*, even Spotted Owlets *Athene brama*. Five to six glossy blue eggs are usual in a clutch, and more than one successive brood may be raised. Flocks commute daily to roost in large trees in urban city parks or reed beds in the countryside.

Bank Myna *Acridotheres ginginianus*

Breeding Resident Common

Bank Myna

Endemic to the Indian subcontinent, resident in its northern half, ranging across the Indo-Gangetic Plain and southwards to Gujarat and Maharashtra. An adaptable, gregarious, commensal species. Monotypic.

These are common mynas in the Delhi area, affecting well-watered floodplains and grazing land about villages and farms, where parties commonly attend grazing cattle. They have adapted very well to congested urban areas as well, frequenting railway and bus stations, town and village markets and refuse dumps, where they live off scraps and other edible rubbish. They breed colonially, excavating holes in earth embankments and in alluvium bluffs left by the river after the monsoons or adopting weep holes in masonry, even down shafts of wells. Their nests are pads of dry grass, feathers, etc., in which three to five eggs are laid. They commute daily to large, noisy roosts in clumps of trees in urban city parks and in reed beds.

Jungle Myna *Acridotheres fuscus*

Casual Non-breeding Visitor Rare

Jungle Myna

A wide-ranging myna of the Himalayan foothills and the eastern parts of the subcontinent, to Assam and east-central India, and disjunctly in peninsular India. Mainly resident, but subject to some local movements. Four subspecies are recognized, of which three occur within Indian limits: nominate *fuscus* in the north, *mahrattensis* in the peninsula and *fumidus* in the Northeast.

Jungle Mynas have been sporadically reported from the area, from Meerut, Okhla and Panipat, and these records have been attributed to escaped or released caged birds. It has, however, been recorded as 'common' (as a casual visitor and/or local migrant?) at the Hastinapur Wildlife Sanctuary on the northeastern side of the area,[239] so the possibility of genuinely wild individuals occurring about Delhi cannot be set aside.

PASSERIFORMES: PASSERINES

TURDIDAE: THRUSHES

Tickell's Thrush, a female at Bhondsi in winter

The family limits among the Turdidae (robins, chats and thrushes) and Muscicapidae (flycatchers), as earlier constituted, had long been confusing, and the morphological diversity in both families had created considerable overlap between them. The robins, chats, etc., have now been shifted to the Muscicapidae on the basis of molecular phylogenetic studies, keeping the thrushes separate.

The Turdidae, as thus reorganized, comprise two main lineages or subfamilies, one, the Turdinae (thrushes), occurring in all continents and well represented in India, the other confined to the Americas. Three genera of thrushes occur about Delhi. *Turdus* are arboreal birds but also feed on the ground, while *Zoothera* and *Geokichla* are mainly terrestrial, living in the undergrowth, and tossing and searching the leaf litter for miscellaneous invertebrates, insects, worms, etc., as well as berries, fruit and other plant items that constitute their diet. Several species are reputed for their songs, mainly heard in the breeding season. Most of Delhi's thrushes, however, are scarce, irregular visitors that breed in the Himalayas and further north, or vagrants driven so far south during cold waves.

India: 33 species, Delhi: 7

Scaly Thrush *Zoothera dauma*

Winter Visitor Rare

Part of a complex of *Zoothera* species from Siberia, East and Southeast Asia, and two in South Asia, *Z. neilgherriensis* of the Western Ghats and *Z. imbricata* in Sri Lanka, which were formerly included in *Z. dauma* as subspecies. They breed throughout the Himalayas eastwards to northern Indochina. These large thrushes are mainly altitudinal migrants along the Himalayan range, but also winter further south to central India, mostly on the eastern side. Rare individuals have wandered westwards, however, recorded as such from Bharatpur and the Aravalli Ranges. Three subspecies are recognized, with nominate *dauma* in India.

Odd individuals have been recorded sporadically in wooded habitats of the Ridge: the first record from Mangar, Faridabad district, on 5 December 2016,[240] then at the South-Central Ridge at Sanjay Van, south Delhi, on 26 October 2018, and at Mangar Bani again on 26 November 2020 and 13 March 2021, these two records possibly pertaining to one wintering individual.

Orange-headed Thrush

Geokichla citrina

Winter Visitor Scarce

Orange-headed Thrush, a scarce winter visitor to woodland patches

Widely distributed across the Oriental region, from India to southern China and southwards through Southeast Asia. The nominate race, which breeds in the Himalayan foothills, disperses southwards over the Gangetic Plain and eastern peninsular India in winter. Twelve subspecies are recognized, of which four are present within Indian limits, including the distinctive resident subspecies *cyanota* in the peninsula, and two other races in the Andaman and Nicobar islands.

These colourful ground thrushes are uncommon winter visitors to the area, affecting shady, damp corners in well-wooded areas. Ganguli mentioned observations in winter months in the 1960s from New Delhi gardens, but a growing number of reports in recent years are mainly from damp depressions with some undergrowth cover in wooded habitats as at the Sultanpur National Park and the Bhindawas Wildlife Sanctuary, Haryana, from Okhla and the Hastinapur Sanctuary, UP, and from the South-Central and Southern Ridge landscapes.

Mistle Thrush *Turdus viscivorus*

Vagrant

Wide breeding distribution from Europe and North Africa to Central Asia, southwards to the western and central Himalayas. Three subspecies are recognized, and the Himalayan populations, race *bonapartei*, show strong seasonal altitudinal movements, though they rarely enter the plains.

A vagrant was observed with Black-throated Thrushes *T. atrogularis* at Sultanpur in February 2002, probably driven southwards with the latter flock by harsh weather.[241]

Song Thrush *Turdus philomelos*

Vagrant

Breeds in Europe and eastwards to central Siberia, wintering to the south in North Africa, West Asia and Arabia. There are very few records from the subcontinent, of birds accidentally straying eastwards of their usual winter range, but one was recorded as close to the Delhi area as the Bharatpur National Park in 1981. Four subspecies are recognized, the few South Asian records presumably pertaining to Siberian breeders, race *nataliae*.

Two birds were recorded at Mangar Valley in the Southern Ridge, Faridabad district, on 11 December 2016,[242] only the third record from India.

A Grey-winged Blackbird, the handsome male of the species

Grey-winged Blackbird

Turdus boulboul

Winter Visitor Rare

Range from the western Himalayas to southern China and northern Indochina, nesting at mid-elevations along the Himalayan range and are altitudinal migrants, dropping to the foothills in winter with a few dispersing deeper into the plains in colder winters. Monotypic.

Grey-winged Blackbirds had not been recorded from the Delhi area historically, but as scattered records in recent years demonstrate, individuals do appear here occasionally between December, when the weather is at its coldest, and early March, when they have been observed in well-wooded city gardens and woodland patches, especially about Okhla and along the Southern Ridge. In most years, however, they are missed.

Tickell's Thrush *Turdus unicolor*

Winter Visitor Rare

A purely Himalayan thrush, nesting from northern Pakistan to Bhutan and wintering mainly in central and eastern India. Monotypic.

Tickell's Thrushes are irregular and rare on passage and winter in the area. Historically, Ganguli had noted just three historical records in February, March and April in the 1960s from Delhi gardens, while A.J. Gaston recorded one on the New Delhi Ridge in late November 1971. There are, however, a number of recent reports from urban gardens, and woodland patches in Ridge habitats or at the Sultanpur National Park and the Bhindawas Sanctuary, Haryana, mainly between late October and March with one report as late as May.

Black-throated Thrush *Turdus atrogularis*

Winter Visitor Uncommon

Breeds in western Siberia and winters South to the Middle East and South Asia along the Himalayas, extending into the northern plains as the winter really sets in. Monotypic.

Black-throated Thrushes are present in most winters in the Delhi area, though rarely before December, and they remain till February–March, when they are most often reported. They may be seen, usually in small parties, in gardens, woodland and scrub on the Ridge, and groves in rural areas, and are regular visitors to wooded habitats in the Sultanpur National Park and the Bhindawas Bird Sanctuary, Haryana.

PASSERIFORMES: PASSERINES

MUSCICAPIDAE: CHATS AND FLYCATCHERS

The Muscicapidae as presently redefined on the basis of molecular phylogenetic studies, group together a number of diverse, mainly insectivorous species in four subfamilies, of which three are represented in India: Muscicapinae (flycatchers and shamas), Niltavinae (blue flycatchers) and Saxicolinae (robins, redstarts, chats, etc.). Some feed by flycatching sallies in the air, others by gleaning or pouncing on prey on the ground from a perch. Some also feed on seeds and berries when insect food is scarce. The Himalayas have a wealth of breeding species of this family which figure on the Delhi list as migrants or accidentals, while the wheatears are mainly arid country species that visit in winter; there are also some familiar residents of the Delhi area amongst the robins and chats.

Verditer Flycatcher, in March, on its way to the Himalayas

India: 104 species, Delhi: 36

Asian Brown Flycatcher *Muscicapa dauurica*

Localized Summer Breeding? Visitor Uncommon

Breeds, albeit very locally, in temperate and subtropical forests of eastern and southern Asia; two subspecies are recognized: nominate *dauurica* breeding in Siberia, Japan, China and India, and *siamensis*, resident in Indochina. In India, birds of the nominate race are summer visitors to the Himalayas and some of the central Indian ranges and are also resident in the Western Ghats; Himalayan breeders winter south of the Vindhyas.[243]

Hutson noted an old July record from the 1940s, but none were reported from the Delhi area thereafter for the next 60 years or so. Since 2015, however, Brown Flycatchers have been observed every year in the summer months from July through September, and with a few reports from April to June as well, in areas of richer, mixed forest on the Southern Ridge – as at the Asola Wildlife Sanctuary in south Delhi, and Mangar Bani, Bhondsi, Manesar and Rewari in Haryana – and woodland patches elsewhere as at Okhla and the Sultanpur National Park. There are three winter records, in end November and December/January. Their breeding season in northern India is between May and July, and juveniles in spotted plumage have been recorded in July and August. Given the pattern of their occurrence, it is not possible to either confirm breeding in the Delhi area or whether these birds are only passing through, having bred in the Himalayan foothills. Other recent reports show their presence in May, June and July in Aravalli habitats in Rajasthan to the south of the Delhi area, and even as far to the west as Jaipur. Whether these records represent an extension of their breeding range in the Aravallis calls for closer study. Unobtrusive birds, they are somewhat crepuscular in habit, and this, along with the fact that these sites have not been watched intensively until recently, may explain the lack of previous records.

Brown-breasted Flycatcher *Muscicapa muttui*

Vagrant

Nest in China and Northeast India and traverse the eastern parts of the peninsula in autumn en route to their wintering areas in the Western Ghats. There are records of birds having drifted as far westwards as Kathmandu, Gujarat and the Satpura Range.

A vagrant was photographed at the Sultanpur National Park, Haryana, on 24 October 2016,[244] the first and only reported occurrence so far in the Delhi area.

Spotted Flycatcher *Muscicapa striata*

Vagrant

Breeding summer visitors across Eurasia, from Europe and North Africa to Central Asia, south to Afghanistan and northwest Pakistan, with the entire population wintering in sub-Saharan Africa. Seven subspecies are recognized over this range; the migration path of race *sarudnyi* between its nesting grounds in Central Asia and its winter quarters in Africa traverses the Indus Valley and the adjoining Indian states of Rajasthan and Gujarat in autumn; in spring, on northward passage, this population takes a more western route, passing mainly west of the Indus.

Migrants have drifted eastwards during passages to reach the Delhi area on four occasions in autumn (September–October) and twice in spring (April–May), recorded thus on the Delhi Ridge in September 1972; the Delhi Zoological Gardens in October 2004; the Aravalli Biodiversity Park, Gurugram, in May 2016; on the Southern Ridge at Raisina, Gurugram district, in September 2016; the Bhondsi Nature Park, Gurugram, in early October 2019; and in central New Delhi in April 2020.

Rufous-tailed Scrub Robin *Cercotrichas galactotes*

Vagrant

Rufous-tailed Scrub Robins are rare vagrants, recorded in Delhi during autumn passage

Summer nesting visitors to North Africa and the Mediterranean, the Middle East and Central Asia, and winter in Africa. Five subspecies are recognized over this range, and birds breeding in its easternmost reaches, of the Central Asian race *familiaris*, traverse western Rajasthan and Gujarat en route to their winter quarters in northeastern Africa in autumn. Their return migration follows a more westerly path.

Vagrants have drifted eastwards from their migration path in autumn to reach Delhi and have been recorded as such on four occasions to date, all in the month of August: single birds on the New Delhi Ridge in 1972, at Surajpur, Greater Noida, on 3 August 2014, and at the Aravalli Biodiversity Park, Gurugram, on 14 August 2017 and 26 August 2018.

Indian Robin *Copsychus fulicatus*

Breeding Resident · Common

Indian Robin, a male of the brown-backed northern race cambaiensis

Endemic to, and widely distributed in, the subcontinent east to Bihar and Bengal, in lightly wooded country, dry cultivation and scrub, often in rather dry, rocky biotopes and commonly near habitations. The Indian Robin, magpie-robins and shamas – formerly placed in separate genera *Saxicoloides*, *Copsychus* and *Kittacincla* – are close enough genetically to be united under one genus, *Copsychus*. Five subspecies are recognized, amongst which the brown-backed race of northern India, *cambaiensis*, is resident in the Delhi area.

These familiar robins are common residents about Delhi in grass-and-bush country, village environs, the rocky scrubland of Ridge landscapes and the larger, more open parks in urban areas. They tend to desert areas if they get too overgrown and lush and are, therefore, much less frequent in central New Delhi than previously. They nest between March and August, when males sing and display, flaunting their white wing patches. Their nests, untidy cups of stems and natural fibres, are usually placed in holes in walls or even other odd places such as some derelict piece of machinery. Usually three eggs are laid in a clutch, and more than one brood may be raised in a season.

Oriental Magpie-Robin *Copsychus saularis*

Breeding Resident · Fairly Common

Common and widespread residents of the Oriental region, appear to be expanding their range westwards. Seven subspecies are recognized, three within Indian limits, with nominate *saularis* extending across the northern Indian subcontinent.

Until the 1950s, Magpie-Robins were considered to be mainly winter visitors about Delhi with only isolated nesting records, and Ganguli categorized them as 'not-too-common' breeders even in the 1960s. Their numbers have increased, they are now common residents, widespread and familiar in better vegetated corners of gardens, groves, plantations and orchards. They begin to sing as the weather starts warming up from February, when their sweet voices are a familiar feature of city parks, and nest through the summer months till July. Nests are untidy cups of twigs, grass, hair and other miscellaneous fibres, placed in holes in trees, walls, a drainpipe or even a nestbox, and hold their clutches of four or five eggs.

Oriental Magpie-Robin, male

Blue-throated Flycatcher *Cyornis rubeculoides*

Vagrant

A Blue Flycatcher of the lower Himalayan ranges and parts of Southeast Asia, with three subspecies, of which nominate *rubeculoides* is a summer nesting visitor to the Himalayas, wintering in peninsular India.

An accidental vagrant on passage. A. J. Gaston recorded "One female or immature trapped on Oct. 22, 1971 (on the New Delhi Ridge) and sent to the BNHS. Identification per A. S. Cheke." The only two later records are a male bird at the Asola Wildlife Sanctuary on the Southern Ridge on 16–20 April 2018 and another (or perhaps two) in woodland at the Sultanpur National Park on 5–8 October 2022.[245]

Tickell's Blue Flycatcher *Cyornis tickelliae*

Localized Breeding Resident? Scarce

Most widespread of the Blue Flycatcher genus *Cyornis* in India, a bird of well-wooded gardens, groves and forest as far west as the Aravalli Ranges in Rajasthan and on the east reaching Bengal. Resident, but shows some localized movements in the non-breeding season, between October and May. Two subspecies are recognized: nominate *tickelliae* in India, *jerdoni* in Sri Lanka.

Tickell's Blue Flycatchers are a breeding resident as close to Delhi as the Sariska National Park in the northern Aravallis. It was rather odd, therefore, that they had not been recorded about Delhi itself, but it was expected they would be sooner or later. They were listed, but without supporting detail, as a passage migrant in the gardens of Rashtrapati Bhavan.[246] A male in an area of native Aravalli woodland at Mangar Bani, Faridabad district, in July 2016 was actually the first documented record for the area, followed by a number of observations in subsequent years during the summer months, February to August, at other sites on the Southern Ridge, with odd winter records as well from Gurugram parks. These records are suggestive of a hitherto overlooked localized resident, or early harbinger of this species establishing itself as a breeder in the Delhi area. A confirmation of nesting is desirable.

Verditer Flycatcher, male

Verditer Flycatcher *Eumyias thalassinus*

Passage Migrant Scarce

Breeds along the Himalayas eastwards to China and spread out in winter over most of the Indian peninsula and Southeast Asia. Two subspecies are recognized: nominate race *thalassinus* in India.

These eye-catching flycatchers are scarce passage migrants in Delhi, with records of single birds or pairs, mainly in March and October, from parks and gardens, woodland patches and the Ridge. This pattern appears to have remained unchanged over the last several decades.

Indian Blue Robin *Larvivora brunnea*

Vagrant

Breeds at mid-elevations 1,800–3,000 m in undergrowth of deciduous and conifer forests in the Himalayas, wintering in the Western Ghats; recorded on passage in the Himalayan foothills and very locally in the peninsula. Rarely reported from the Gangetic Plain, but has been recorded at Lucknow. Monotypic.

A vagrant was photographed in woodland at the Sultanpur National Park on 1 October 2022.[247]

Bluethroat *Luscinia svecica*

Winter Visitor Fairly Common

A male Bluethroat, assuming his breeding colours

Breeds widely across the northern and temperate Eurasia south to Turkey, Afghanistan and Ladakh, wintering in northern Africa and southern Asia. Up to 12 subspecies have been named over this range, of which at least four winter in India: nominate *svecica* which breeds in northern Asia, *pallidogularis* and *tianshanica* of the Central Asian plains and mountains, respectively, and *abbotti* of Afghanistan and Ladakh.

Bluethroats are fairly common visitors to the Delhi area from late August/September to April, frequenting damp corners in parks and gardens, reed beds and waterbodies of all types, small or large, as long as there is nearby cover into which they can retreat. By March, many males are in fresh breeding plumage with glistening blue throats. Race *pallidogularis* has been identified at this time; some of the other subspecies may be expected to winter as well.

Blue Whistling Thrush *Myophonus caeruleus*

Vagrant

Large whistling thrushes of mountain streams and rivers running through forested ravines and open country, often close to human habitations, largely resident from southern Central Asia, Afghanistan and the Himalayas eastwards to China and Southeast Asia. Six subspecies are recognized, of which *temminckii* is a common and familiar resident of the Himalayas. This population shows some altitudinal movements down to the foothills in winter but is known to wander into adjacent plains on rare occasions, having been recorded as far to the south as Neemrana and Alwar, Rajasthan.

Blue Whistling Thrushes have been recorded as vagrants with just two records from the area. One bird, in poor plumage, was reported from the Delhi Cantonment area on 22 March 2010 and another individual was photographed at Rithal, Gohana, near Sonipat, Haryana, in January 2018.[248]

Siberian Rubythroat *Calliope calliope*

Vagrant

Breeds in Siberia east of the Urals and south to Mongolia, and winters in eastern India and Southeast Asia, but have wandered westwards as far west as Bharatpur, Rajasthan. Monotypic.

There are a mere handful of records of vagrant individuals of these striking rubythroats from the Delhi area. Frome mentioned two historical records in November in the 1940s and one in 1952. Later records include single males on the New Delhi Ridge in end October 1978 and Okhla in March 1979,[249] and at Sultanpur in late February 2014.

Himalayan Bush Robin *Tarsiger rufilatus*

Vagrant

Formerly lumped with the migratory Orange-flanked Bush Robin or Siberian Bluetail *T. cyanurus* but now split;[250] breeds in the Himalayas in open *Abies-Rhododendron* forest at 3,000–4,400 m and is an altitudinal migrant, wintering down to 1,300 m. Monotypic.

These bush robins do not normally enter the plains in winter, yet a vagrant immature male (presumed to be Himalayan, though field separation from its sister Siberian species could not be confirmed – there are only two records of the latter from Northeast India) was recorded in woodland on the Southern Ridge at Mangar Bani, Faridabad, in the last week of November 2020. A small party of Grey-hooded Warblers *Phylloscopus xanthoschistos* and a Whistler's Warbler *P. whistleri* were reported at the same time and location, an indication that they may have all been driven southwards by harsh weather conditions in the Himalayan foothills. The bird was photographed again at the same site on 7 March 2021, and it seems likely that this vagrant individual had spent the entire 2020–21 winter there.[251]

A male Ultramarine Flycatcher on passage through the area

Ultramarine Flycatcher *Ficedula superciliaris*

Passage Migrant · Rare

Common Himalayan species, nesting in broadleaf forest at mid-elevations and wintering in central India. Two subspecies of this handsome flycatcher are recognized: nominate *superciliaris* in the western and central Himalayas, and race *aestigma* from the eastern Himalayas to southern China.

There are scattered records from the Delhi area of straggling passage migrants of the west Himalayan race, most in March and October: historically, in the 1960s and 1970s,[252] and in recent years, from Delhi gardens and parks, woodland on the Ridge and wooded areas in cultivated country, as at the Sultanpur National Park or the Bhindawas Sanctuary, Haryana. Individuals may even winter rarely, with one at Sunder Nursery, New Delhi, in January–February 2015 and another on the South-Central Ridge at Sanjay Van in January 2017.

Rusty-tailed Flycatcher *Ficedula ruficauda*

Passage Migrant Rare

Largely endemic to the subcontinent, breeding in coniferous and mixed forests from Tajikistan and Afghanistan to the western and central Himalayas, and wintering in southwestern India. Monotypic.

These flycatchers have been recorded in the area as straggling passage migrants in both autumn and spring. Ganguli had mentioned just two historical records in April 1957 and January 1970, but there is a smattering of reports of birds on passage in recent years, mainly in April and between late September and mid-October, from Delhi gardens, woodland on the Ridge (including several reports from Mangar Bani, Faridabad district) and from wooded areas at Okhla and the Sultanpur National Park.

Taiga Flycatcher *Ficedula albicilla*

Passage Migrant Uncommon

Replace Red-breasted Flycatchers *F. parva* as a breeding species in the eastern Palaearctic with their winter range correspondingly more towards the south and east in the subcontinent and Southeast Asia. The two were formerly considered conspecific, but they are genetically divergent, diagnosably distinct with differences in moult patterns and appear to represent independent evolutionary lineages.[253] Monotypic.

Taiga and Red-breasted Flycatchers were not subspecifically distinguished in the past, so their status in the area is difficult to ascertain with any degree of confidence. Taiga Flycatchers appear to be primarily passage migrants about Delhi, en route to parts of India to the immediate south and southwest and are certainly much scarcer about Delhi than their Red-breasted siblings in winter. Many reports from the area, moreover, seem to have been misidentifications.

Taiga Flycatcher, a male in breeding plumage

Kashmir Flycatcher *Ficedula subrubra*

Vagrant

Restricted as a summer breeding visitor to Kashmir; records of Kashmir Flycatchers during passage to and from their winter quarters in the Nilgiris and Sri Lanka are rare, most birds apparently covering the distance in a single hop. Monotypic.

There is one definite record from the Delhi area, a male photographed at the Sultanpur National Park on 14 September 2015.[254] A couple of later claims from the same location cannot be validated.

Red-breasted Flycatcher, male

Red-breasted Flycatcher

Ficedula parva

Winter Visitor Common

Breeding summer visitors to Eastern Europe and Russia west of the Urals, and winter mainly in India, common in the west and centre and progressively scarcer towards the eastern and southern parts of the country. Towards the northwest of the subcontinent, they are mainly passage migrants, both in spring and autumn. Monotypic.

Red-breasted Flycatchers are common passage migrants through the area, with fair numbers remaining through the winter as well. Autumn passage begins in October and continues into November, but they are especially conspicuous in spring, from mid-March through April, when males are in showy breeding dress. They are birds of open woodland, parks and gardens, which can teem with these little flycatchers on peak passage days.

Plumbeous Water Redstart *Phoenicurus fuliginosus*

Vagrant

Breeds along the Himalayan range, west to east, and in southern China. They are altitudinal migrants in the Himalayas and regularly descend to the foothills in winter, dispersing further southwards into the plains during cold spells. Two subspecies are recognized; the Himalayan population is of the nominate race, *fuliginosus*.

There are three records to date of vagrant individuals reaching the Delhi area: from Okhla in the 1990s, one near Sonipat that stayed from December 2001 to February 2002[255] and a female/juvenile again at Okhla on 17 February 2018.

White-capped Redstart *Phoenicurus leucocephalus*

Vagrant

Altitudinal migrants, descending to the foothills in winter, when they frequent rivers as they enter the *terai* and often their associated canal systems. As a species, their range extends westwards to Uzbekistan and eastwards to much of China. Like the Plumbeous Redstarts *P. fuliginosus* that share their habitat, individuals may wander further southwards into the northern plains in severe winter weather and have been reported as such from the Sariska National Park and even as far as the Indira Gandhi Canal in Rajasthan. Monotypic.

These lovely redstarts of fast flowing Himalayan streams are mere vagrants in the Delhi area, individuals having wintered at least twice near Sonipat, remaining at the same stretch of canal from December 2001 to February 2002,[256] and another that stayed from mid-November 2016 till January 2017. A third was reported from Bandhwari, Gurugram district, on 21 December 2018.

Blue-capped Redstart *Phoenicurus coeruleocephala*

Vagrant

Breeds from Central Asia to Afghanistan and the western Himalayas; are altitudinal migrants, wintering at mid- to low-elevations in the foothills in open pine forest and scrub-covered slopes. Monotypic.

Untypical redstarts, the only member of the genus lacking red in male plumage, these robins do not normally venture further southward into the plains in their winter displacements, yet an adult male was observed closely on the stone edge of the dam at Badkhal Lake, Faridabad, in November 2000,[257] an unusual record of vagrancy for the Delhi area. There are, however, two other records from the wider neighbourhood a little beyond the limits of the area defined for this Checklist: a first winter male at Bharatpur, Rajasthan, in early October 2017, and another from the Haiderpur Wetland, UP, on the Ganga northeast of Delhi, on 1 February 2020.

Black Redstart *Phoenicurus ochruros*

Winter Visitor Common

Distributed from Europe across to Central Asia, the Himalayas and China, wintering to the south of this breeding range in North Africa, Arabia and South Asia. Seven rather distinct races are recognized, two of which winter in India, the western Himalayan *phoenicuroides* and the eastern *rufiventris*.

The greyer-crowned race *phoenicuroides* is a common winter visitor to the area between September and April/early May, when it may be seen in city gardens, parks and woodland, and generally in lightly wooded country. There are rare records of the black-crowned eastern race *rufiventris* from about Delhi as well.

The Black Redstart, a common visitor to parks and woodland alike

Blue-capped Rock Thrush *Monticola cinclorhynchus*

Passage Migrant Rare

Summer breeding visitors in open mixed pine and broadleaf forest and sparsely wooded rocky slopes at mid-elevations along the Himalayas, and winter across much of central and southern India. Monotypic.

These colourful rock thrushes are rare, irregular passage migrants through the area. Historically, there are just three mentions of this Rock Thrush from Delhi: Ganguli noted two observations in September from Delhi gardens, while A.J. Gaston recorded a female on the New Delhi Ridge in mid-April 1974. In spring it seems particularly scarce, with just two recent spring records in April 2000 and 2021 but seems rather less rare in autumn with scattered reports in recent years, mainly in October, from woodland areas, including the Sultanpur National Park, the river *khadar* and a few sites on the Southern Ridge.

Rufous-tailed Rock Thrush *Monticola saxatilis*

Vagrant

Breeds from southern Europe and the Middle East to Central Asia, and winters in East Africa. Their migration path in autumn cuts across the westernmost parts of the subcontinent, when they have been recorded in small numbers in Pakistan and more rarely from trans-Himalayan Ladakh and the northwestern Indian states of Himachal Pradesh, Rajasthan, Haryana, Gujarat and elsewhere in peninsular India, mainly between August and October. Very scarce in the subcontinent in spring, when their migratory path is further to the west; occasional nesting has, however, been recorded in open rocky slopes in Baluchistan, Pakistan. Monotypic.

A vagrant was recorded at the Aravalli Biodiversity Park, Gurugram, on 15 September 2019.

Blue Rock Thrush *Monticola solitarius*

Winter Visitor Scarce

Distributed from the Mediterranean region to the Himalayas, China and Japan, nesting across this wide range in stony mountain cliffs and gorges and wintering in North Africa, Arabia and throughout southern Asia. Five subspecies are recognized, two of which, *longirostris* and *pandoo*, breed within Indian limits in the northwestern and central Himalayas, and winter to the south across the country, in broken rocky terrain, quarries ancient forts and even occupied structures.

Blue Rock Thrushes are scarce winter visitors to the Delhi area between October and April, perhaps more frequently reported during passages, affecting historical monuments and ruins, abandoned mines and rock formations, and similar habitats on the Southern Ridge. These presumably belong to the Himalayan race *pandoo*.

Blue Rock Thrush, a female

Stoliczka's Bushchat *Saxicola macrorhynchus*

Localized Winter Visitor? Rare

A Stoliczka's Bushchat performing its characteristic 'puff-and-roll' display

An Indian endemic species, rare and localized in arid semi-desert with sparse bushes in Rajasthan and Kachchh, Gujarat. There are century-old records from Haryana too, mostly in winter, and have being reported since then from near Hisar between late January and March,[258] most recently in December 2018. The records suggest that these bushchats perform short-distance seasonal movements, but these are poorly understood and documented. Monotypic.

The status of Stoliczka's, or White-browed, Bushchat in the Delhi area remains unclear, with sparse reports scattered over the winter months between November and April, and two in August. The first records were one from Tughlaqabad in November 1993 and three observed in open country near the Sultanpur National Park in early February 2001 that seemed to be establishing territories but disappeared by early April.[259] After over a decade, during which there were no further reports (possibly because these areas were not investigated thoroughly), there were a few wintering records from the Sultanpur National Park neighbourhood: in November 2014; one (or perhaps two) that appear to have spent the 2017–18 and 2018–19 winters there, between mid-October and mid-March; and one in the same general area in January 2020. Summer records included one in reed beds in the Yamuna *khadar* in north Delhi on 25 August 2016[260] in seemingly untypical habitat, and one in open ploughed fields at Manjhawali, Faridabad district, in mid-August 2021, both possibly being dispersing juveniles of the year. The breeding biology of this bushchat is poorly known, but juveniles have been reported in August–September, and evidence suggests that the breeding season may extend from March to July. The records are suggestive of passage movements through the Delhi area and localized wintering in the neighbourhood of the Sultanpur National Park, which, as part of the *dabar* area of southwest Haryana, shows marked desert influences in its avifauna.

Siberian Stonechat *Saxicola maurus*

Winter Visitor Fairly Common

Widespread as a breeding species across Asia, from the Caspian Sea to Siberia, and southwards to the Himalayas, and winter through much of southern Asia, from South to Southeast Asia. These stonechats were formerly considered conspecific with European and African Stonechats *S. rubicola* and *torquatus*. Five subspecies are recognized, of which at least three winter in India: the Siberian nesting *maurus*, Himalayan *indicus* and Tibetan *przewalskii*.

Siberian Stonechats are fairly common visitors to the area (races *maurus* and *indicus*) from late August to April–May, though numbers increase during passage periods, and in spring they can be common in the Yamuna floodplain which offers the ideal mix of cultivation, dry grass-and-bush country and rank reedy patches that suit their requirements. They also affect reed beds and marshes alongside White-tailed Stonechats *S. leucurus*.

White-tailed Stonechat *Saxicola leucurus*

Localized Breeding Resident Uncommon

Restricted to the floodplains of the Indus, Ganga and Brahmaputra rivers across the northern subcontinent, affecting areas of tall grass and reeds interspersed with open marshy ground liable to seasonal inundation. Monotypic.

These resident stonechats were not listed for Delhi prior to the 1970s and were first recorded in the Yamuna floodplain only in 1978, around the same time that two other species which occupy the same grassland/reed bed habitat, Striated Grassbird *Megalurus palustris* and Yellow-bellied Prinia *Prinia flaviventris*, were also recorded for the first time. Also resident in the Haiderpur Wetland and adjacent Hastinapur Sanctuary, the first records again go back to around the same time, late 1970s and early 1980s.[261] White-tailed Stonechats remain limited to remnant patches of such marshy habitat, resident and breeding in the mosaic of reeds and open damp ground in the Yamuna *khadar* at Wazirabad and Okhla. Males begin singing by late January with song flights in March; by the end of that month, some pairs are building, a few already feeding young. Their nests are cups of grass and other fibres, tucked into a tussock or at the foot of a bush, with clutches of usually three eggs. Over the last couple of decades, with the changes in the character of the habitat at Okhla, its numbers appear to have declined at this site.

Pied Bushchat *Saxicola caprata*

Breeding Resident Fairly Common

Widespread and largely resident except for the northernmost populations, from Central Asia and Iran across the subcontinent to Southeast Asia. No less than 16 subspecies have been named over this range, three in India, with race *bicolor* in northern India.

These striking bushchats are quite common residents about Delhi in grass-and-bush country, often near wetlands and marshes, along canal embankments, fields and thorn hedges near villages. They normally nest between February and July, building a nest of grass and rootlets in a depression in the ground under a bush or earth bank, in which three to five eggs are laid, but males have been observed singing and performing song flights again in September (a late second brood?). Males are pugnacious towards conspecifics in their territory, announcing their ownership by song and with slow flapping flights when their white wing patches are prominently displayed.

Grey Bushchat *Saxicola ferreus*

Winter Visitor Scarce

Mainly resident in open scrub- and bush-covered hillsides, at mid-elevations in the Himalayas, eastwards to China and parts of Southeast Asia. Himalayan breeders are altitudinal migrants, wintering to the foothills and extending further into the plains at times, during cold spells. Two subspecies are recognized, the nominate in India.

Ganguli considered them accidental in Delhi and mentioned only two 'doubtful' records by Hutson, presumably in the 1940s. They may have been overlooked, as there has been a smattering of reports from across the area in recent years, between October and February, from gardens, the river floodplains, canal embankments, woodland patches in cultivated areas and Ridge habitats, often near water.

Isabelline Wheatear *Oenanthe isabellina*

Winter Visitor Uncommon

Like most of our wheatears, breeds in desert habitats to the west and north of the subcontinent and winters in the drier parts of northwestern India. Isabelline Wheatears are summer breeding visitors to Turkey and across Central Asia, south as far as Afghanistan and Baluchistan, and winter in northern Africa, Arabia and northwestern India. Monotypic.

The dry bare flats favoured by Isabelline Wheatears are being increasingly lost to expanding agriculture and urbanization in the Delhi area, and wintering numbers have come down. Ganguli noted pre-1970 winter records from Tughlaqabad and Sultanpur. Recent reports, all between end September and February, are mainly from open ground near the Sultanpur National Park and the Bhindawas Sanctuary in Haryana. They are the most terrestrial of our wheatears, with a rather distinctive habit of running, rather than hopping, on the ground.

Desert Wheatear *Oenanthe deserti*

Winter Visitor Uncommon

Widespread from North Africa to Central Asia as far south as Iran, Afghanistan and Ladakh, and winter in the Sahara, Arabia to western and northern India. Four subspecies are recognized over their range, with race *oreophila*, breeding in West and Central Asia, wintering in India.

A winter visitor in small numbers to the area between October and March, perhaps most frequent in the latter month during spring passage. They affect fallow ground in the floodplain and countryside, and bare sandy flats and overgrazed ground near villages.

Desert Wheatear, male in breeding colours

Brown Rock Chat *Oenanthe fusca*

Breeding Resident Common

Endemic to northern and central India. They were formerly placed in the mainly African genus Cercomela, but recent molecular studies have concluded that it actually belongs to the wheatear genus *Oenanthe*. Indeed, as Hugh Whistler, writing in his *Popular Handbook of Indian Birds*, pertinently observed years ago, 'It is a regular wheatear in its habits ...' Monotypic.

Common residents in the area, and here they seem to have largely abandoned their natural habitat of rocky cliffs and similar features for human-modified environments, living and nesting in buildings, old and new, archaeological monuments, forts, etc., and are tame, familiar birds that even enter inhabited houses on occasion. Males begin singing from March, and they nest between April and August in ledges and crevices, both natural and otherwise, building a rough cup of grass, feathers and other fibres, often with a rampart of earth around it, which holds a clutch of three or four eggs.

Variable Wheatear *Oenanthe picata*

Winter Visitor Scarce

Breeds from Central Asia to Pakistan, wintering in the Indus Plain and western India, east to the Delhi neighbourhood. Monotypic, but occurs as three morphs: *picata*, *opistholeuca* and *capistrata*, which are usually not treated as valid taxa, and intergrades are common.

Variable Wheatears are scarce winter visitors to the area between late October and March, affecting dry flats and stony waste ground. With such habitat becoming increasingly restricted around Delhi, they are now scarcer and less widespread since the 1960s when Ganguli considered them 'uncommon'. Recent reports are mostly from the Sultanpur neighbourhood and the barer parts of the Southern Ridge. All three morphs have been recorded in the area, with *picata* the most frequently reported.

Red-tailed Wheatear *Oenanthe chrysopygia*

Winter Visitor Rare

Nest in Iran and Afghanistan, wintering in Arabia, the Persian Gulf, Pakistan and northwestern India, mainly in the Thar desert and Kachchh. Monotypic.

Red-tailed, or Persian, Wheatears have straggled eastwards as far as Delhi on rare occasions in winter. There is just one historical, pre-1920, record from Delhi and scattered observations in later, post-1975, years from arid habitats in Haryana, in the Sultanpur National Park environs, near the Bhindawas Wildlife Sanctuary and dry, open ground on the Southern Ridge.[262]

PASSERIFORMES: PASSERINES

DICAEIDAE: FLOWERPECKERS

An Oriental–Australasian family of small, often brightly coloured, birds with tubular tongues addicted to nectar and berries of parasitic mistletoes of the families Loranthaceae and Viscaceae, with which they exist in a symbiotic relationship. In spite of their name, they are mainly berry (especially mistletoe) eaters but may be seen pecking at other fruit as well, especially guavas *Psidium* sp. They live in wooded and open habitats, including orchards and city gardens, but of the several species on the Indian list, only one reaches the Delhi area as a rare, irregular visitor.

India: 9 species, Delhi: 1

Thick-billed Flowerpecker *Dicaeum agile*

Winter Visitor Scarce

Mainly resident, though subject to local movements, in much of India, Sri Lanka and Southeast Asia, with up to eleven subspecies named over this range; nominate *agile* in most of India with *pallescens* in the Northeast.

Delhi is on the western fringes of the flowerpecker's normal residential range in the northern plains and it is only an occasional visitor here, mainly in winter between October and February. There are historical records from the Delhi Ridge and almost annual records in recent years from Delhi's parks and gardens; they are possibly more regular than these scattered records suggest but are overlooked.

PASSERIFORMES: PASSERINES

NECTARINIIDAE: SUNBIRDS

A family of small birds with a large number of species in Africa and southern Asia, and with long decurved bills adapted for feeding on flower nectar, of which they are inordinately fond. Small insects and spiders are also an important part of their diet. Dressed in bright shades of red, yellow, green, blue and purple, many have brilliantly iridescent plumage. They live in varied habitats, from dry woodland, gardens and parkland to tropical rainforest. Of the several species in India, only one is resident in the Delhi area.

India: 15 species, Delhi: 1

▲ *A male Purple Sunbird, in his breeding colours*

Purple Sunbird *Cinnyris asiaticus*

Breeding Resident · Common

Common residents from the Persian Gulf to the Indian subcontinent and mainland Southeast Asia. Three subspecies are recognized: *brevirostris* ranges from the Persian Gulf to western India, nominate *asiaticus* in most of the country and *intermedius* from eastern India to Indochina.

Familiar birds of urban gardens, woodland on the Ridge and generally anywhere that there are flowering trees and shrubs in the countryside. Males begin to moult into their breeding finery from November and are singing and displaying as soon as the weather begins to warm up in February. Their nests are remarkable structures: a soft oblong purse with a side entrance and a covering porch, the whole constructed of vegetable fibres and cobwebs, and camouflaged with dried leaves, social spider nests, bark, etc. to resemble a mass of miscellaneous rubbish, and suspended from a twig or a hanging rope or chain, even in a balcony or veranda. The clutch is usually of two eggs. The breeding season continues till May and June, and by the rainy season males assume their eclipse plumage again, with many juveniles in fresh and bright female-type plumage around at that time.

PASSERIFORMES: PASSERINES
PLOCEIDAE: WEAVERS

A large Afro-Asian family of sparrow-like seed-eaters of tropical grasslands and open woodlands, famed for their intricately knotted and finely woven nests of leaf fibre. Just as remarkable is their polygynous breeding biology, in which males alone weave the nests and then use them as display platforms to lure successive females. After the eggs are laid, typically in clutches of three eggs per nest, incubation and care of the chicks is by the females alone. They feed predominantly on seeds, but insects and flower nectar also figure in their diet.

The family is highly diversified in Africa, both in genera and species, but Asia hosts only five species in the genus *Ploceus*; four are found in India, one of which weaves perhaps the most remarkable pendulate nest structures in the family.

India: 4 species, Delhi: 4

◄ *A male Black-breasted Weaver in breeding plumage*

Streaked Weaver *Ploceus manyar*

Breeding Resident · Fairly Common

Resident across the Indian subcontinent and much of Southeast Asia in wetland habitats with reeds and tall grasses and have been introduced successfully in the Nile delta and other locations in the Arabian peninsula. Four subspecies are recognized, of which two occur within Indian limits, *flaviceps* in most of the country and *peguensis* in northeastern India.

Locally common in the area. Both Streaked- and Black-breasted Weavers may breed in close proximity to each other as at Okhla or Basai, but the former also occupies another, rather different ecological niche, smaller channels and ponds with overhanging grass and reeds rather than expanses of such habitat. They nest in small, loose colonies or sometimes singly, and males are conspicuous and noisy from April to September with a similar successively polygynous breeding biology as its two congeners. After breeding, they disperse but do not seem to associate with parties of Baya Weavers in mixed flocks in cultivation as the Black-breasteds sometimes do.

Baya Weaver *Ploceus philippinus*

Breeding Resident Common

Baya Weaver, male, building nest

Widespread residents in the Indian subcontinent and Southeast Asia, these are weavers of openly wooded countryside, often near water, but not associated with wetlands as closely as the others on the Delhi list. Five subspecies are recognized over their range: three in India, the nominate in most of the country, *travancoreensis* in the southwest and *burmanicus* in the Northeast.

Bayas are common birds in cultivation and parkland about Delhi, and their colonies with nests suspended from branches are a familiar rural sight – *babools* and *khajur* palms overhanging ponds, ditches, and canals are especially favoured. Completed nests are the most exquisite, finely woven structures with pensile suspensions and long entrance tubes, often weighted by mud packs daubed in the 'chamber'. Males practice successive polygyny; they begin building by late April, but females do not arrive at the colonies until May or later, and eggs are laid only by July after the rains break. Outside the breeding season, they roost communally in reed beds, often in association with other weavers, starlings, wagtails, etc. There is concern about apparently declining Baya numbers in the Delhi area as elsewhere in India, possibly because of indiscriminate pesticide use in cultivation.

Finn's Weaver *Ploceus megarhynchus*

Vagrant, probable escapes

Endemic to India with a restricted range in seasonally flooded *terai* grasslands, with two separate populations localized in Uttarakhand in the west, and in north Bengal and Assam in the east, both severely threatened by loss of habitat and illegal trapping. The species is categorized in the Global IUCN (2021) Red List as Endangered. Two subspecies are recognized: the nominate in the west and *salimalii* in the east.

Breeding colonies of this rare weaver, with a maximum of 35 nests, were reported in June and July in early 1980s at Hastinapur near Meerut. Until 1990, trappers from Meerut reported that they used to catch 20–30 Finn's Weavers from Hastinapur each year.[263] This colony has disappeared since then. However, one bird was reported from Sultanpur in 1985 and two males were observed building nests at Okhla in June/July 1993. These nests were destroyed by rising water levels and there have been no further records.[264] These are likely to have been escaped or released birds, or more remotely, stragglers from the Hastinapur marshes whence, however, they have not been reported again either.

Black-breasted Weaver *Ploceus benghalensis*

Breeding Resident Common

Endemic to the Indo-Gangetic floodplains and are locally common across the northern subcontinent in areas of tall grass and reeds near water, spreading more widely as monsoon rains create areas of suitable habitat. Usually considered monotypic, but results from morphological and (limited) vocalization studies indicate that this weaver may be polytypic with two breeding morphs, black-faced and white-faced, belonging to allopatric populations and potentially representing different taxa.[265]

This is a common weaver of the Gangetic floodplains, spreading to wetlands and wet grasslands in the countryside in the monsoon months. Ganguli had called them 'very uncommon' in the 1960s, but they have evidently increased since then. Males assume their striking and distinctive breeding plumage from mid-April, the birds in the Delhi area being all of the 'black-faced' morph; their nests, often in small, loose colonies, are anchored in a canopy of reeds or tall grasses pulled together, and nest-building may be delayed till monsoon grasses grow tall enough. Females arrive at the colony by May/June, choose mates and lay eggs; the male begins another nest to attract a second female, practising successive polygyny till September. Post-breeding flocks feed on flowering *Saccharum* grass heads along with Baya Weavers *P. philippinus* and munias in winter.

PASSERIFORMES: PASSERINES

ESTRILDIDAE: MUNIAS

An African, southern Asian and Australasian family of small, lively, finch-like birds adapted for a granivorous diet. Highly diversified in Africa, in both genera and species, these are often boldly coloured and patterned birds of grasslands, wetlands, cultivation and lightly wooded open country in general. Popularity in the pet trade and illegal collection and trafficking has seriously impacted at least one Indian species, the Green Munia *Amandava formosa*. The family is divided into two main subfamilies, Estrildinae and Lonchurinae, both represented in India.

India: 8 species, Delhi: 6

◄ *The male Red Munia, denizen of wet grassy areas*

Red Munia *Amandava amandava*

Breeding Resident · Fairly Common

Native to South and Southeast Asia, but which have been introduced and naturalized in a number of other countries as well, in southern Europe, Japan and some Pacific islands. Three subspecies are recognized; only the nominate in South Asia.

Bright, confiding little birds, Red Munias are locally common in the Delhi area, seemingly more so today than in the past when Ganguli, in her *Guide* deemed them uncommon. They keep to clumps and patches of tall grass about marshes, reed beds and wet cultivation, breeding freely wherever such habitats occur between July and October. The Yamuna floodplains about Okhla, seepage marshes along canals, rice and sugarcane cultivation and wetlands in general hold numbers of breeding pairs. Red Munias' nests are typically balls of coarse grass, the chamber lined with fine fibres and with a side entrance hole, tucked into grass clumps or anchored in reeds.

Green Munia *Amandava formosa*

Vagrant, probable escape

An endemic munia of grass-and-bush country and open woodland in central India, from the Aravallis in Rajasthan, Madhya Pradesh, Maharashtra and northern Andhra Pradesh, this vulnerable species has suffered greatly from illegal trapping for the pet trade. Monotypic.

One observed at Sunder Nursery, New Delhi, in October 1978[266] with a party of Red Munias was almost certainly an escaped caged bird, possibly from the adjacent Delhi Zoo gardens.

Indian Silverbill *Euodice malabarica*

Breeding Resident · Common

Resident from the Persian Gulf to much of the Indian subcontinent. Monotypic.

Widespread about Delhi, this dry country species is the commonest of the family in arid habitats: grass-and-thorn scrub of the Ridge, tall grass and waste ground around cultivation and villages, vacant plots on the outskirts of towns, open *babool*/*kikar* woodland and so on. It keeps in small parties, even during the breeding season that extends over much of the year, mainly during the summer and monsoon months, from April to September. They have an interesting nesting economy, males bringing nesting material while females build the nest which is usually an untidy ball of dry grass in a thorn tree; several birds have been seen to enter one nest and the number of eggs is often more than the normal clutch of four to eight; as many as 25 have been recorded,[267] suggesting that several females deposit their eggs in one nest. They are often seen near Baya Weaver colonies and heronries, where they may appropriate abandoned weaver nests or build in the large stick nests of herons and egrets.

Indian Silverbill carrying nesting material

The Scaly-breasted Munia, a handsome little finch

Scaly-breasted Munia *Lonchura punctulata*

Breeding Resident Uncommon

Widespread and resident in its native range across the Indian subcontinent and Southeast Asia, but widely introduced and naturalized elsewhere, in the Indian Ocean islands and the Caribbean. Twelve subspecies are recognized, with two in India: nominate *punctulata* in most of the country and *subundulata* in the Northeast.

Delhi is towards the western borders of their range in the northern plains, and these munias tend to be rather uncommon here. They frequent gardens, the Ridge and long grass areas in the floodplain, preferring lusher areas than the Silverbill and appear to have increased in numbers about New Delhi over the last few decades with increasing greenery (and releases?). They breed in July–September, the males flying to their nest sites holding long green strips of grass that trail behind them like a long green tail. The nests are loose balls of grass, that hold typically five to seven eggs with larger broods possibly the result of two or more females laying in same nest. Usually in small parties, but flocks of up to 25 to 30 birds, both adults and juveniles, may also be seen.

Tricoloured Munias now nest in Delhi, but many are probably escaped or released birds.

Tricoloured Munia *Lonchura malacca*

Breeding Resident, probable escapes or releases Scarce

A resident munia of peninsular India south from Gujarat, Madhya Pradesh and Odisha, and in Sri Lanka. It has expanded its range in northern India in the Indo-Gangetic Plain as far as Jammu in recent decades, both through human agency but also apparently naturally. Introduced and naturalized in the Hawaiian Islands, the West Indies and Central America. Formerly lumped with Chestnut Munia *L. atricapilla*. Monotypic.

Delhi is to the north of the Tricoloured Munia's native range, but small parties, rarely of as many as 40 birds with juveniles, are not infrequently reported from marshy and rank grass areas in the floodplains of the Yamuna and Ganga rivers, and about wetlands. Such records go back several decades, and Ganguli had noted them as uncommon but nesting in the Delhi area even in the 1960s. Their nests are balls of grass with a side entrance, placed in a bush or in clumps of grass or reeds, with clutches of five or six eggs. Many birds artificially coloured (by bird dealers), a pointer that these have escaped from captivity or been deliberately released for religious reasons. Though they are breeding about Delhi, it is uncertain whether the population would be able to sustain itself were it not for the continuing accretion through fresh releases. Rare individuals with cinnamon-tinged underparts in the Delhi area have been claimed to be hybrids with Chestnut Munias, but other authorities consider them plumage morphs of Tricoloured; such intermediate-looking individuals also occur in Sri Lanka, well outside the range of *L. atricapilla*,[268] and in naturalized populations in central America. In a mixed captive group, these two species associated and paired assortatively which supports the view that these are colour aberrations rather than hybrids.

Chestnut Munia *Lonchura atricapilla*

Vagrant, probable escapes?

Widely distributed in northern India along the Himalayan foothills, Bengal and Assam, eastwards to Southeast Asia. Ten subspecies are recognized, with two in India: nominate *atricapilla* in eastern India and race *rubronigra* extending westwards along *terai* grasslands and wetlands as far as the Yamuna river. The range of this latter race appears to be curiously contracting and is now almost restricted to Nepal with a few localized reports from northern UP as well.

Chestnut Munias have been reported from the area far less often than their Tricoloured sibling, with very few records in recent decades (these are mostly from the Ganga *khadar* at the Hastinapur Wildlife Sanctuary, the only recent report from Delhi's immediate neighbourhood was in July 2021 at the Dhanauri wetlands, UP). It is quite possible, however, that these munias may wander here from the *terai* without human agency. There are two historical records, a pair in a New Delhi garden in July 1945 with a party of Silverbills and one in June–July with Streaked Weavers, probably also in the 1940s. Interestingly, until Hutson's time, there were only these two records from around Delhi with no mention at all of the Tricoloured; since the latter is more often kept in captivity or for release, it is arguable that these earlier Chestnut Munia records were of genuine vagrants rather than released birds.

PASSERIFORMES: PASSERINES

PASSERIDAE: SPARROWS

Too familiar to need a description, this Old World family is now represented worldwide through human-aided introductions. Primarily granivores, sparrows also supplement their diet with insects, especially for their chicks. This diet also enables them to survive in diverse habitats, from cultivation and grasslands, desert, and open woodland, and as commensals in towns and villages. Some, such as the Spanish Sparrow, are gregarious and migrate in flocks, even breeding communally in large groups, while others like the Yellow-throated Sparrow *Gymnoris xanthocollis* nest individually in tree holes. House Sparrows *Passer domesticus* nest in holes in walls, monuments and buildings. Macdonald (1960) has some engaging paragraphs in his book *Birds in My Indian Garden* on a pair of House Sparrows that nested in a bookshelf in his library in New Delhi.

India: 12 species, Delhi: 4

The Yellow-throated Sparrow, a woodland bird

House Sparrow *Passer domesticus*

Breeding Resident Common

Native to Europe, Central and South Asia, House Sparrows are now present in all continents through deliberate introduction. Up to 12 subspecies are recognized, with three occurring within Indian limits: *indicus* as a resident across the country, *parkini* an altitudinal migrant in the western Himalayas and *bactrianus* as a winter visitor from Central Asia to the northwestern subcontinent.

House Sparrow, male

Adaptable and commensal, they are, or perhaps were, familiar birds of urban Delhi and village precincts, breeding freely in houses and buildings from April to August by when insect life is abundant to feed their brood of four to six chicks. Migratory populations in Asia also breed in holes in cliffs. Multi-brooded, and when not breeding, they tend to gather in flocks to feed and roost. Local declines in urban House Sparrow populations have attracted attention, including about Delhi where these birds are no longer such a feature of urban areas as they were earlier, in parks, bazaars, bus and railway stations, etc. But they remain locally common where they do occur on the outskirts of the city and rural areas; the House Sparrow is Delhi's State Bird.

Spanish Sparrow *Passer hispaniolensis*

Winter Visitor Uncommon

Breed from the Mediterranean region and North Africa to Kazakhstan and Afghanistan, and locally, erratically, elsewhere, and tend to be nomadic, wintering in the Middle East and West Asia as far as western India. Multi-brooded, they breed in large colonies, building globular nests of grass on trees. Two subspecies are recognized; the birds wintering in India belong to race *transcaspicus* that breeds in Central Asia.

Delhi is near the eastern limits of their wintering quarters and their appearances and numbers here tend to be irregular, mainly between November and March; it is, however, regularly reported from other rural areas in Haryana. Migrating Spanish Sparrows travel in huge flocks, often along with House Sparrows of the migratory Central Asian race *P. d. bactrianus*, but about Delhi they will usually be encountered in small parties feeding with Short-toed Larks *Calandrella* sp., doves *Streptopelia* sp., starlings *Sturnus* sp. and other sparrows in fallow fields.

Sind Sparrow

Passer pyrrhonotus

Breeding Resident

Uncommon

Sind Sparrow, male

Formerly restricted to the Indus's floodplains in Pakistan, just extending across the border into India along the Sutlej river into the Punjab in riverine or swampy areas, occurring in wet tall grass, *Tamarix* and *Acacia* jungle. Over the last two decades, they have colonized and are now naturalized in several districts of Punjab, Haryana and Delhi.

In an extraordinary natural eastward expansion of its range, they were first reported in *babool* plantations along canals at Sonipat in early 2001,[270] followed by discovery of the first nests in May that year. They then spread rapidly across the Haryana plains following the network of irrigation canals and waterways and within a couple of years were recorded nesting in north Delhi, Bhindawas, Sultanpur and Manesar. They breed in *babool* woodland near water in small, loose colonies from February to July, building untidy globular nests of grass in small trees and roost in reed beds in the non-breeding season; unlike House Sparrows, they are not commensal with humans. At Bhindawas, a small colony nested in the bases of old egret nests in the heronry.[271]

Yellow-throated Sparrow *Gymnoris xanthocollis*

Breeding Resident   Fairly Common

Distributed from southeastern Turkey and the Persian Gulf region to the subcontinent east to Bengal, Yellow-throated Sparrows (also known by an awkward alias, Chestnut-shouldered Petronia) are residents of open woodland habitats in India, though populations to the west, including in Pakistan and northwestern India, are mainly summer breeding visitors there. Two races: *transfuga* in the west of its range, nominate *xanthocollis* in India.

Widespread and quite common locally as breeding residents in the Delhi area. In the 1960s, Ganguli saw them in parks and gardens, but they seem to be commonest today in *babool*/*kikar* woodland on the Ridge and around villages and rural groves. They feed on grain and seeds, berries, insects and, commonly, nectar in season. Males become vocal as the weather begins to warm up in February and they breed from April to June in tree holes in such habitat, laying clutches of three or four eggs. Larger parties, of 40 or more, gather in shrubbery in the non-breeding season to roost.

PASSERIFORMES: PASSERINES

MOTACILLIDAE: WAGTAILS AND PIPITS

Represented on all continents, this family of elegant, mostly long-tailed, insectivorous ground-living birds is most diversified in the Old World–the wagtails almost entirely so, while pipits are more widely distributed. They are structurally similar, but very different in plumage; wagtails' patterns and colours tend to be bright and contrasting, and with their vivacity and confiding demeanour they are a delight to watch. Indian pipit species on the other hand are soberly coloured, but phenotypic characters, structure, size, pattern, voice and behaviour are useful identification tools.

Wagtails and pipits represent separate clades in the family, which are sometimes treated as different subfamilies, Motacillinae and Anthinae. Both are well represented in the open wetland and open dryland habitats of the Delhi area, which suit them admirably, but most are migrants and only one of each group breeds with us.

India: 21 species, Delhi: 16

▲ *The Western Yellow Wagtail, a male of the black-headed race*

Forest Wagtail *Dendronanthus indicus*

Vagrant

Breeds in northeastern Asia and winters in southwestern India, Sri Lanka and Southeast Asia. Unusually for wagtails, they nest in trees in these northern forests often along tracks and streams, and in winter, frequent woodland and plantations in southern Asia. They are known to wander west from their usual migration path in autumn, having even reached the Persian Gulf fairly regularly. Monotypic.

There are only four records of vagrants of these aberrant wagtails from Delhi, including three from the Okhla area: one historical, by H. Alexander in September 1949, and two recent, on 16 October 2016 and 28 February 2020; and one from woodland at the Sultanpur National Park on 2–3 October 2022.[272]

Grey Wagtail *Motacilla cinerea*

Passage Migrant Uncommon

Breeds by fast flowing woodland streams in Europe, North Africa and Asia, eastwards to Japan and southwards to the Himalayas, and winters in parts of Africa and southern Asia. Three subspecies of this graceful wagtail are recognized; the nominate race ranging across Europe and Asia.

Grey Wagtails are mainly passage migrants through Delhi in small numbers, passing through in September/October and in March/April. A few spend the winter here as well, in ones or twos by artificial ponds and channels in the Delhi Zoo and in parks such as Lodhi Gardens but are noticeably fewer than during passage periods. They prefer shaded streams and running water instead of the open marshes beloved of the Yellow Wagtails.

Western Yellow Wagtail *Motacilla flava*

Winter Visitor Common

Diverse subspecies of Western Yellow Wagtails breed in marshes and meadows across temperate Europe and Asia and winter in sub-Saharan Africa and India, in similar habitats, including wetland margins and damp grassland. Ten subspecies are recognized, of which at least four winter in India: *thunbergi* (that breeds in northwestern Siberia), *beema* (Central Asia), *feldegg* (Turkestan) and *leucocephala* (Mongolia). Subspecies *lutea* (that breeds in southwestern Russia and winters mainly in East Africa) is also claimed to occur occasionally.

Yellow Wagtails are common visitors to the area between late August and May, swarming at times during passage in spring, and affect damp grazing grounds and short-grass areas, wet mudflats, marshes and irrigated fields. They often feed around cattle enclosures, where flushed insects offer rich pickings. Three races occur regularly, the males identifiable as they assume breeding dress in spring: the grey-headed *thunbergi* and blue-headed *beema* are the commonest, while the black-headed *feldegg* is the last to depart in spring–most birds in late April/ May are of this striking race. The white-headed *leucocephala* is a rare migrant with just three records from our area: in April 1965 near Okhla, January 2003 at Sonipat, and 12 April 2009 at Okhla again. Spectacular numbers of Yellow Wagtails gather to roost in winter in reed beds at Okhla and elsewhere.[273]

Citrine Wagtail *Motacilla citreola*

Winter Visitor Common

Breeds in Eurasia, Central Asia and the western Himalayas, wintering in southern Asia. Three subspecies are recognized, all winter mainly in India: the nominate of northern Siberia, *werae* of Eastern Europe and Russia and *calcarata* which breeds in Central Asia south to the Himalayas.

At least two races winter commonly in the area, arriving by late August and remaining till April/May, and are very distinct in their breeding plumages that they assume before emigration: the northern, grey-backed *citreola* leaves by March/April, while the showy black-backed Himalayan breeding *calcarata* is most in evidence between mid-March and late May. They are more aquatic in their habitat preference than the Yellow Wagtails and prefer wet muddy margins, even wading in shallow water on occasion, and roost in large numbers with other wagtail species in reed beds at Okhla.

White-browed Wagtail

White-browed Wagtail

Motacilla maderaspatensis

Breeding Resident · Fairly Common

Subcontinental endemics, with pairs resident around a variety of waterbodies, both large and small, including rivers and canals, tanks, even artificial ponds and garden pools. Monotypic.

Fairly common breeding residents about Delhi, White-browed Wagtails are not birds of squelchy ground or marshes, preferring flowing watercourses with rocky banks; or in their absence, bridges and canal sluices, weirs, ghats and even artificial waterbodies in the larger gardens, where they breed between March and June. Nests, built by both sexes, are cups of soft vegetable fibres tucked in a hole or ledge under a bridge or similar structure near water and hold a clutch of usually four eggs. Males advertise territories through their spirited songs which may also be heard through much of the year.

White Wagtail *Motacilla alba*

Winter Visitor · Common

Very widespread birds, breeding across all of Europe and Asia south to the Himalayas, and wintering in Africa and throughout southern Asia. Ten distinct subspecies with marked variations in plumage colour and pattern have been named over this range, of which one, *M.a. alboides*, is an altitudinal migrant breeding along the Himalayan range, and four other races, *dukhunensis* (breeding extralimitally in Siberia), *personata* (Central Asia), *baicalensis* (southern Siberia) and *leucopsis* (East Asia), winter in India.

Common winter visitors from late August till May, affecting open ground both near water and away from it and, very often, irrigated lawns and garden paths. Two grey-backed subspecies, the black-faced *personata* and the white-faced *dukhunensis* winter in large numbers in the Delhi area, while two black-backed subspecies, the black-faced *alboides* and the white-faced *leucopsis*, are both far scarcer – there are just a handful of reports of the former from Delhi (though *alboides* has been termed a regular winter migrant in the Meerut region), while *leucopsis* is more often met with. White Wagtails, like their congeners, gather in large numbers at communal roosts in reed beds at Okhla and elsewhere.

Richard's Pipit *Anthus richardi*

Winter Visitor · Scarce

Summer breeding visitors to open steppe grasslands of the eastern Palaearctic and winter from the Indian subcontinent – mainly its southern and eastern parts – to Southeast Asia but appear to wander west of their usual wintering areas fairly frequently. Five subspecies are recognized; Indian wintering birds are probably mainly of the nominate race *richardi* that breeds in Siberia and northern Central Asia.

These large, relatively long-legged and long-tailed pipits have been reported from the Delhi area occasionally in winter, often on damp ground near wetlands and seepage marshes. Identification issues confound claimed records, and the few reliable reports are mainly from the Sultanpur neighbourhood and Greater Noida wetlands from as early as September.

Paddyfield Pipit *Anthus rufulus*

Breeding Resident Common

The area's only resident pipit, a bird of open country habitats and stubble fields throughout its range; found in the Indian subcontinent and Southeast Asia. Six subspecies are recognized, three of which occur within Indian limits: *waitei* in the northwest, nominate *rufulus* in most of the country and the dark race *malayensis* in southwestern peninsular hills.

Generally common in the countryside, these pipits will usually be seen in pairs or singly, never in flocks, in areas of dry short grass and overgrazed pastures about villages and bare sandy tracts in the floodplain. The race here is presumably *waitei*. Males begin to show territorial behaviour with aerial song displays from March; they breed between April and July building nests on the ground, a cup of grass, etc. lined with finer material, in a hoofprint or depression, to hold a clutch of three or four eggs.

Long-billed Pipit *Anthus similis*

Winter Visitor Uncommon

Range widely but patchily over much of Africa and West Asia to India and Myanmar. As many as twenty subspecies recognized, with four occurring within Indian limits: the greyer race *decaptus* of Iran and Pakistan winters in northwestern India, *jerdoni* breeds in the Himalayas and winters in the northern Indian plains, while nominate *similis* and the darkest race *travancoriensis* are resident in the peninsular hills.

These large pipits visit Delhi in small numbers (race *decaptus* and/or *jerdoni*) between October and April, favouring the broken stony ground, abandoned mines and low scrub of the Southern Ridge, which suit their requirements perfectly. They may also be found at times in sandy flats, as about Sultanpur, or stone-clad embankments by the river and canals. They tend to be locally distributed in view of their habitat preference.

A Long-billed Pipit in its preferred rocky scrub habitat

Blyth's Pipit *Anthus godlewskii*

Winter Visitor Uncommon

Breeds in the Mongolian steppes and winters mainly in eastern and southern India. Field identification can be a challenge, and most records are difficult to pin down with certainty. Monotypic.

Ganguli did not include Blyth's Pipit from Delhi in her *Guide to the Birds of the Delhi Area*, but as identification criteria have developed and are now more clearly understood, a growing number of convincing reports from the field suggest that small numbers probably visit the area fairly regularly, mainly between mid-September and February, preferring grassy ground which is less bare and closely grazed than Paddyfield's typical habitat.

Tawny Pipit *Anthus campestris*

Winter Visitor Uncommon

Summer breeding visitors to Eurasia, from Europe and North Africa to Central Asia, wintering southwards to African Sahel, Arabia and the northwestern subcontinent. Three subspecies are recognized across this range, with mainly the nominate wintering in India.

Tawny Pipits are uncommon winter visitors to the area from October to April, usually in scattered parties in open scrub country, bare sandy flats and overgrazed ground.

Rosy Pipit *Anthus roseatus*

Winter Visitor Fairly Common

Breeds abundantly in soggy alpine meadows and boulder-strewn grassy slopes of the high Himalayas, Tibet and southern China, wintering in the northern Indian subcontinent, usually near water and in marshy areas. Monotypic.

Rosy Pipits are locally common in the area from October to April, singly or in small loose parties about margins of lakes and marshes, canal banks and bunds by the river. Large numbers collect to roost in marshes, milling about in the darkening sky at sunset and suddenly dropping like rain to disappear in the vegetation. They begin to moult into breeding plumage in March, and many are in fine breeding plumage by the time they leave.

Tree Pipit *Anthus trivialis*

Winter Visitor Fairly Common

Breeds widely in open woodland across temperate Eurasia southwards to the western Himalayas, migrating in winter to Africa and India. Two subspecies are recognized: nominate *trivialis* of Eurasia and *haringtoni* breeding in the western Himalayas, Kashmir to Garhwal, both winter in the Indian peninsula.

Fairly common about Delhi from September to April, Tree Pipits tend to be localized in suitable habitats, preferring grassy areas with some tree and shrub growth. Passage movements are most in evidence in September–October and from March to mid-April, when numbers may collect in the evenings to roost in reed beds, as at Okhla, or grassy woodland as at the Aravalli Biodiversity Park, Gurugram.

Olive-backed Pipit *Anthus hodgsoni*

Winter Visitor · Uncommon

The Olive-backed Pipit, a pipit of open woodland

Essentially Asian in distribution, breeding in grassy and bracken covered clearings in woodland, from Siberia and north-eastern China southwards to the higher Himalayas, wintering in India, southern China and Southeast Asia. Two subspecies are recognized: *yunnanensis* breeding in Siberia and nominate *hodgsoni* in the Himalayas, China and Japan.

Olive-backed Pipits are regular visitors to the area between October and March in open *Acacia*/*Prosopis* parkland or orchards and groves with short grass understorey but no undergrowth; regular sites include woodland patches at the Sultanpur National Park and other sites in Haryana including the Bhindawas Wildlife Sanctuary, Asola–Bhatti and Surajkund on the Southern Ridge, Delhi parks such as Hauz Khas, Mehrauli and Lodhi Gardens, and the Delhi Zoo. Most birds that winter here appear to be of the plainer-backed Siberian race *yunnanensis*.

Red-throated Pipit *Anthus cervinus*

Vagrant

The bulk of the Red-throated Pipit populations that breed mostly north of the Arctic Circle across the Palaearctic, winter to the west and east of the Indian subcontinent in Africa and Southeast Asia, and only a few straggle into the subcontinent itself. Monotypic.

Ganguli deemed these pipits rare passage migrants through the Delhi area, but noted only two historical records from Okhla, in May 1949 and April 1965, of birds in breeding plumage. Their preferred wintering habitat is mainly damp grassland and marshes, and several claimed sightings from the Yamuna floodplains in 1970s are vitiated by confusion with Rosy Pipits. The very few validated reports in recent years include singles at Okhla in December 1998 and two individuals in their distinctive breeding dress at the Basai wetlands, Haryana, in early April 2016.[275]

Water Pipit *Anthus spinoletta*

Winter Visitor · Scarce

Water Pipits that breed in the mountains of southern Europe, the Caucasus and Central Asia, winter southwards as far as North Africa, West Asia to northwestern India, and southern China. Three subspecies are recognized: the nominate nesting in Europe, *coutellii* in Turkey and the Caucasus, and *blakistoni* in Central Asia, with the latter wintering in the subcontinent.

Delhi is near the eastern edge of their winter range in India, and they are usually rather scarce here. Ganguli recorded a few historical observations between October and April, from the 1960s and earlier. There have been almost annual records in recent years, however, of single birds or small parties between November and February, mainly from the neighbourhood of wetlands in Haryana.

PASSERIFORMES: PASSERINES
FRINGILLIDAE: FINCHES

The Fringillidae are a widespread family comprising three lineages or subfamilies: Fringillinae (with four species in one genus, *Fringilla*, the chaffinches and Brambling), Carduelinae with 183 species in 49 genera and Euphoniinae, which is restricted to the New World.

These small- to mid-sized, often colourful, songbirds, with conical 'seedcracker' bills, are well diversified in the Himalayas, but only one species, the Common Rosefinch *Carpodacus erythrinus*, winters widely in India.

India: 48 species, Delhi: 2

Brambling *Fringilla montifringilla*

Vagrant

Widespread and abundant as a nesting species across the northern Palaearctic from Europe to Siberia, wintering between October and March–April in Europe to the Mediterranean region, and in Asia south to Iran, Central Asia and China. Large numbers of these finches may show random dispersive movements in years of periodic abundance following successful breeding seasons. They are locally common in winter in the mountain ranges of Afghanistan, Pakistan, Kashmir and Ladakh, and in smaller numbers in the western Himalayas east to Nepal, vagrants having reached Bhutan. Monotypic.

One individual, a male moulting into breeding plumage, was reported from the Bhondsi Nature Park, Gurugram district, in mid-March 2022, is the first record from the plains of the subcontinent south of the Himalayas; the bird remained for several days at the site. A vagrant, whose genetically programmed 'compass' may have misled it to migrate south rather than north.

Common Rosefinch *Carpodacus erythrinus*

Passage Migrant and Winter Visitor   Fairly Common

Wide breeding range in Central Europe across temperate Asia, nesting in forests, woodland and orchards in mid- to high elevations in the Himalayas, wintering in India, southern China and Southeast Asia. Five subspecies are recognized, distinguished mainly by tone and extent of red in breeding males: race *erythrinus* breeds from Europe to central Siberia, *grebnitskii* in eastern Siberia, *kubanensis* in Turkey and Iran, *ferghanensis* in Central Asia south to Ladakh and *roseatus* in the Himalayas and China. All except *grebnitskii* winter in India.

These rosefinches are fairly common passage migrants through Delhi in both autumn and spring, with some remaining all winter. Spring passage is conspicuous from mid-March to early May, when parties of Common Rosefinches frequent city gardens, the Ridge and groves and thickets in rural areas. At this time, they are especially fond of mulberries and flower nectar. At least two races, if not more, occur in the Delhi area; wintering birds that frequent woodland on the Ridge are duller, with restricted pink in the males and diffusely streaked females, and are presumably of the nominate race, while spring passage is dominated by the brilliant Himalayan-breeding *roseatus* with carmine males and darker, well-marked females, which apparently winters further south.

PASSERIFORMES: PASSERINES

EMBERIZIDAE: BUNTINGS

These are a family of small seed-eating birds with typically conical, sharp-tipped bills and longish tails, widespread in open habitats in Europe, Asia and Africa. Feed on the ground or from flowering grass heads, gleaning seeds and grain, and insects as well when feeding chicks. Genetic distances between the heterogeneous bunting species on our list are small enough for all to be placed in a single genus, *Emberiza*, including the very different-looking, and our only resident species, the Crested Bunting, which was hitherto placed in its own genus *Melophus*. Buntings are well-known wanderers; of the other buntings on the Delhi list, only the Himalayan-nesting White-capped Bunting *E. stewarti* is a regular winter visitor, three that breed to the northwest of the subcontinent migrate through the Delhi area to their wintering grounds in the peninsula, and five others are vagrants.

India: 20 species, Delhi: 10

▲ *The striking Crested Bunting male*

Crested Bunting *Emberiza lathami*

Breeding Resident · Uncommon

Largely resident, though subject to local movements, in dry, stony hillsides and scrub jungle, in the lower Himalayas and central India, eastwards to China. Monotypic.

These perky buntings had formerly been considered scarce passage migrants about Delhi with scattered records from the Yamuna floodplain in September–October and March–May,[277] and odd reports in the winter months as well, believed to be Himalayan breeders that had dispersed into the plains in the non-breeding season. However, they are breeding residents in the Aravalli Ranges (e.g. at the Sariska National Park, 150 km south of Delhi), and in recent years have also been found to nest in small numbers at the northernmost extensions of this range as it enters our limits, in the open scrub landscapes of the Southern Ridge, with males in song between May and July. A monogamous species, the nest is typically a small cup of fine grasses tucked low under a stone or in a wall crevice, that holds a clutch of three or four eggs. They are rarely in parties like other buntings and are usually seen singly.

Black-headed Bunting *Emberiza melanocephala*

Passage Migrant Rare

Buntings of lightly wooded country and cultivation, nesting in groves, vineyards and orchards from southeastern Europe and Turkey east to Iran and Baluchistan, the entire population wintering in west-central India. Monotypic.

Delhi is on the fringe of their migration path through Gujarat, Rajasthan and Sindh, and their main passage movement passes just south of our area. They are rare here: Ganguli mentioned just two historical records of single birds in flocks of Red-headed Buntings, in April 1944 and 1962. Because of difficulties in separating this species from Red-headed Buntings in female-type plumage, considerable confusion prevails in records from the Delhi area. In recent years, there are only two verifiable records in autumn, from the Basai marshes, Haryana, in September 2001 and November 2002; and two in April, from Okhla in 2013 and Basai in 2016. There are two mid-winter records as well, from the Sultanpur National Park in December 2012 and the Yamuna *khadar*, Faridabad, in December 2016.

Red-headed Bunting, male in breeding colours

Red-headed Bunting *Emberiza bruniceps*

Passage Migrant Fairly Common

Summer visitors to the steppes and cultivated areas of trans-Caspian Central Asia, to the north and east of its black-headed sibling's breeding range, these buntings spend the winter in west-central India, from Rajasthan and Madhya Pradesh southwards to Karnataka. Monotypic.

Fairly common on both passages through Delhi, in autumn between August and November and in spring between mid-March and April, when they feed on ripening grain in rural areas and the floodplain. They are usually in parties of 20–50 but there are reports of larger flocks too which are, in fact, more typical of these buntings on migration. There are scattered records in mid-winter as well.

Chestnut-eared Bunting *Emberiza fucata*

Vagrant

Breeds in three areas in Asia: the Himalayas, in northeastern Asia and in eastern China. Three subspecies are recognized over this range: *arcuata* nesting in the western Himalayas and wintering in the *terai* grasslands in the foothills; nominate *fucata* in the Amur basin, Korea and Japan and wintering in northeastern India and southern China; and *kuatunensis* in eastern China, wintering in Southeast Asia.

Some Himalayan breeders occasionally wander further south of their usual winter quarters along the rivers of the floodplains and vagrants have thus reached the Delhi area at least thrice: a small party at Garhi Bala village, Sonipat, Haryana, on 2 November 2002; one by the Yamuna at Okhla, Delhi, on 2 November 2005;[278] and a juvenile at Dilshad Garden, east Delhi, on 27 October 2020.

Rock Bunting *Emberiza cia*

Vagrant

Breeds from the Mediterranean to Central Asia and the Himalayas, with four subspecies, of which *stracheyi* and *flemingorum* are altitudinal migrants nesting in the Himalayas, and *par*, breeding Central Asia to Afghanistan with small numbers straggling southwards in winter into the plains of the northwestern subcontinent.

Ganguli, in her *Guide to the Birds of the Delhi Area*, mentioned only one historical record, a bird collected in late February 1925 in north Delhi, and 'doubtful records' from the Ridge in December and from Tughlaqabad in November 1945. The sole recent record from within the Delhi area is one in a patch of thorn scrub woodland at Maroudi Jatan, Rohtak district, on 27 February 2014, but a party of four birds has also been reported from Kaithal, Haryana, which is just beyond its limits, on 24 February 2018.

White-capped Bunting *Emberiza stewarti*

Winter Visitor Uncommon

Nests on lightly wooded mountain slopes in Central Asia southwards to Afghanistan and the western Himalayas, wintering in scrub jungle and wasteland in northern India, to UP, Gujarat and Maharashtra. Monotypic.

These buntings are uncommon but widespread winter visitors to the Delhi area, in the drier thorn scrub habitats of the Ridge, along canal embankments and patches of native *babool* woodland. Gaston[279] considered them 'numerous winter visitors' to the New Delhi Ridge between October and May in the early 1970s, in 'flocks of up to 20', though they have more usually been recorded in small parties of two to five birds in recent decades. They are most frequent and more widely distributed during passages in October–November and especially March–April, when the males are in their striking breeding dress.

A male White-capped Bunting in breeding plumage

Grey-necked Bunting *Emberiza buchanani*

Passage Migrant Scarce

Breeds on dry, rocky hillsides in Central Asia southwards to Iran and Afghanistan, and winters in similar habitats in Gujarat, central India and the Deccan Plateau. Three races recognized but variation is slight and clinal. All three presumably mix in their winter quarters in India.

These buntings are passage migrants through the Delhi area on their way to their usual wintering quarters further south in the peninsula. Ganguli recorded just three observations of small parties in stony scrubland till the 1960s, while Gaston[280] considered it 'a not uncommon spring passage migrant' with one autumn record in October 1971. There are a number of recent reports from old quarries and stony scrub habitats of the Ridge in south Delhi and Haryana, and a few from the Yamuna floodplains, mainly in spring on northward passage in March–April, but with some autumn records in September–November as well.

Ortolan Bunting *Emberiza hortulana*

Vagrant / Reconfirmation desirable

Breeds in Europe and Russia east to central Siberia and winters mainly in Arabia and Africa. Monotypic.

They are vagrants in India, a mega-rarity with just three or four records of stray migrants in Kashmir, Lahaul-Spiti and Gujarat. Ganguli documented an observation of a party of five seen on 18–19 April 1970 at Mehrauli, south Delhi, by Peter Jackson, A.J. Gaston and herself.[281] Grey-necked Bunting is the main confusion species, and her detailed comparison with BNHS skins of Grey-necked is suggestive but still leaves scope for doubt in identification. There have been no further reports, and the record is retained here as requiring reconfirmation.

Striolated Bunting *Emberiza striolata*

Vagrant

Resident, subject to local movements, in dry, rocky, often hilly scrub country from northeastern Africa and Arabia to India, east to Madhya Pradesh and south as far as northern Maharashtra. Three subspecies are recognized, the nominate in India.

The Delhi area is on the fringes of their range in the subcontinent, but they have been reported on a few occasions from just beyond our limits in Haryana, from appropriate habitats in the Hisar, Bhiwani and Mahendragarh districts, and also from dry sandy ground at the Haiderpur Wetland in UP. From within the Delhi area, Ganguli mentions two historical records, both in March, Mehrauli in 1952 and Tughlaqabad in 1958. There are just three later reports of single birds, from Jhajjar district on 4 December 2014, from Faridabad, Haryana, on 22 April 2017, and from lightly wooded 'Aravalli' habitat at Manethi, Rewari district, on 27 August 2021.

Eurasian Reed Bunting *Emberiza schoeniclus*

Vagrant

Breeds in wetlands and reed beds across the temperate Palaearctic and is partially migratory, a rare visitor to northwestern India with a few records from Kashmir, Punjab and Haryana.

The first, and only, record to date of these buntings was a female or first winter bird seen at Sultanpur in November 2003:[282] As the observer, Bill Harvey, has noted, 'it was observed from very close quarters in a grazed clearing in the tall grass/reeds at Sultanpur. I am very familiar with the species in Europe.'[283]

5 Historical and Expunged Records

Historical Records (Pre-1975)

Twenty-two species that had been reported from the Delhi neighbourhood in the past have not been recorded from the area again since 1975. The NCR has seen enormous, accelerating ecological change over the past half-century or so with dramatic expansion of urban areas, altered patterns of irrigation and agriculture, loss of wetlands, changes in river hydrology and deterioration of air and water quality. This has, naturally, resulted in the local extirpation of some birds that figure in this list, while several vagrants recorded in the past may or may not reappear in Delhi's neighbourhood. In addition, some old sight records are unverifiable today in the absence of details – often, even the year of the sighting is not mentioned in the primary source. The risk of misidentification remains in the absence of sufficient documentation.

Tundra Swan *Cygnus columbianus*

Swans of the far north whose usual wintering range in Asia includes the Caspian Sea, China and Japan, but which have been recorded as very rare vagrants from the subcontinent. There is a record from 'Sutana, about 45 miles north of Delhi' (near Panipat) from December 1937,[284] but the specimen itself in the BNHS museum is labelled as from 'Rajpur' near Delhi.[285]

Common Golden-eye *Bucephala clangula*

Golden-eyes breed in lakes and rivers in boreal forests of North America and northern Eurasia; the latter populations winter along coasts and inland waters across Eurasia, Japan and China, straggling into the northern subcontinent in winter. Ganguli records two observations from Delhi, in early March 1957 (by V. Martin, at Dasna, Ghaziabad district) and 24 November 1963 (by herself, on the Yamuna River).

White-headed Duck *Oxyura leucocephala*

This endangered species breeds very locally from the Mediterranean to Central Asia, a few wintering to Pakistan, very rarely further into India. Three were obtained in 1882–83 near Delhi.[286]

Pin-tailed Sandgrouse *Pterocles alchata*

Pin-tailed Sandgrouse breed from Spain and across North Africa to Central Asia, the latter populations migrating to the northwestern part of the subcontinent, mainly Pakistan, in winter. They are now the least numerous and most localized of the subcontinent's sandgrouse species. Blewitt, in 1875, mentions large flocks, exceeding 300, near Najafgarh, with a specimen deposited in the Hume Collection (BMNH), while Chill records having obtained specimens from flocks of 200 and 500 in early February 1883 near Farrukhnagar. There have been no further reports from the Delhi area, and these nineteenth-century records may have been episodic influxes of this species which is known to exhibit nomadic behaviour. A detailed note on these unusual records is posted on eBird.[287]

Black-bellied Sandgrouse *Pterocles orientalis*

This large sandgrouse ranges from Spain to Central Asia and winters eastwards to western Rajasthan. Its wintering patterns in India may have changed over recent decades and it seems to have become far less common and more localized than in the past. Ganguli calls it an uncommon winter visitor in southern and southwestern Delhi and mentions mid-winter records, the last report, however, being an unusually late bird in early May 1968.[288]

Slaty-breasted Rail *Lewinia striata*

Hutson[289] recorded observations from reedy pools about Delhi 'in March, April and May'; the years were not specified but presumably from the 1940s. There has been no subsequent record.

Siberian Crane *Leucogeranus leucogeranus*

The western population of the Siberian Crane, whose last winter refuge in India used to be the Keoladeo National Park, Bharatpur, Rajasthan, stands effectively extirpated now. Wintering numbers at Bharatpur fell from around 75 in the early 1970s to two by 1998, with the last birds recorded there in 2001–02. They were also reported from elsewhere in northern India in the past, and Hume & Marshall[290] stated that it was recorded 'at least once' in Najafgarh about 1875.

Great Indian Bustard *Ardeotis nigriceps*

The fate of this spectacular endemic bustard hangs by a slender thread today. There are nineteenth-century records from near Delhi and Gurugram,[291] and Basil-Edwards in the 1920s was informed that 'it used to be found about Delhi'. There is no habitat left today to support this species in the Delhi area, even if it were not so rare.

Macqueen's Bustard *Chlamydotis macqueenii*

This West and Central Asian species, split from the North African Houbara, is a winter visitor to the drier northwestern parts of the subcontinent. There are anecdotal reports of its presence about Delhi in the early 1900s, and Ganguli records that 'one [was] obtained about 60 miles from Delhi on the road to Jaipur' about 1950.

Lesser Adjutant *Leptoptilos javanicus*

Lesser Adjutants range from the eastern and southern subcontinent across Southeast Asia. Vagrants have been reported from Delhi historically[292] 'occasionally in company with Greater Adjutants'.

Spot-billed Pelican *Pelecanus philippinus*

Spot-billed Pelicans occasionally wander from their localized breeding colonies in eastern and southern India, apparently much less often now than formerly, and Ganguli recorded observations at the Sultanpur lakes in the winter of 1969–70.[293] It has not been reported since.

Cinereous Vulture *Aegypius monachus*

These huge vultures breed locally from Spain to Central Asia, the eastern populations wintering in West Asia to northwestern India and the Himalayas. They were recorded as uncommon winter visitors to refuse dumps about Delhi with observations from the 1940s till 1969,[294] but none have been reported since.

Saker Falcon *Falco cherrug*

These large falcons of Eastern Europe and Central Asia winter mainly, but sparsely, in the north and west of the subcontinent. Frome[295] collected a specimen on 21 November (no year mentioned!), and Ganguli refers to one seen by H. Alexander in January (again no year!).

Rosy Minivet *Pericrocotus roseus*

Rosy Minivets breed in the Himalayan foothills and locally in the Eastern Ghats, straggling southwards into the peninsula in winter. There are two old observations from north Delhi: Frome[296] saw one in December (no year mentioned! 1940s?), and Ganguli mentions 'a few' seen by Alexander in "1946 or 1947."

Black-crowned Sparrow Lark *Eremopterix nigriceps*

A common sparrow lark of deserts from the Sahara across the Arabian Peninsula and West Asia eastwards to the Thar, that appears to have been recorded once in the Delhi area, a single bird in February at Najafgarh.[297]

White-browed Bulbul *Pycnonotus luteolus*

A bulbul of peninsular India, and Delhi is much beyond the range of this bird, but Ganguli mentions sight records on 16 September (no location or year – 1950s?) and at the Begumpuri Mosque, south Delhi on 2 January 1952.[298] The possibility of escapes cannot be ruled out.

Black Bulbul *Hypsipetes leucocephalus*

These are common Himalayan bulbuls that descend to the foothills and adjoining plains in winter but would be very unusual as far south as Delhi. The only records are by Gaston,[299] who saw two or three on the New Delhi Ridge on 12 and 24 February and 7 March 1972. He notes that these records followed heavy and unseasonal late snowfalls in the hills of Himachal Pradesh and UP, and large numbers of Black-throated Thrushes were also seen at the same time. (Note: There are a number of records of these bulbuls wintering, in parties of up to 35, from just beyond the limits of the Delhi area at Haiderpur Wetland – which is much closer to the Himalayan foothills – between mid-January and mid-March.)

Plain Leaf Warbler *Phylloscopus neglectus*

These tiny warblers breed in Central Asia and winter in the Persian Gulf and riverine thorn forest in the Indus Plains, mainly in Sindh. A few have been recorded in the desert areas of western Rajasthan. Ganguli mentions observations by H. Alexander in March 1928, November 1946 and January 1950,[300] but without further detail. It has been claimed since, especially from Tughlaqabad, but in the absence of more detailed data these historical records can only be retained here tentatively, subject to reconfirmation.

Tawny-bellied Babbler *Dumetia hyperythra*

A South Asian endemic, placed in the redefined babbler family Timaliidae, widespread in undergrowth and forest habitats in India including the Aravallis, northwards at least to the Sariska National Park. Ganguli mentions a party seen at the New Delhi Golf Links Colony on 22 September 1942 and 'about a dozen birds seen singly and in small parties of 2 or 3 in the shrubbery in Lodhi gardens on 12 November', probably in 1961.[301] (Note: This babbler has also been reported from just beyond the limits of the Delhi area as defined for this book, from Haiderpur Wetland, the Hastinapur Wildlife Sanctuary, in the third week of June 2021.)

Bar-tailed Treecreeper *Certhia himalayana*

A west Himalayan treecreeper that is known to extend short distances into the adjoining plains in winter. Hutson[302] is credited with seeing one in light woodland about Delhi, on 4 January (presumably in the 1940s, the date misquoted by Ganguli as 4 June). There are no further details. (Note: There are recent records of this treecreeper from the Haiderpur Wetland, just outside the limits of the Delhi area but adjacent to the Hastinapur Wildlife Sanctuary, in early February and March 2020, and January 2022.)

Gold-fronted Chloropsis *Chloropsis aurifrons*

A bird that stayed for much of August 1944 at the Alipur Nursery near Safdarjung, New Delhi, was evidently an escape.[303]

Trumpeter Finch *Bucanetes githaginea*

A desert finch of North Africa and West Asia, ranging southwards to Rajasthan in winter. Hume[304] records one obtained at Hattin in Gurugram district, Haryana, on 16 December 1877.

Expunged Records

Some of the records in this list are by experienced observers but have been set aside for the moment, for the reasons mentioned here. Checklists are not frozen in time, however, and these records may be reassessed in the light of additional information as it becomes available. Many experienced observers have questioned their own identifications of past years.

Orange-breasted Green Pigeon *Treron bicincta*

A red-legged green pigeon, in extremely poor plumage, was photographed on 29 September 2016 at Okhla and identified as this species. This is far from the species' closest known range in the UP *terai* (although it has wandered as far south in the Gangetic Plains as Bareilly and Lucknow); further, the angle and quality of the photograph does not show features that could convincingly eliminate female Ashy-headed Green Pigeon *T. phayrei*, which has occurred as a vagrant in Delhi. Moreover, the state of this unfortunate bird's plumage suggests that it escaped either from captivity or from a raptor attack.

White-bellied Heron *Ardea insignis*

These magnificent herons are Critically Endangered today with tiny remnant populations in the far eastern Himalayas. Ganguli wrote that several (eight or nine) of these herons were seen by Peter Jackson, Roger Holmes and herself in May/June 1968 at the Shamspur Jheel west of Najafgarh. Jackson[305] recorded: 'a few years ago, Dr Sálim Ali had raised the possibility that some herons we saw at Shamaspur ... might be the Great White-bellied Heron ... this year, Mrs Ganguli, Mr Holmes and I saw several of these herons again ... and from the bright white of the breast and belly and the distinctly larger size than the Grey Heron which was also to be seen, we concluded that they were almost certainly *A. insignis*.' Ganguli added that Jackson had seen these herons again in 1970. These records are quite extraordinary given the range of the species, its extreme rarity compared to the numbers seen and unsuitability of the habitat. Jackson himself never reconfirmed these observations, and in their comprehensive *Handbook of the Birds of India and Pakistan*, Ali & Ripley[306] make no mention of the record. It can only be considered hypothetical, a case of mistaken identity.

Eurasian Woodcock *Scolopax rusticola*

Woodcocks breed widely across temperate Europe and Asia, southwards to the Himalayas, and in India they winter in the southern Indian hills. They are rarely reported in between on passage, which probably takes place in a single hop over the Indian mainland. Ganguli[307] mentions a sight record by H. Alexander, but with no further details. It is not proposed to permit such vague records in the Annotated Checklist.

Bridled Tern *Onychoprion anaethetus*

A pelagic tern of tropical seas, it is only exceptionally found inland having been blown astray by cyclonic weather or ocean storms and never far from the coast. A sight record of one over the Yamuna in June 1999 by Clive Harris[308] is quite extraordinary and is excluded from the Annotated Checklist.

Changeable Hawk Eagle *Nisaetus cirrhatus*

There are two races of this species on mainland India, the crestless *limnaetus* of the Himalayan foothills and the long-crested nominate of peninsular India. Ganguli[309] mentions one seen at New Delhi by H. Alexander 'prior to 1954'. No further details or, surprisingly, not even the race were mentioned, and this seems typical of the casual reporting of observations by some otherwise accomplished observers of the time. (Note: there is a record of a vagrant individual, apparently race *limnaetus*, from Haiderpur Wetland, just outside the limits of the Delhi area, in early August 2020.)

Golden Eagle *Aquila chrysaetos*

A very wide-ranging species of Europe, Asia, North Africa and North America, Golden Eagles are resident in the higher Himalayas, making only small altitudinal movements, and are not known to enter the Gangetic Plains even during severe winters. This may not hold true, however, in the western parts of the subcontinent, in Pakistan, Rajasthan and Gujarat, from where odd wintering birds have been reported. Hutson[310] recorded 'a female near at Qadipur on 31 December', no year mentioned, 'on the ground with two King Vultures, considerably larger than the latter and almost uniformly dark brown with "trousered" legs and a pale grey beak'; almost certainly misidentified.

Upland Buzzard *Buteo hemilasius*

Upland Buzzards breed north of the Himalayas and winter on their southern slopes, but do not normally enter the plains, and identification difficulties confound the odd reports of their occurrence here. In addition, hybrids within the genus *Buteo* are not rare, in particular between *rufinus* and *hemilasius*. Many buzzard phenotypes from India suggest hybrids in appearance, though genetic confirmation would be desirable.[311] There are five claimed records for Delhi: Ganguli mentions two dubious sightings in 1951; Gaston reports one over the New Delhi Ridge in March 1973; one photographed at the Madanpur *khadar* in December 2002 is misidentified;[312] and finally, one was photographed at the Yamuna Biodiversity Park, north Delhi, on 1 March 2015.[313] The thinly and sparsely feathered tarsi of the Yamuna Park bird were the basis for identification, but its jizz is Long-legged rather than Upland, whose tarsi feathering is also usually denser and more obvious; a hybrid with Long-legged Buzzard seems to be the best explanation. In the absence of any conclusive record, Upland is excluded from the Annotated Checklist.

Crested Goshawk *Accipiter trivirgatus*

There is a perplexing record from Hastinapur[314] in the late 1970s, with the observer's comment, 'This bird of prey is most conspicuous in the non-breeding months of winter. It has been observed for a long time since 1978. It prefers to stay at a particular spot at the edge of the forest overlooking the marshes, offering a good view, often perched atop a disused electric pole or on a scanty-leaved branch of a tree', apparently referring to one individual bird. Crested Goshawks, subspecies *indicus*, are breeding residents along the Himalayan foothills, *bhabar* and *terai* from Uttarakhand eastwards and not known to disperse further south into the Gangetic Plains in winter. There have been no reports from the Delhi area since that time, and this record is retained here pending fresh information emerging.

Mottled Wood Owl *Strix ocellata*

Ganguli attributed a record from Delhi to Hutson, but the latter did not include it in his own list,[315] except for a passing mention of crows tormenting Mottled Wood Owls amongst other species. The Delhi area is, however, within the residential range of this owl in India.

Streak-throated Woodpecker *Picus xanthopygaeus*

This sedentary, non-migratory woodpecker of forest edges in the Himalayan foothills and adjacent plains, and peninsular India as well, is known from the southernmost Aravallis (Mount Abu) but not further northwards along the range. Dispersal abilities of woodpeckers are low, and a female reported from the Asola Wildlife Sanctuary on the Southern Ridge in June 2015 would be a most unlikely vagrant from the Himalayan foothills. Though a photograph was put out by the observer, accompanying details remain ambiguous, and the record is excluded from the Annotated Checklist for the present.

Ashy Woodswallow *Artamus fuscus*

Lav Kumar Khachar had withdrawn his own records of 1950s,[316] referred to in Ganguli's *Guide to the Birds of the Delhi Area.*

Maroon Oriole *Oriolus traillii*

A Himalayan species that shows limited seasonal altitudinal movements; an 'immature male' reported from the Lodhi Gardens, New Delhi, in April 1980 without any supporting detail was termed as a probable escape.[317] There are no further reports from the Delhi area, but eBird checklists record Maroon Orioles from Roorkee in March 2021 and from the 'Haiderpur Wetland' in February and November 2021, both locations close to, but outside the limits of the area and much closer to the *terai*, where these orioles do winter occasionally.

Bar-winged Flycatcher-shrike *Hemipus picatus*

Two subspecies of this flycatcher-shrike are resident, with only minor altitudinal movements, along the Himalayas and, disjunctly, in central and southwestern India. Yado Mohan Rai, who had studied the birds of the Meerut and Hastinapur region since 1976 and later years, wrote,[318] 'Two of these birds, never known to occur in the north of peninsular India, were observed in Hastinapur in Dec. 1980 in a small wooded patch overlapping the marshes. Their status in this region could at best be only occasional or stray.' There has been no further record from the Delhi area, or from the Gangetic Plains away from the Himalayan foothills, and this perplexing record is retained here, excluded from the Checklist for the time being.

Grey-backed Shrike *Lanius tephronotus*

A Himalayan breeding shrike that winters down to the foothills and rarely into the adjacent plains in severe weather. Ganguli[319] mentions 'two or three records' by H. Alexander, with no further details. Confusion vis-à-vis *L. s. caniceps* is, however, evident from her further remarks.

Rook *Corvus frugilegus*

The Rook is a winter visitor to the northwestern parts of the subcontinent, mainly in northern Pakistan but straggling eastwards to the Punjab. A report of five birds in flight at Okhla by H. Alexander was termed "doubtful" by the observer himself.

Sykes's Lark *Galerida deva*

This distinctive endemic lark of dry landscapes of the Deccan Plateau, Gujarat and central India east of the Aravallis has been listed for the Delhi area with no supporting details, and a photographic record from Basai can be immediately dismissed. Excluded from the Annotated Checklist.

Lanceolated Warbler *Locustella lanceolata*

This elusive warbler winters mainly in Southeast Asia and in smaller numbers as far westwards as the northeastern subcontinent. J. Donahue's specimen, collected at New Delhi on 17 August 1962, was re-examined and confirmed as Grasshopper Warbler.[320] It is excluded from the Annotated Checklist.

Dusky Warbler *Phylloscopus fuscatus*

The Dusky Warbler, similar in appearance to the Smoky Warbler *P. fuligiventer* and closely related to that species, nests in eastern Siberia and China and winters on damp ground along the Himalayan foothills from Nepal eastwards, in northeastern India and mainland Southeast Asia. Vagrants have been recorded from peninsular India west to Harike in Punjab and Bharatpur in Rajasthan; it has also been reported from the Delhi area at Okhla and from the Haiderpur Wetland, but these records cannot be validated.

Rufous-vented Grass Babbler *Laticilla burnesii*

This bird, formerly called Rufous-vented Prinia, is now placed in the babbler family Pellorneidae. This species is a local inhabitant of the Indus floodplains east to Harike on the Sutlej, with scattered records from Punjab and Rajasthan, and has been recorded as a resident from as close to our area as Ottu Barrage, Sirsa district, Haryana. Ganguli[321] wrote that Alexander had recorded one at a marsh along Mathura Road, Delhi, in January. Although other species with similar distributions, e.g. Sind Sparrow, are known to have extended their ranges to Delhi, this record is unverifiable.

Slaty-blue Flycatcher *Ficedula tricolor*

Nests at rather high elevations in subalpine shrubbery and forest undergrowth in the Himalayas, dropping to the foothills and *terai* in winter, when it prefers tangled vegetation, tall grass and reeds. One, at Nehru Park, New Delhi, in end-January 2015, a 'probable male'[322] seen without optical equipment from a jogging track, is the only report from the Delhi area. There are no records of these birds wintering south of the *terai* in the northern plains, except from the Bharatpur National Park, where they have been reported in eBird checklists on rare occasions between late November and early February in 1999–2001 and in 2015. It is possible, therefore, that this flycatcher could visit in Delhi in winter, but the single report has been set aside here, given the circumstances of the sighting.

Pale-billed Flowerpecker *Dicaeum erythrorhynchos*

A South Asian endemic, Delhi is beyond the normal range of this flowerpecker, which is known, however, from as close as 250 km about Kalesar, Haridwar and Bareilly. Ganguli[323] mentions observations by H. Alexander in 'Qudsia Park, North Delhi, Lodhi Gardens and possibly in other localities' but with no details. It has not been reported since from the Delhi area, and Alexander's records are unverifiable today from such vague statements. (Note: There are recent records from just outside the limits of the Delhi area, from Haiderpur Wetland, the Hastinapur Wildlife Sanctuary, Meerut district, in late February–March 2020.)

6 Birding Hotspots

WHERE TO WATCH BIRDS ABOUT DELHI

BRIEF NOTES, INDICATIVE MAPS, REPRESENTATIVE SPECIES

There are birds – common, uncommon or rare, resident or migrant – anywhere and everywhere in the Delhi area. One just has to keep one's eyes and ears open wherever one goes!

The three sectoral maps in the following pages point to some particularly bird-rich and frequently visited options in Delhi's neighbourhood. The city itself offers some fine birding opportunities with its tree-lined avenues, historical sites, 'city forests' and many large and wooded parks and gardens. An added advantage is that these are all easily accessible.

A few spaces deserve special mention here. The Delhi Zoological Park with its waterbodies and waterbird enclosures, for example, hosts a large heronry, with breeding Painted Storks, ibis and diverse heron and cormorant species, and its wooded nooks and corners attract many passerine migrants in season.

The green, wooded remnants of the famed Delhi Ridge: the Old Delhi Ridge near Delhi University, now designated as the Kamla Nehru Biodiversity Park, and the New Delhi Ridge behind Rashtrapati Bhavan, that includes the spacious Buddha Jayanti and Mahavir Jayanti Parks, the polo grounds and Malcha Mahal, both offer birding opportunities all through the year, attracting several species of woodland and scrub, both breeding birds and migrants in season.

Other bird-rich areas within city limits include the Humayun's Tomb area with the adjacent Sunder Nursery, Lodhi Gardens, Hauz Khas and Sanjay Van city forests, the historical sites and gardens at Mehrauli, Tughlaqabad, Moth Masjid and several other open, green spaces; a longer list would include the city's biodiversity parks, which are dealt with in the following pages.

Mandothi Jheel landscape

The River Floodplains

The floodplains of the Ganga and Yamuna, with their mosaic of oxbow lakes, woodland, grassland and marsh, offer fine birdwatching opportunities. Annually flooded with a fertile overlay of alluvium, they are heavily encroached by cultivators and only remnants of bird-rich natural habitat remain.

The Yamuna Biodiversity Park, north of Wazirabad, Delhi, comprises 185 ha of restored floodplain habitats and has developed into a fine refuge for forest, grassland, marsh and waterside birds. It hosts a heronry with breeding night herons, egrets, darters and cormorants, and wintering flocks of waterfowl; the braided river channels upstream hold diverse waders, pratincoles, plovers and terns along the sandbanks and islets.

Further south, the Kalindi Biodiversity Park aims to restore over 115 ha of floodplain ecosystems and will form a valuable adjunct to the Okhla Bird Sanctuary.

The Okhla area has long been famed as a rich bird habitat. It has changed character over the decades, since the Okhla barrage was constructed in the 1980s, with its large impoundment area protected as a sanctuary. Its extensive reed beds interspersed with water channels hold a diversity of waterfowl in winter. Three species of bitterns breed in the reed beds as do watercocks, jacanas and other marsh species as well. The surrounding woodland attracts a variety of passerine migrants and winter visitors, and the sanctuary is renowned for being home to several raptors, with rarities such as Lesser Fish Eagles and White-tailed Eagles recorded here. It hosts a heronry with breeding Night Herons, darters and cormorants.

Away from the river and increasingly hemmed in by the expanding urban sprawl of Greater Noida, the Surajpur Bird Sanctuary also hosts a heronry with openbills, ibis, various herons and cormorants and many species of waterfowl in winter. The surrounding bands of woodland have produced interesting records of owls and raptors, and patches of grassland hold breeding Bristled Grassbirds in the summer season.

The Dhanauri *Jheel* south of Greater Noida, measuring about 100 ha, is a first-class bird habitat, being considered as a potential Sarus sanctuary and a Ramsar site. It hosts a diversity of waterfowl and waders in winter, and raptors, including Steppe, Eastern Imperial, Bonelli's, Greater Spotted and Indian Spotted Eagles, and falcons, including Red-necked and Peregrines. It is famed for the large numbers of Sarus Cranes that breed and roost here.

At the northeastern edge of the area, the Hastinapur Wildlife Sanctuary (which includes the 1,200 ha Haiderpur Wetland created by the impoundment of the Bijnor barrage just upstream, albeit just beyond the borders of the area covered in this book) has productive floodplain habitats. Parts of the sanctuary are heavily encroached, but the Haidarpur Wetland is protected as a Ramsar site for their diverse mammal and bird fauna. These areas have only recently begun to be explored ornithologically, but they host some remarkable bird species, both residents and winter visitors, because of their location closer to the Himalayan foothills and *terai*, and with vastly improved access by road from Delhi, would reward an observer with interesting records, perhaps even new additions to the Delhi area checklist.

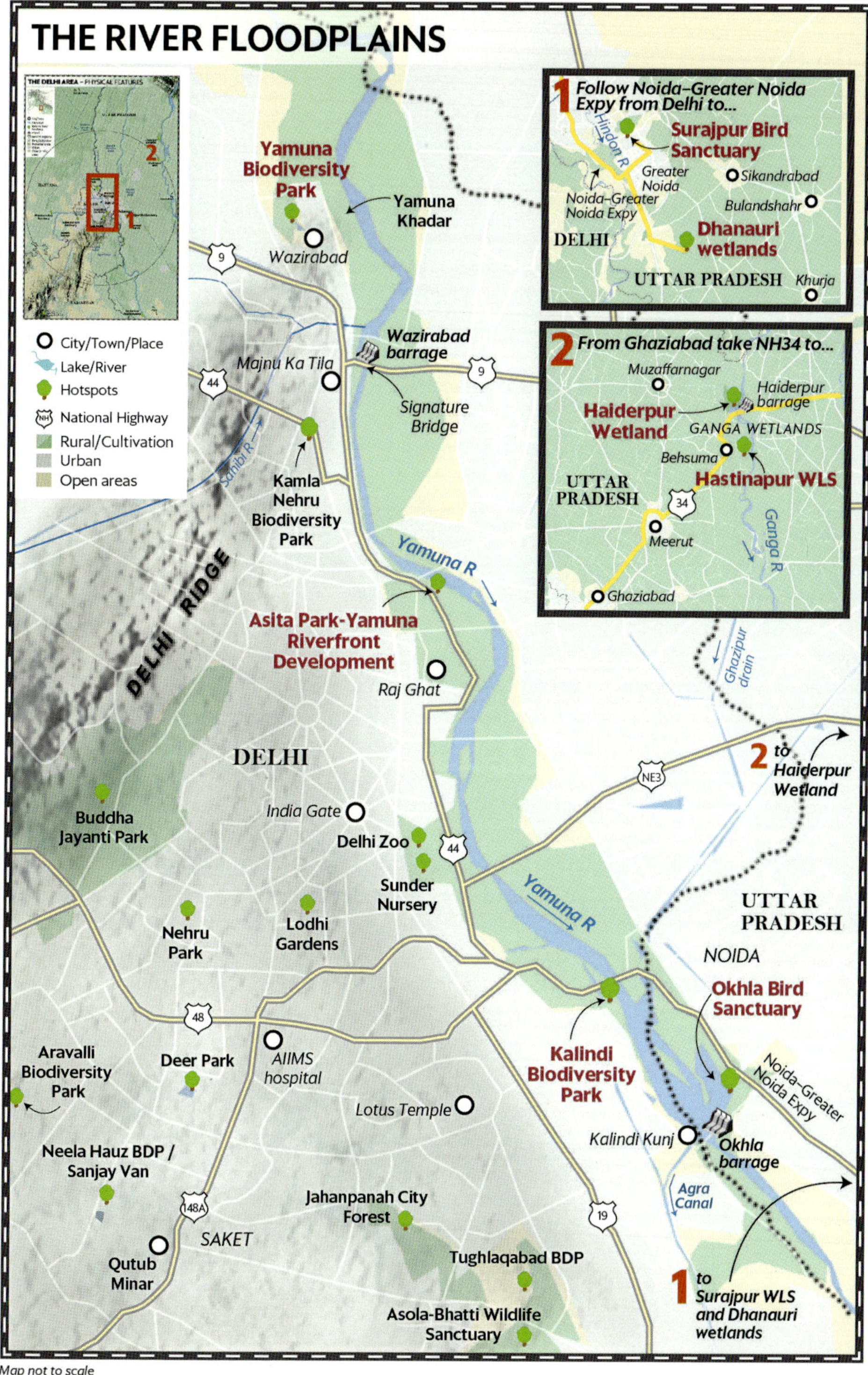

Map not to scale

The Haryana Wetlands

The Sultanpur National Park, extending over 143 ha and about 16 km by road southwest of Gurugram, is very well known as a refuge for waterfowl. It presently hosts perhaps the largest heronry in the Delhi area with a wide range of storks, ibis, herons, darters and cormorants breeding annually here. A diversity of waterfowl, including waders, fill the waterbodies of the sanctuary each winter and have consistently produced rarities. Being a wooded patch in an otherwise intensely cultivated area, the surrounding woodlands offer a refuge for migrant and wintering passerines, including warblers, flycatchers, thrushes, etc. The large numbers of waterfowl attract a range of raptors that gather here, especially in winter.

The Sultanpur neighbourhood offers some fine birding areas; remnant patches of grassland, as at Jhanjraula to the north of the sanctuary, attract a number of breeding grassland species, including at least two bustard-quails, Rain Quails, Black Francolins, Indian Coursers, Singing Bush Larks and the rare Stoliczka's Bushchats. Grassland-haunting harriers, Short-eared Owls and open country falcons also are regularly recorded.

In spite of its presently polluted status, the Najafgarh drain hosts a heronry with herons, darters and cormorants breeding along its tree-planted embankments. This is the channel of the former Sahibi river, which, in the past, used to flood to form a vast shallow lake, the Najafgarh *Jheel*, of which only remnants remain. These remnant wetlands and associated flooded fields, located between the Najafgarh drain and the road to Sultanpur near the village of Chandu-Budhera, can offer outstanding birding opportunities, including large flocks of flamingos, pelicans, cranes, storks, herons, waders and waterfowl with an impressive range of raptors as well. The Basai marshes, also located along the Sultanpur road, were another outstanding birdwatching site in past decades, but have now been partly drained, lost in the rampant urban expansion that now smothers them.

Further afield in Haryana, notably in the Sonipat, Rohtak and Jhajjar districts, a number of wetlands, lakes and spillage marshes along the distributary network of the Western Yamuna Canal, offer fine birding. These host a range of waterfowl and waders, owls, including Dusky Horned and Brown Fish Owls, and raptors, such as Osprey, Eastern Imperial, Greater Spotted and Indian Spotted Eagles. Of these, the Bhindawas Wildlife Sanctuary, a 410 ha reservoir in Jhajjar district, the Dighal wetlands and the Mandothi wetlands near Bahadurgarh are of special interest and have regularly recorded several rare species.

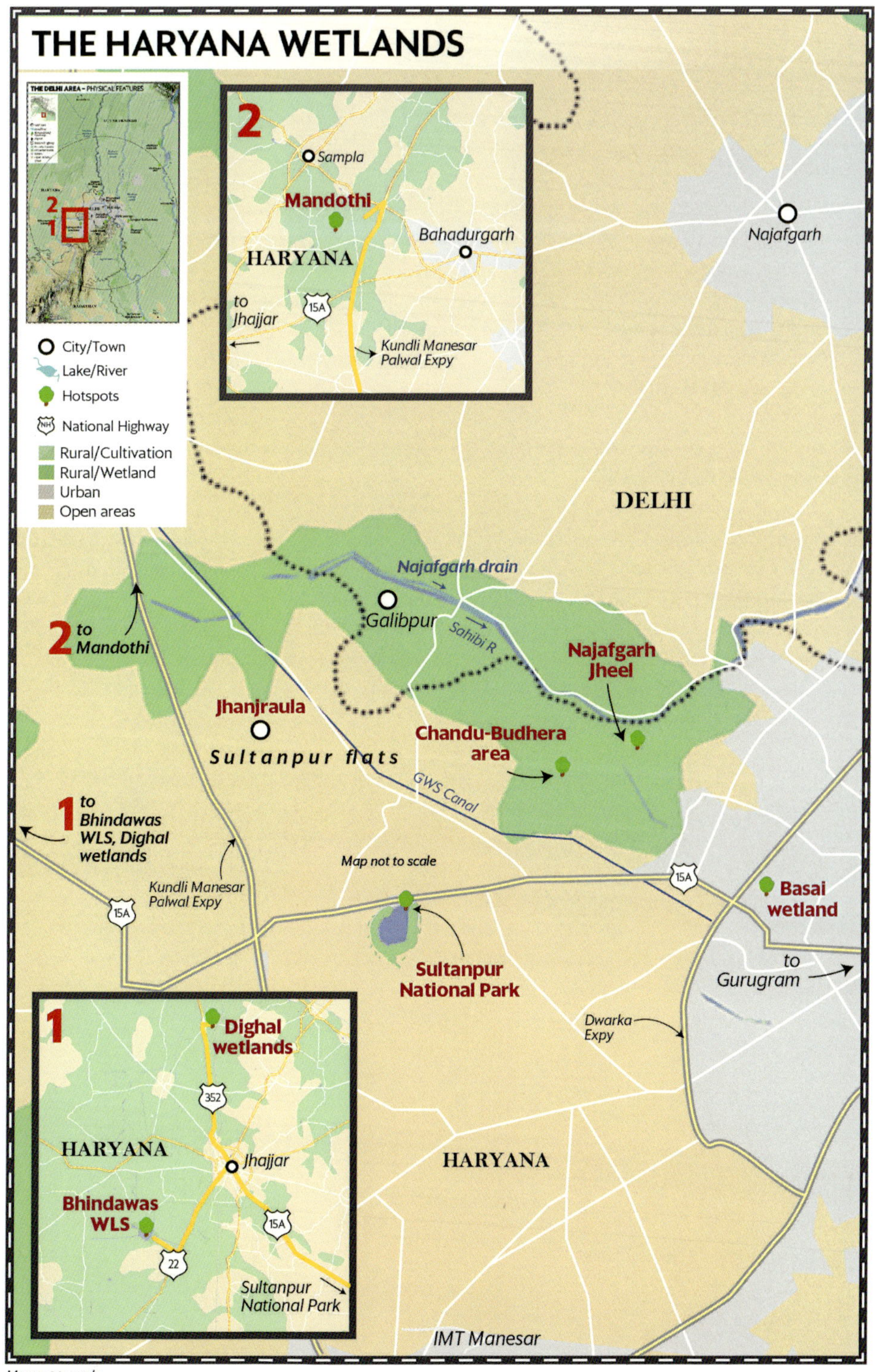

Map not to scale

The Aravallis

The Aravalli habitats that extend into the area from the south are of special interest as they hold a number of breeding species not present elsewhere in the area.

The Aravalli Biodiversity Park, Vasant Vihar (283 ha) and Neela Hauz Biodiversity Park (part of the Sanjay Van Reserve Forest, 317 ha) in the Mehrauli neighbourhood together protect over 600 ha on the South-Central Ridge; and the Tilpath Valley (70 ha) and Tughlaqabad biodiversity parks (81 ha) on the Southern Ridge. These, along with the Aravalli Biodiversity Park in Gurugram (154 ha), are hilly areas deforested and scarred by mining that have been restored by human effort and are beginning to regain their original floral and faunal diversity; they now form an extensive, albeit disjointed, complex of woodland-and-scrub 'Aravalli' habitats.

The Southern Ridge extends southwards from Tughlaqabad and the Asola-Bhatti Wildlife Sanctuary into the Faridabad and Gurugram districts.

Of particular ornithological interest are the diverse habitats protected by the Asola-Bhatti Wildlife Sanctuary, an area of woodland and abandoned stone mines that now form deep lakes, with visitor and educational facilities and a butterfly park. Further to the south, Mangar Bani, a 266 ha sacred grove in a sheltered ravine protected by religious sentiment, still retains much of its vegetation in its natural state. The area and its surrounding rocky hills and wooded valleys offer some first-class birding in a small, easily accessible area and host breeding populations of birds such the Indian Pitta, Common Cuckoo, Black-headed Cuckooshrike and White-bellied Drongo, typical species of the Aravalli Range.

The Bhondsi Nature Park, just off the Gurugram-Sohna Road, is another easily assessed site with a similar birdlife. It seems to hold a particular attraction for passerine migrants, such as Fire-capped Tits, which appear regularly at this site every spring, but with very few records elsewhere in the area.

The Aravalli habitats continue southwards, to the Damdama Lake, with its Openbill Storks, and Sohna; the Sakatpur area with its Jungle Bush Quails and Sirkeer Cuckoos; to Manesar on the Delhi-Jaipur highway. A little further to the west are wooded outliers of the Aravallis, at Manethi near Rewari, special habitats for some rare avian desiderata of the Delhi area, including the White-bellied Minivet, White-naped Tit and Indian Spotted Creeper.

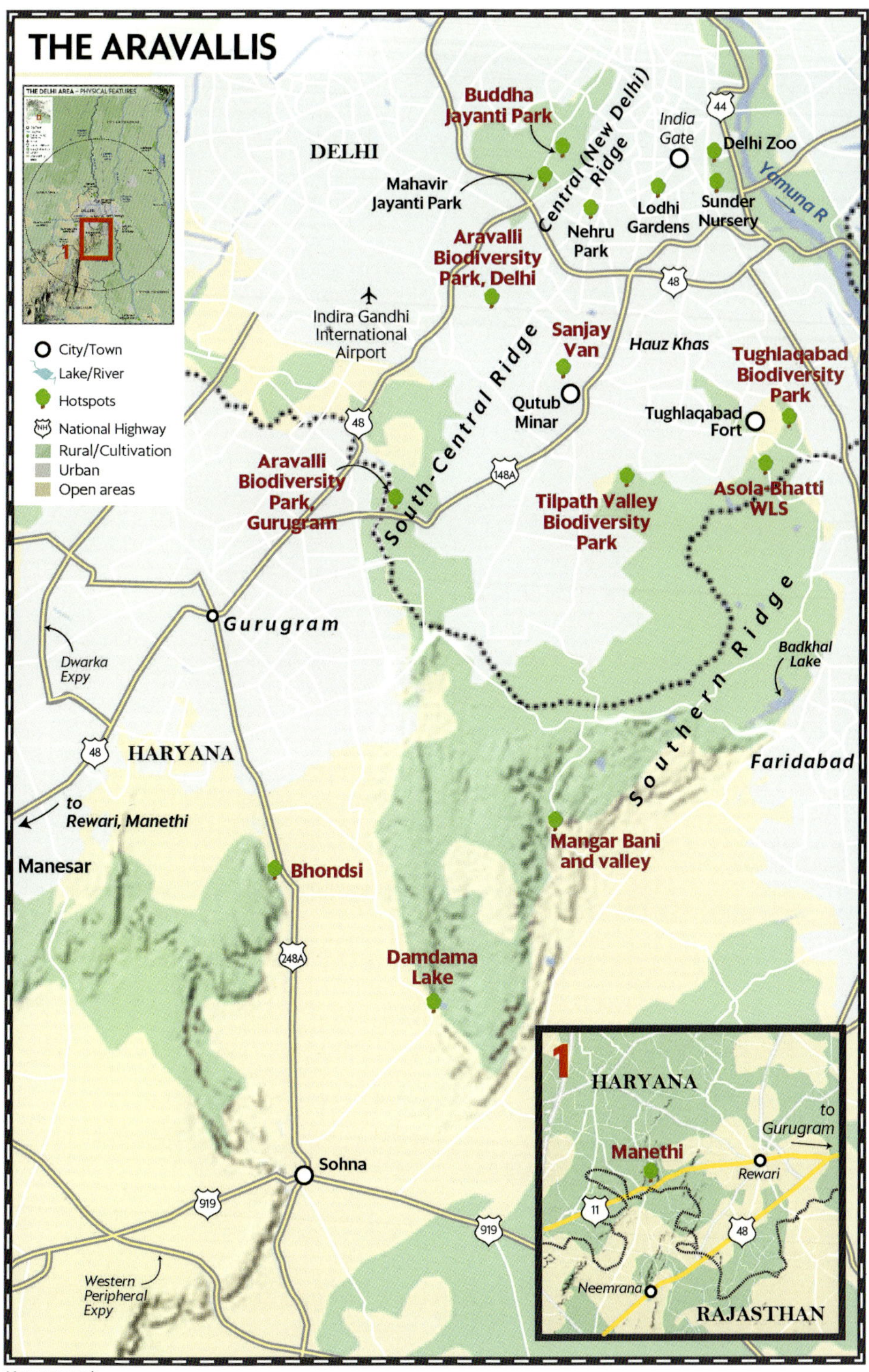

Map not to scale

Glossary

Acacia: Common name for flowering, thorny trees of the family Fabaceae, also called mimosas, widespread in arid tropical habitats, which were all formerly placed in the genus *Acacia*, now split into various genera with *Acacia* itself largely restricted to Australian species. Several species are common in the Delhi area, including babool *Vachellia nilotica*, ronjh *V. leucophloea*, Phulai *Senegalia modesta*, Kumttha *S. senegal*, Khair *S. catechu*, etc., which continue to be generally referred to by their former, familiar name acacias, or *babools*, as they are in this book

Accidental: An extremely rare transient visitor, usually synonymous with vagrant (as in this book). Refers to the fact that the occurrence of the bird is by accident, for a variety of reasons

Allopatric: Two or more related species whose geographical ranges do not overlap

Altricial: A bird that hatches helpless and requires significant parental care

Amaltas (Hindi): *Cassia fistula*, or Indian laburnum; a showy species planted in gardens and thinly scattered in remnant woodland areas on the Ridge and elsewhere

Axillaries: Underwing feathers at the base of the wings, the axilla or 'armpits' of a bird

Babool (Hindi): *Vachellia nilotica*, widely referred to as acacia, native to Africa, the Middle East and the Indian subcontinent. Still placed in its former genus *Acacia* by several authorities, it is referred to in this work by this long-familiar name, as *Acacia nilotica*, syn. *Acacia arabica*.

Bagar (Hindi): The local term for dry, sandy plains, showing desert influences, mainly in northwestern India

Bangar (Hindi): The local term for cultivated plains of old alluvium drained by the major rivers of northern India

Banyan tree: *Ficus benghalensis*, bargad or barh in Hindi, whose figs are very popular with frugivorous birds. There are a few large specimens in Delhi city itself, but it is more common in the countryside, being widely planted in village squares or near temples (because of their religious significance) and commonly spread, self-sown in rock crevices and elsewhere through seeds dispersed by birds.

Ber (Hindi): *Zizyphus mauritiana*, or Indian jujube; a dense, spiny tree with edible berries, fairly common in arid habitats and also widely planted commercially in groves near villages

Bhabar (Hindi): Local term for the low foothills of the Himalayas, composed of well-drained, coarse gravel deposits supporting sal *Shorea robusta* forests, which drain into the swampy, low-lying *terai*

Biogeographic realms: The broadest biogeographic divisions of the Earth's land surface and characterized by the shared evolutionary history of the flora and fauna they sustain. Divided into bioregions or major habitat types, which are further subdivided into ecoregions, or habitats with uniform environmental conditions. There are eight such realms.

Bistendu (Hindi): *Diospyros cordifolia*, a low, lush evergreen tree common in remnant areas of native vegetation on the Ridge

Brood parasitism: A bird species, like several cuckoos, laying eggs in the nest of another species, and leaving it to the foster parents to incubate the eggs and raise the chicks till they fledge

Clade: A monophyletic group, i.e. a group of organisms evolved from a common ancestor, composed of the common ancestor and all its lineal descendants

Class (in biology): A taxonomic rank used in the classification of organisms, below phylum and above order. Birds form the class Aves.

Crepuscular: Active at dusk and dawn

Dabar (Hindi): Local term for low-lying, poorly drained terrain, often sandy and saline, waterlogged after rain

Dhak/Palash (Hindi): *Butea monosperma*, or flame of the forest, with its show of orange-red flowers when its branches are leafless in March, is widespread but scattered on the Ridge and in uncultivated areas near villages of the area. A favourite of nectarivorous birds when in bloom.

Dhau (Hindi): *Anogeissus pendula*; a short, drought-resistant tree that can grow in dense monospecific stands on hilly ground and native to undisturbed valleys of the Aravallis in the area

Dhauldhak (Hindi): *Erythrina suberosa*, and *E. variegata*, also called coral trees or just erythrinas, planted in gardens in the area and an unfailing magnet for nectar-feeding birds when in flower

Diurnal: Active during daytime

Duars: Local term for the eastern portion of the *terai* in West Bengal and Assam

Eclipse plumage: Dull, female-like plumage worn by drakes after breeding, in waterfowl

Ecological niche: The role or position occupied by a particular species in its ecosystem

Endemic: Indigenous, limited to a specific area

Eurasia: The land mass of Europe and Asia together

Eucalyptus: *Eucalyptus tereticornis*, or Red Gum, introduced from Australia, now common in India in commercial plantations, and in the Delhi area especially for strip afforestation along canal embankments and railway lines

Extinct: No longer in existence

Extirpated: Locally extinct

Fallow: Cultivated land, prior to ploughing for the next crop

Family: A taxonomic rank used in the classification of organisms; a clade, or monophyletic (related) group of genera, which ranks below order but above subfamily/genus

Feral: A once-domesticated species that has reverted to a wild state

Fledgling: A young chick that has just acquired feathers

Genus: A clade or monophyletic group of related species, which ranks below family/subfamily but above species

Holarctic realm: One of the six major terrestrial biogeographic realms of the world, a composite of the Palearctic (Old world) and Nearctic (North American) realms, thus comprising the diversity of terrestrial ecosystems and habitats in the Northern Hemisphere

Irruption: The mass movement of a population from one place to another, usually linked to food supplies; sometimes cyclical, but typically irregular or nomadic

Jamun (Hindi): *Syzigium cumini*; a large native tree of monsoon forests in India, commonly planted as an avenue tree in Delhi and in gardens; its deep purple berries are edible and harvested commercially

Jhau (Hindi): *Tamarix aphylla*, or tamarisk; a small tree or bush with small pink flowers, adapted to sandy and saline conditions, often present in the *khadar* along the large rivers

Jheel (Hindi): A shallow lake or wetland

Jizz (in bird identification): The general impression of a bird that takes a holistic view of its characteristics (colour, pattern, behaviour, voice, etc.) which are unique to a species and assists in field identification

Juvenile: The stage of a bird's development after fledging but prior to adulthood. In species that take longer than a year to reach maturity, later stages are often termed as immature.

Kans (Hindi): *Saccharum spontaneum* or wild sugarcane; a native grass of the Gangetic plains, forming almost pure stands in the *terai* and some *khadar* areas

Kareel/akh (Hindi): *Capparis decidua*, or capparis; spiny bush widespread in arid areas especially on the Ridge; leafless for much of the year but covered with showy pink and red clusters of flowers in the summer and after the monsoon

Khadar (Hindi): Local term for the low-lying, flood-prone land of new alluvium, enriched by fresh deposits of silt every year along the course of the great rivers of the Gangetic Plains

Khajur (Hindi): *Phoenix sylvestris*, or wild date palm; planted in gardens and in groves near villages

Khus (Hindi): *Vetiveria zizanioides*; a fragrant tall grass, common in *khadar* habitats along the course of the larger rivers

Kikar (Hindi): Two species of these thorny trees with feathery foliage, in the genus *Prosopis* occur in the area, the kikar or khejri *P. cineraria*, a native of Delhi's original arid zone flora but now restricted to remnant patches, though more frequent in rural areas; and vilayati kikar *P. juliflora* or mesquite, introduced from Central America for afforestation of the Ridge a century ago, has proved invasive, now widespread and common in Delhi to the detriment of native flora

Lek: A gathering of males of certain polygynous species, including some birds, where they engage in competitive displays and courtship rituals to entice visiting females ('lekking'), or the space or plot utilized by the displaying males for this purpose

Lobes: Fleshy extensions on the sides of toes of some aquatic birds that aid in underwater propulsion

Lotus: *Nelumbo nucifera*, the sacred lotus or kamal in Hindi; aquatic plants with large, spectacular pink flowers, India's national flower

Mango: *Mangifera indica*, or aam in Hindi; a popular fruit tree commonly planted in city gardens and groves in rural areas

Migration: The regular, usually seasonal, movement of all or part of an animal population to and from a given area, usually following a regular route

Monophyletic: A clade; a group of organisms classified within the same taxonomic rank and share a most common recent ancestor

Monotypic: Having a single form with no subspecies

Morph: A distinctive colour or pattern variant within the population of a species, which is not geographically isolated (as a subspecies would be) yet maintains its distinctiveness and interbreeds freely with 'normally' plumaged birds

Mulberry: *Morus alba*, or toot in Hindi; naturalized in the area, sometimes cultivated but often self-sown, spread by frugivorous birds that are very fond of its fruit

Nearctic realm/region: One of the eight biogeographic realms constituting the Earth's land surface, covering most of North America

Neem (Hindi): *Azadirachta indica*; a large tree with a dense canopy, widespread, planted along avenues in Delhi and other towns in the area, cultivated for its reputed medical and disinfectant properties, and naturalized in many areas. Its drupes are favoured by many frugivorous birds.

Nidicolous: Chicks that remain in nests for a long duration after hatching

Nidifugous: Chicks that leave the nests shortly after hatching

Nidification: Relating to nests or nesting

Nocturnal: Active during the night time

Nominate subspecies: First subspecies to be formally named in a species, and it repeats the same subspecific scientific name as the species itself

Order: A taxonomic rank used in the classification of organisms between class and family; a group of related, monophyletic families. There are 41 orders of birds extant in the world.

Oriental realm/region: One of the eight biogeographic realms, comprising the Indian subcontinent, southern China and Southeast Asia

Palearctic realm/region: The largest of the eight biogeographic realms, encompassing all of Eurasia north of the Himalayas, and North Africa

Palmyra palm: *Borassus flabellifer*, toddy palm or tad/tal in Hindi, native palm of South and Southeast Asia, with large fan-shaped leaves and producing woody fruits with sweet edible pulp; the fermented sap is consumed as an alcoholic drink, toddy

Parapatric: Populations of a species or closely related species that occupy geographic areas that are partially overlapping or have a partial barrier between them

Peelu (Hindi): *Salvadora persica*; a native tree of desert areas from Arabia to northwestern India, present in Delhi in many historic gardens such as the Qutb Minar, near villages and on the Aravallis

Peepal (Hindi): *Ficus religiosa*, best known and commonest of the area's *Ficus* spp.; a large tree with distinctive leaves and small but profuse figs which are a favourite of many frugivorous birds. The other species include the bargad *F. benghalensis*, gular *F. racemosa*, pilkhan *F. virens*, etc.

Phragmites: *Phragmites karka*, or nal/narkul in Hindi; tall, flowering perennial reeds with many leaf blades, common in dense stands on flooded ground

Polyandrous: The female of a species having more than one mate

Polygamous: Either sex of a species having more than one mate

Polygynous: The male of a species having more than one mate

Polymorphic: Occurring in several forms

Precocial: Chicks that hatch with open eyes, covered with down and capable of independent locomotion. However, not all precocial birds leave the nest; some may stay in the nest and be fed by parents there and are, thus, considered nidicolous rather than nidifugous.

Primaries: Outer flight feathers or 'hand' of a bird's wing. Typically nine to ten in number, depends on the species.

Race: Subspecies

Reedmace: *Typha angustifolia, T. elephantina*, aquatic reeds with fleshy leaves that form dense stands (or reed beds) in shallow water

Record: A published, or otherwise broadcast, report of occurrence, ideally retrievable for reconfirmation

Rectrices: Tail feathers

Remiges: The primary and secondary feathers of a bird's wing taken together

Riparian: Inhabiting streams, rivers, and waterways

Ronjh (Hindi): *Vachellia leucophloea*, a small thorn tree with a crooked trunk and light-yellowish-grey bark, native to South Asia; still widely placed in its former genus *Acacia*, and is referred to here by this long familiar name *Acacia leucophloea*.

Sarkanda (Hindi): *Tripidium bengalense*, syn. *Saccharum bengalense*; a tall flowering grass species native to northern South Asia, harvested for thatching and cane furniture in rural areas

Semal (Hindi): *Bombax ceiba* or silk cotton; a tall, dominating, deciduous tree native of monsoon grassland and forests in India, commonly planted in the Delhi area. Its large, showy scarlet flowers with waxy petals produce abundant nectar, which is a favourite of frugivorous bird.

Shisham (Hindi): *Dalbergia sissoo* or Indian rosewood; large tree native to the Himalayan foothills, in gravelly or alluvium soils deposited along banks and islets of rivers where these debouch from the hills into the plains; also planted along commercially for its valuable wood

Secondaries: Flight feathers growing from the inner part, or 'arm', of the wing, adjacent to the primaries. The number of feathers varies from 9–25 depending on the species.

Species: (abbreviated as sp., plural spp.) A taxonomic rank and the basic unit of classification, below genus, and most often defined as the largest group of organisms within which any two male and female individuals can produce fertile offspring by sexual reproduction. Other criteria may also be employed to define species, including DNA sequences, morphology, behaviour or ecological niche.

Speculum: The colourful, usually iridescent, patch on the secondary feathers of the wing of some ducks

Subfamily: An auxiliary (intermediate) taxonomic rank, next below family but more inclusive than genus

Subspecies: (Abbreviated as ssp.) A taxonomic category that ranks below species, usually a distinct form or population living in a different geographical part of the species' range at least in the breeding season and varying in some morphological characteristics. The differences between subspecies are usually less distinct than the differences between species. It is synonymous with race.

Symbiotic: Close and interdependent relationship between distinct species

Sympatric: Related species or populations occurring within the same geographical area; overlapping in distribution

Taxon (pl. taxa): A group of organisms considered by taxonomists to form a unit and given a taxonomic rank

Taxonomy: The grouping and naming of organisms based on a scientific system of classification

Taxonomic rank: The relative level of a group of organisms (a taxon) in the taxonomic hierarchy. The main ranks in descending order, as relevant to ornithology, are kingdom (Animalia), phylum (Vertebrata), class

(Aves), order, family, genus and species. A given rank subsumes under it the less general categories, i.e. more specific descriptions of life forms.

Terai: The strip of low-lying, poorly drained country running west to east along the foothills of the Himalayas, from the Yamuna in the west to the Brahmaputra in the east. Marshy ground, grasslands and forests extend along its length.

Tertials: The innermost feathers of a bird's wing, covering bases of some of the secondaries

Totipalmate: All four toes of an aquatic bird's foot connected by webs, which aids in swimming and diving

Transitional plumage: The plumage of a bird which is in the process of moulting its body feathers, usually from immature to adult, or non-breeding to breeding plumage (or vice versa)

Ulu (Hindi): *Imperata cylindrica*, or cogon grass; common native grass with fluffy, silky white flowering spikes, quick to colonize fallow fields and clearings

Vagrant: A bird that appears outside the normally recorded distribution of the species. It is usually synonymous with accidental, but a fine distinction is sometimes made: a vagrant would be a bird recorded far from its usual distribution, a European species appearing in far eastern Asia for instance, whereas accidental may refer to a very rarely and irregularly occurring individual within a species' overall range, for example, a long-distance migrant that normally overflies an area non-stop, but could be compelled to land en route because of illness, inclement weather or any other reason.

Water chestnut: *Trapa bicornis*, or singhada in Hindi; floating aquatic plants bearing three spined nuts cultivated in village ponds for their edible, crisp-fleshed corms

Water hyacinth: *Pontederia crassipes* (formerly in genus *Eichhornia*) or jal kumbhi in Hindi; an aquatic plant with showy purple-pink flowers, native to South America but introduced, mostly accidentally, across the world. Now naturalized and invasive, it is a highly aggressive weed, forming dense floating colonies that block sunlight and oxygen from the water and crowd out native species.

Water lily: Several species of flowering aquatic plants of the genus *Nymphaea*, kumud in Hindi, and *Euryale ferox* (makhana in Hindi, with white edible seeds) occur in India, both as wild species and in water gardens for ornamental purposes

Wing coverts: The feathers which cover the bases of the primary and secondary feathers of a bird's wing, on both its upper and lower surfaces, forming the upper wing coverts and underwing coverts (or 'wing lining') respectively

Zygodactyl: Having two toes directed forward and two backwards, rather than three forward and one (hallux) backwards

Notes

1. See Pradip Krishen, *Trees of Delhi: A Field Guide* (Delhi: Dorling Kindersley, 2006), 33.
2. See Sydney Basil-Edwardes, 'A contribution to ornithology of Delhi, Part 1,' *JBNHS 31 (1926a), 262.*
3. See Krishen, *Trees of Delhi*, 32.
4. 'Ecoregions' as defined by the Worldwide Fund for Nature (WWF) scientists (Olson et al. 2001) is, to date, one of the best researched efforts to map habitat types worldwide.
5. Krishen, *Trees of Delhi*, 19–23, has succinct but very readable accounts of these ecotones or micro-habitats of the area.
6. See Krishen, *Trees of Delhi*, 24–25, for a description of Mangar Bani and its native flora.
7. *DDA Biodiversity Parks of Delhi* (2021) by Delhi Biodiversity Foundation D.D.A. elucidates the concept behind the initiative.
8. See *Landscape and Environmental Planning Department – Biodiversity Park* (2021) by Delhi Biodiversity Authority for brief write-ups on biodiversity parks.
9. See also A. J. Urfi, 'Wetlands of ornithological significance in the Delhi region,' *Oriental Bird Club (OBC) Bulletin* 22 (1995): 38–41.
10. From an interesting study by Sundar et al., 'Does the stork bring home the owl? Dusky Eagle-Owls *Bubo coromandus* breeding in Woolly-necked Stork *Ciconia episcopus* nests,' *BioTROPICA* 54, no. 2, 1–5.
11. *Namami Gange* is an integrated conservation mission, approved as a flagship programme by the Government of India in June 2014 to accomplish the objectives of effective abatement of pollution, conservation and rejuvenation of the Ganga.
12. See Rahmani et al. (2016) for definitions and complete listing of IBAs in India.
13. See also Anthony Gaston, 'Some comments on the "revival" of Sultanpur Lake,' *Oriental Bird Club Bulletin* 20 (1994): 49–50.
14. See Urfi, *The Painted Stork: Ecology and Conservation* (New York: Springer-Verlag, 2011).
15. Khera et al. (2009), though dated, is an interesting study of the subject.
16. See *Gazetteer of the Delhi District 1883–84*. (1884): 20–21.
17. See G.N. Sinha. ed., *An Introduction to the Delhi Ridge*, (New Delhi: Department of Forests and Wildlife, Government of NCT of Delhi, 2014).
18. Editors (1950), 811–812.
19. See A.O. Hume, *The Nests and Eggs of Indian Birds Vol I–III*, 2nd ed (London: R.H. Porter, 1889–90).
20. Basil-Edwardes, 'A contribution to the ornithology of Delhi. Part 1,' *Journal of the Bombay Natural History Society* 31, no. 2 (1926a), 261–273 and 'A contribution to the ornithology of Delhi. Part 2,' *Journal of the Bombay Natural History Society* 31, no. 3 (1926b): 567–78.
21. Norman Fredrick Frome, 'The Birds of Delhi and District,' *Journal of the Bombay Natural History Society* 47, no. 2 (1948): 277–300.
22. Anonymous, 'Announcements: The Delhi Bird-watching Society,' *Journal of the Bombay Natural History Society* 49, no. 3 (1950): 595.
23. Praveen and Jayapal, 'Checklist of the Birds of India v7.0' (28 February 2023); periodically revised, this is perhaps the most up-to-date checklist of the birds of India.
24. Clements et al., 'eBird/ Clements Checklist of the Birds of the World,' v. 2022 (August 2022), now in its 7th edition.
25. Gill, Donsker, and Rasmussen (eds.), IOC World Bird List, v13.1 (2023).
26. The 'Tobias criteria', proposed and elucidated in Tobias et al. (2010), addresses the issue of delimiting species and determining their taxonomic status by developing quantitative methods for assessing levels of divergence between related bird populations.
27. See Cicero et al., 'Integrative taxonomy and geographic sampling underlie successful species delimitation,' *Ornithology* 138, no. 2 (2021) for an elucidation of this approach.
28. See Aasheesh Pittie, 'Bibliography of South Asian Ornithology,' (2018), a valuable resource for ornithological research in South Asia.
29. See 'IUCN Red List of Threatened Species' at https://www.iucnredlist.org/.
30. See 'The flyways concept can help coordinate global efforts to conserve migratory birds,' 2010, http://datazone.birdlife.org/sowb/casestudy/the-flyways-concept-can-help-coordinate-global-efforts-to-conserve-migratory-birds.
31. See *Central Asia/South Asia. BirdLife Flyway Factsheet 7* (Cambridge: Birdlife International, 2010b).
32. See Alexander Lees and J. Gilroy, *Vagrancy in Birds* (Princeton: Princeton University Press, 2022) for a detailed exploration of vagrancy in birds.
33. See *East Asia/East Africa. BirdLife Flyway Factsheet 6* (Cambridge: Birdlife International, 2010) for details of this migratory route.
34. See *East Asia/Australasia. BirdLife Flyway Factsheet 8* (Cambridge: Birdlife International, 2010c) for details of this migratory route.
35. See Usha Ganguli, *A Guide to the Birds of the Delhi Area* (New Delhi: Indian Council of Agricultural Research, 1975), 46.
36. Clutch sizes and some other breeding details quoted in this Checklist are henceforth either from personal

observations or sourced from Sálim Ali and Sidney Dillion Ripley, *Handbook of the birds of India and Pakistan together with those of Nepal, Sikkim, Bhutan and Ceylon Vols. 1–10* (Bombay: OUP, 1968–1974), unless noted specifically for a species.

37. Kalpavriksh (Group Account), eBird checklist, 6 November 1993, https://ebird.org/view/checklist/S52219186.
38. From Hume and Marshall, *The Game Birds of India, Burmah, and Ceylon, Vol III,* (Calcutta, published by the authors, 1881), 77; Praveen, Jayapal and Pittie, 'Notes on Indian rarities-2: Waterfowl, diving waterbirds, gulls and terns,' *Indian BIRDS* 9, nos. 5 and 6 (2014). Specimens deposited at the British Museum of Natural History, London, numbered BMNH 1894.6.1.75–77.
39. See Harvey, Devadar and Grewal, *Atlas of the Birds of Delhi and Haryana* (New Delhi: Rupa and Co, 2006), 298. Later records are from eBird checklists.
40. From A.O. Hume, 'Notes,' *Stray Feathers* 8, nos. 2–5 (1879): 411. Also collected by W. M. Chill from Sultanpur, as subsequently clarified by Hume in *The Game Birds* (Calcutta, published by the authors, 1881, 231).
41. Frome, 'The Birds of Delhi and District,' 299, described it as 'occasional winter visitor'; specimen in the BNHS collection.
42. These later records are from Sudhir Vyas, 'Checklist of the Birds of the Delhi Region: An Update,' *Journal of the Bombay Natural History Society* 93, no. 2 (1996): 223; Abhinav and Dhadwal, 'Falcated Duck *Mareca falcata*: An addition to the avifauna of Himachal Pradesh,' *Indian BIRDS* 13, no. 6 (2017): 154–156.
43. From A.O. Hume, 'Notes,' *Stray Feathers* 10, nos. 1–3 (1887): 174; Ganguli, *A Guide to the Birds of the Delhi Area*, 48, respectively.
44. Harvey et al., *Atlas of the Birds of Delhi and Haryana*, 299. Baer's remnant wild populations are believed to be additionally threatened by hybridization with the Ferruginous Duck *A. nyroca* in particular, and these vagrant records would not have been scrutinized from this angle as comparison criteria for hybrids had not been developed.
45. From A.O. Hume, 'Notes,' *Stray Feathers* 10, nos. 1–3 (1887): 174.
46. From E.C. Benthall, 'The Birds of Delhi and District,' *Journal of the Bombay Natural History Society* 48, no. 2 (1949): 369.
47. See Mehta and Dogra, Snapshot sightings: Jungle Bush Quail from Faridabad, Haryana', *Indian BIRDS* 13, no. 5 (2017): 140A, recording presence of Jungle Bush Quail in the area after a long hiatus.
48. From Ganguli, *A Guide to the Birds of the Delhi Area*, 44.
49. See Tom J. Roberts, *The Birds of Pakistan: Regional Studies and Non-Passeriformes, Vol 1* (Karachi: OUP, 1991), 52–53.
50. See Rakesh Ahlawat, 'Horned Grebe *Podiceps auritus* at Dighal, Haryana, India,' *Indian BIRDS* 14, no. 4 (2018): 121; Praveen, Jayapal and Pittie, 'Notes on Indian rarities-2: Waterfowl, diving waterbirds, gulls and terns'; only the third record of this grebe from India.
51. These numbers are from Ganguli, *A Guide to the Birds of the Delhi Area*, 22.
52. See Sondhi et al., 'First record of the Pompadour ('Ashy-headed') Green Pigeon *Treron pompadora conoveri*/phayrei from Uttarakhand, India,' *Indian BIRDS* 11, no. 1 (2016): 21–22.
53. Harvey et al., *Atlas of the Birds of Delhi and Haryana*, 344.
54. It is also one of the few birds that relishes the poisonous fruit of the yellow oleander *Cascabela thevetica*.
55. Anthony Gaston and J. Mackrell, 'Green Munia Estrilda formosa at Delhi, and other interesting records for 1978,' *Journal of the Bombay Natural History Society* 77, no. 1 (1980): 44; Vyas, 'Some interesting bird records from the Delhi area,' *Journal of the Bombay Natural History Society* 99, no. 2 (2002): 328.
56. See Rasmussen and Anderton, *Birds of South Asia: the Ripley Guide, Vols I and II* (Washington, D.C. and Barcelona: Smithsonian Institution and Lynx Edicions, 2012), 228, for details of the split.
57. From Vyas, 'Checklist of the birds of the Delhi region: an update,' *Journal of the Bombay Natural History Society* 93, no. 2 (1996): 229.
58. From Frome, 'The Birds of Delhi and District,' 287.
59. From Ganguli, *A Guide to the Birds of the Delhi Area*, 152-153; Harvey et al., *Atlas of the Birds of Delhi and Haryana*, 303.
60. See Julian P. Donahue, 'Notes on a collection of Indian birds, mostly from Delhi,' *Journal of the Bombay Natural History Society* 64, no. 3 (1968): 418.
61. From Anthony J. Gaston, 'The seasonal occurrence of birds on the New Delhi Ridge,' *Journal of the Bombay Natural History Society* 75, no. 1 (1978a): 123.
62. These records are from Hutson, *The Birds About Delhi, Together with a Complete List of Birds Observed in Delhi and the Surrounding Country* (Delhi: The Delhi Bird Watching Society, 1954), 110; Nigel Redman in Harvey et al., *Atlas of the Birds of Delhi and Haryana*, 306; Kavi Nanda, 'A Sykes's Nightjar *Caprimulgus mahrattensis*, and a summer record of Eurasian Skylark *Alauda arvensis* from Delhi-NCR Region,' *Indian BIRDS* 15, no. 1 (2018): 27, respectively.
63. Donahue, 'Notes on a collection of Indian birds, mostly from Delhi,' 418.
64. Gaston, 'The seasonal occurrence of birds on the New Delhi Ridge,' 123.
65. From Frome, 'The Birds of Delhi and District,' (1948), 752.
66. Gaston, 'The seasonal occurrence of birds on the New Delhi Ridge,' 123; Ganguli, *A Guide to the Birds of the Delhi Area*, 166. These former summer records of

Indian Nightjar, including calling birds, are of interest compared with the apparent lack of reports of this species in recent years.

67. A recent study by Sangster, Mohan and Pandey (2021) concludes that three lineages can be distinguished within the 'Savanna Nightjar' complex based on detailed comparisons of their songs, and these are best considered as species: *C. monticolus* (Franklin's Nightjar; India to Thailand and southern China), *C. affinis* (Savanna Nightjar; Indonesia) and *C. griseatus* (Philippines) – an example of three valid species that have long been treated as a single species without a solid scientific basis.
68. Yogesh Parashar, personal communication with the author, e-mail dated 23 April 2018.
69. See, e.g., Hugh Whistler, 'Some notes on the Genus Caprimulgus (Nightjars) in the Punjab ["With a note on the nightjars of Sind by Dr C.B. Ticehurst"],' *Journal of the Bombay Natural History Society* 27 no. 2 (1920): 368–69; Frome, 'The Birds of Delhi and District,' 289; Julian Donahue, 'Notes on a collection of Indian birds, mostly from Delhi,' 418.
70. See Tom J. Roberts, *The Birds of Pakistan: Regional Studies and Non-Passeriformes,* 492; Prasad Ganpule, 'Roosting behaviour of Franklin's Nightjar Caprimulgus affinis,' *Indian BIRDS* 6, nos. 4 and 5 (2010): 92–94, for such flocking behaviour of Savanna Nightjars.
71. See Harvey et al., *Atlas of the Birds of Delhi and Haryana*, 304.
72. From Ganguli, *A Guide to the Birds of the Delhi Area*, 169; Vyas, 'Checklist of the birds of the Delhi region: an update,' 229.
73. From Vyas, 'Checklist of the birds of the Delhi region: an update,' 226.
74. From Julian Donahue, 'Notes on a collection of Indian birds, mostly from Delhi,' 414; Sudhir Vyas, 'Checklist of the birds of the Delhi region: an update,' 226.
75. From Ali and Ripley, *Handbook of the birds of India and Pakistan together with those of Nepal, Sikkim, Bhutan and Ceylon. Stone Curlews to Owls* (Bombay: OUP, 1969b), 160–61 and Rasmussen and Anderton, *Birds of South Asia: the Ripley Guide, Vols I and II*, 143.
76. Harvey et al., *Atlas of the Birds of Delhi and Haryana*, 99.
77. Anand Arya, photograph, in a personal communication with the author; also see Clive Harris, eBird checklist, 29 September 2007, https://ebird.org/view/checklist/S39926331.
78. From Ganguli, *A Guide to the Birds of the Delhi Area*, 866 and Vyas, 'Checklist of the birds of the Delhi region: an update,' 225.
79. These numbers are quoted from Ganguli, *A Guide to the Birds of the Delhi Area*, 125–126 and H.P.W Hutson, *The Birds About Delhi*, 158.
80. Personal observation by the author; Harvey et al., *Atlas of the Birds of Delhi and Haryana*, 310.
81. From Ganguli, *A Guide to the Birds of the Delhi Area*, 120–21.
82. From Bill Harvey et al., *Atlas of the Birds of Delhi and Haryana*, 123.
83. See Frome, 'The Birds of Delhi and District,' 295.
84. Harvey et al., *Atlas of the Birds of Delhi and Haryana*, 129.
85. See Graham Appleton, 'Following Sociable Lapwings,' *Wadertales* (blog), 3 January 2021, for an interesting study of Sociable Lapwings' migration routes.
86. These records, respectively, are from Frome, 'The Birds of Delhi and District,' 295; Hutson (1954), 164; Ganguli, *A Guide to the Birds of the Delhi Area*, 98; and Harvey et al., *Atlas of the Birds of Delhi and Haryana*, 311.
87. From S. Sridhar, eBird checklist, 5 March 2012, https://ebird.org/view/checklist/S96332519.
88. See Harkirat Singh Sangha, *Waders of the Indian Subcontinent* (Jaipur: Privately published, 2021), 148–152.
89. See Harvey et al., *Atlas of the Birds of Delhi and Haryana*, 311. The 1965 record is from Ganguli, *A Guide to the Birds of the Delhi Area*, 102.
90. For full account of field observations, see Vyas,' Checklist of the birds of the Delhi region: an update,' 227.
91. Personal observation by the author; see 'Checklist of the birds of the Delhi region: an update,' 226.
92. An unusual large concentration of more than 125 Painted Snipes together in four or five loosely scattered groups, mainly adults but including some juveniles, was reported by S. C. Sharma at Mandothi Jheel, Jhajjar district, on 10 June 2022, either a post-breeding gathering or birds concentrating in the few remaining pools in a desiccating habitat.
93. Tom J. Roberts (1991, 284) details the westward expansion of this species in Pakistan.
94. Ganguli, *A Guide to the Birds of the Delhi Area*, 113–114.
95. Maitreya Sukumar in a personal communication to the author.
96. From Harvey et al., *Atlas of the Birds of Delhi and Haryana*, 117.
97. From Harvey et al. *Atlas of the Birds of Delhi and Haryana*, 309.
98. For references to these records, see Ali and Ripley (1969, 307); Reyan Sofi, eBird checklist, 10 August 2021, https://ebird.org/view/checklist/S93032571; and Rohit Kumar Baldodia, eBird checklist, 24 September 2019, https://ebird.org/view/checklist/S60832608 [Accessed 30 June 2022].
99. Harvey et al., *Atlas of the Birds of Delhi and Haryana*, 309.
100. From Sharma et al., 'Long-billed-billed Dowitcher *Limnodromus scolopaceus* at Sultanpur National Park, Haryana, India,' *Indian BIRDS* 8, no. 4 (2013): 101–03; and Suresh C. Sharma, eBird checklist, 17 July 2020, https://ebird.org/view/checklist/S71568825.
101. See Harvey et al., *Atlas of the Birds of Delhi and Haryana*, 308; and the re-confirmation is by the observer, Bill Harvey in a personal communication.

102. These records are from Ganguli (1975, 112); Vyas (1996, 227); and Harvey et al. (2006, 309) respectively; the others are from eBird checklists.
103. From Sangha, *Waders of the Indian Subcontinent*, 427-430.
104. These two records are from Ganguli (1975, 119) and Harvey et al. (2006, 309), respectively.
105. These records are from Mridul Anand, eBird checklist, 23 July 2016, https://ebird.org/view/checklist/S30815969; Rakesh Ahlawat, eBird checklist, 27 June 2021, https://ebird.org/view/checklist/S90847293; and Kamal Sahansi, photograph in Facebook, 9 June 2022, https://www.facebook.com/photo/?fbid=10228224755272309&set=pcb.5283484911727078; also Vinod Gupta, eBird checklist, 19 June 2020, https://ebird.org/view/checklist/S113245286 [All accessed 30 June 2022].
106. From Harvey et al., *Atlas of the Birds of Delhi and Haryana*, 311.
107. See Ganguli, *A Guide to the Birds of the Delhi Area*, 128.
108. See Ganguli, *A Guide to the Birds of the Delhi Area*, 131.
109. See Harvey et al., *Atlas of the Birds of Delhi and Haryana*, 312; Praveen et al., 'Notes on Indian rarities–2: Waterfowl, diving waterbirds, gulls and terns,' (2014) for an analysis of this record.
110. From James Lambert, 'Snapshot sightings: Little Gull at Okhla, Uttar Pradesh,' *Indian BIRDS* 10, no. 1 (2015): 28A.
111. From 'Common Gull *Larus canus* Linnaeus recorded in India,' *Journal of the Bombay Natural History Society* 90, no. 3 (1994): 509–10; Praveen et al., 'Notes on Indian rarities–2: Waterfowl, diving waterbirds, gulls and terns,' (2014).
112. See Ganguli, *A Guide to the Birds of the Delhi Area*, 134; Vyas, 'Checklist of the birds of the Delhi region: an update,' 228.
113. These records are, respectively, from Ganguli (1975, 134–35); R.S.E. Swanqvist, eBird checklist, 2 February 1984, https://ebird.org/view/checklist/S57135231; Devasar and Suvarna (2018, 133); and Amit Sharma, eBird checklist, 2 April 2022, https://ebird.org/view/checklist/S106067843.
114. For first-hand accounts of the probable occurrence of this rarity in Delhi, see Alexander (1950, 120–21); Vyas (1996, 328); and Harvey et al. (2006, 314). Praveen et al., 'Notes on Indian rarities–2: Waterfowl, diving waterbirds, gulls and terns,' 113–136, analyses these reports.
115. From J.S. Waraich in a personal communication, and Ritwik Singh, eBird checklist, 5 July 2018, https://ebird.org/view/checklist/S46999465 respectively.
116. See Ganguli, *A Guide to the Birds of the Delhi Area*, 137.
117. From Frome (1947), 298.
118. See K.S. Gopi Sundar et al., 'Does the stork bring home the owl? Dusky Eagle-Owls *Bubo coromandus* breeding in Woolly-necked Stork *Ciconia episcopus* nests,' 1–5.
119. From Ganguli, *A Guide to the Birds of the Delhi Area*, 41.
120. These records may be accessed, respectively, from Ganguli (1975, 41); Peter Kaestner, eBird checklist, 9 July 1981, https://ebird.org/view/checklist/S66951729; Harvey et al. (2006, 321); Vyas (1996, 222).
121. From Urfi, *The Painted Stork: Ecology and Conservation*; Harvey et al., *Atlas of the Birds of Delhi and Haryana*, 187.
122. These details are from Frome (1948, 297); Ganguli (1975, 24); Vyas (1996, 221).
123. From Vyas, 'Checklist of the birds of the Delhi region: an update,' 221; Urfi (2003).
124. From Dinesh K. Singal, eBird checklist, 8 July 2018, https://ebird.org/view/checklist/S47063875 [Accessed 30 June 2022].
125. From Rajni Trivedi and B.M. Parasharya, 'Probable breeding of Little Bittern *Ixobrychus minutus* at Nalsarovar Bird Sanctuary, Gujarat, western India, with notes on identification of juveniles,' *Indian BIRDS* 15, no. 1 (2019): 17–20.
126. See S.C. Sharma, in Harvey et al., *Atlas of the Birds of Delhi and Haryana*, 320; Narender S. Khaira, eBird checklist, 26 June 2021, https://ebird.org/view/checklist/S90802352; and various observers' eBird checklists from Mandothi wetland, e.g., Ritwik Singh 1 September 2021, https://ebird.org/view/checklist/S94024069 [Accessed 30 June 2022].
127. Harvey et al., *Atlas of the Birds of Delhi and Haryana*, 172.
128. Personal observation by author.
129. From Koparde, Pankaj and Pierre Yesou, 'Interrelationship of birds and habitat features in urban greenspaces in Delhi, India,' *Urban Forestry & Urban Greening* 8, no. 3 (2009): 238–246.
130. From Ganguli, *A Guide to the Birds of the Delhi Area*, 34–35, and Harvey et al., *Atlas of the Birds of Delhi and Haryana*, 319.
131. See e.g. Ganguli, *A Guide to the Birds of the Delhi Area*, 43; Vyas, 'Checklist of the birds of the Delhi region: an update,' 222; Harvey et al., *Atlas of the Birds of Delhi and Haryana*, 185.
132. From Winkler et al., 'Hawks, Eagles, and Kites (Accipitridae)', version 1.0., in *Birds of the World*, ed. Billerman et al. (Ithaca, NY: Cornell Lab of Ornithology, 2022)
133. See Richard Cuthbert et al., 'Rapid population declines of Egyptian Vulture *Neophron percnopterus* and Red-headed Vulture *Sarcogyps calvus* in India,' *Animal Conservation* 9, no. 3 (2006): 349–354; and Galligan et al., 'Have population declines in Egyptian Vulture and Red-headed Vulture in India slowed since the 2006 ban on veterinary diclofenac?' *Bird Conservation International* 24, no. 3 (2014): 272–281, for useful studies on the Egyptian Vulture's decline in India.
134. See Vibhu Prakash et al., 'The population decline of *Gyps* vultures in India and Nepal has slowed since veterinary use of Diclofenac was banned,' *PloS ONE*

7, no. 11 (2012): 1–10, and Vibhu Prakash et al., 'Recent changes in populations of Critically Endangered *Gyps* vultures in India,' *Bird Conservation International*, 29, no. 1 (2017): 1–16, for this assessment.

135. From Richard Cuthbert et al., 'Rapid population declines of Egyptian Vulture *Neophron percnopterus* and Red-headed Vulture *Sarcogyps calvus* in India,' 349–354 and Galligan et al., 'Have population declines in Egyptian Vulture and Red-headed Vulture in India slowed since the 2006 ban on veterinary diclofenac?' 272–281.
136. See Frome (1947), 290; Hutson (1954), 124–125.
137. Surya Prakash in a personal communication with the author.
138. See Vibhu Prakash et al., 'Recent changes in populations of Critically Endangered *Gyps* vultures in India,' 1–16.
139. V.M. Galushin, 'A huge urban population of birds of prey in Delhi, India (Preliminary note),' *Ibis* 113, no. 4 (1971): 522.
140. Personal observation by the author.
141. See Vibhu Prakash et al., 'Recent changes in populations of Critically Endangered *Gyps* vultures in India,' 1–16.
142. From Suresh C. Sharma in a personal communication.
143. From Pankaj Gupta in a personal communication.
144. Devojit Das et al., 'Diclofenac is toxic to the Himalayan Vulture *Gyps himalayensis*,' *Bird Conservation International* 21, no. 1 (2011): 72–75.
145. Photographs posted by Ritwik Singh, Google Groups [Delhibirdpix], 03 December 2016, https://groups.google.com/forum/#!searchin/delhibirdpix/snake$20eagle%7Csort:date/delhibirdpix/F6QcYZbeA3w/w3AVX8F_DAAJ and https://groups.google.com/forum/#!searchin/delhibirdpix/snake$20eagle%7Csort:date/delhibirdpix/Gdt7e4OEUxg/ar-V7r9_DAAJ [Accessed on 30 June 2022].
146. Sydney Basil-Edwardes, 'A contribution to the ornithology of Delhi. Part 2.' *Journal of the Bombay Natural History Society* 31, no. 3 (1926b): 571.
147. To provide a sense of the physical and temporal spread of records of Black Eagles from the Delhi area, these include: single birds at the Aravalli Biodiversity Park, Delhi, in February 2012 and February 2022; near Mehrauli, south Delhi, in early November 2015 and November 2021; the Aravalli Biodiversity Park, Gurugram, in October 2018; from Southern Ridge landscapes at the Asola-Bhatti Wildlife Sanctuary in January 2019, December 2020, January–March 2022, and March 2023; Mangar Bani, Faridabad, in January 2013, September 2015, October 2019 and two together there in late January 2021; the Bhondsi Nature Park in February–March 2019; and Rewari, Haryana in October 2016. These are all from 'Aravalli' habitats in south Delhi and Haryana, but there are two reports from wooded wetlands in the river floodplain as well – at Surajpur in early November 2014, and the Yamuna khadar, Noida, in mid-November 2019.
148. From Suresh Sharma and Jaideep Chanda, 'Indian Spotted Eagle *Aquila hastata* nesting in Sonepat, Haryana, India,' *Indian BIRDS* 6 no. 1 (2010): 18; Rakesh Ahlawat in a personal communication.
149. From Sant et al., 'Breeding ecology of the Indian Spotted Eagle *Clanga hastata* around Belgaum, India,' *Indian BIRDS* 16, no. 6 (2020): 176–84.
150. From Ali and Ripley (1968), 276–278.
151. A photograph of a Tawny Eagle on a nest near the Red Fort was published in MacDonald (1960).
152. For references, see Frome, 'The Birds of Delhi and District,' 290; Hutson (1954), 128–129; 'Checklist of the birds of the Delhi region: an update,' 224; Harvey et al., *Atlas of the Birds of Delhi and Haryana*, 318.
153. The two nesting records are from Koshy, in a email dated 23 April 2018; and S. Rajagopal, report posted on Google Groups [Delhibirdpix], 20 November 2017, https://groups.google.com/forum/#!searchin/delhibirdpix/bonelli$27s$20eagle%7Csort:date/delhibirdpix/q9tqvNo9eDM/ZfGfMFjlAQAJ.
154. From Sangha et al., (2011), 139–140.
155. Records of Besra from the area include single birds at the Delhi Zoo in December 1997; on the South-Central Ridge in February 2017 and December 2018; and the Southern Ridge in November 2016 and October–November 2021; several from the Sultanpur National Park neighbourhood in November 2001, September 2010, December 2019 and February 2021; and from Sonipat, Haryana, in January 2020.
156. From Ali and Ripley, *Handbook of the birds of India and Pakistan together with those of Nepal, Sikkim, Bhutan and Ceylon. Divers to Hawks*, 233–34; Rishad Naoroji, *Birds of Prey of the Indian Subcontinent* (New Delhi: Om Books, 2006), 392–394.
157. From Sohail Madan, eBird checklist, 26 February 2022, https://ebird.org/view/checklist/S103733252. Identification since confirmed.
158. Videographed by Suresh C. Sharma, https://youtu.be/t5A5QxxV_0Q and posted on the Delhibirdpix Google Groups on 21 June 2022, with the comment 'Have been seeing Brahminy Kite pair carrying nesting material to their half-built nest, flying in circles around the trees, but never seen any evidence of their successful breeding in Haryana during nearly five decades of my birdwatching. This morning while returning from Kalesar/Hathni Kund ... we observed an adult Brahminy Kite feeding two juvenile birds which were demanding food making shrill cries near Indri (Karnal).'
159. For photographs of this wintering juvenile, see Rohit Sharma, eBird checklists, 27 February and 4 March 2021, https://ebird.org/view/checklist/S82424063 and https://ebird.org/view/checklist/S82795175 respectively.
160. From BirdLife International, 'Species factsheet: *Haliaeetus leucoryphus*,' http://www.birdlife.org on 29 November 2022.

161. See Hutson, *The Birds About Delhi*, 134, and Ganguli, *A Guide to the Birds of the Delhi Area*, 68–69, for these nesting records.
162. These records of Pallas' Fish Eagle in the 1980s are from: Vyas, (1996, 224–25 [at Okhla]); Debashish Chakravarti, provided in a personal communication, (at Sultanpur); and Harvey et al. (2006, 314) (at Bhindawas, reported by S. C. Sharma).
163. The data used here were accessed from Movebank (movebank.org, study name 'Pallas's Fish-eagle (Mongolia & Bangladesh, 2020–2022)', study ID 1233565100) on 30 June 2022.
164. Yogesh Parashar, in an e-mail dated 23 April 2018.
165. See Rohit Sharma, eBird checklists, 27 February and 4 March 2021, https://ebird.org/view/checklist/S82424063 and https://ebird.org/view/checklist/S82795175 respectively, for photographs of both these individuals. The immature bird was initially misidentified as Grey-headed Fish Eagle on account of streaked head, nape and breast, but identity confirmed by D. Karuthedathu, an expert on this species.
166. Suresh C. Sharma in a personal communication, and Harvey et al., *Atlas of the Birds of Delhi and Haryana*, 318.
167. Personal observation by the author; also see Purbasha Banerjee and Vibhu Prakash, 'Record Of Grey-Headed Fish-Eagle *Ichthyophaga ichthyaetus* from Bhindawas Wildlife Sanctuary, Haryana, India,' *Journal of the Bombay Natural History Society* 114 (2017) for a note recording details of the same bird.
168. From Eugene McCarthy, *Handbook of Avian Hybrids of the World*, (New York: OUP, 2006) 183; also Hans Peeters in an e-mail dated 26 October 2018.
169. From Tom J. Roberts, *The birds of Pakistan: Regional Studies and Non-Passeriformes, Vol 1*, 192.
170. From K.S. Gopi Sundar et al., 'Does the stork bring home the owl? Dusky Eagle-Owls *Bubo coromandus* breeding in Woolly-necked Stork *Ciconia episcopus* nests,' 1–5.
171. Abhinand Chandran et al., 'The Pallid Scops Owl *Otus brucei* in southwestern India, with notes on its identification,' *Indian BIRDS* 12, nos. 2 and 3 (2016): 56–63.
172. Records of Pallid Scops Owl from the area include single birds at their roosts along the Najafgarh Canal (January 2015) at the Asola Wildlife Sanctuary (December 2015 and late November 2016); at Okhla through much of January 2016; from the Aravalli outliers at Manethi, Rewari district (November–December 2020, and again October–December 2021); near Bhindawas, Jhajjar district (late November–December 2021) and from Delhi (December 2021, where an injured bird was received at Wildlife Rescue Centre, Wazirabad, Delhi, but did not survive).
173. These references are, respectively, from Basil-Edwardes (1926b, 570); Gaston (1978a, 123); Turaga (2014, 76–77); and from eBird checklists for Asola.
174. Anthony Gaston, 'The seasonal occurrence of birds on the New Delhi Ridge,' 125.
175. See K.S. Gopi Sundar et al., 'Does the stork bring home the owl? Dusky Eagle-Owls *Bubo coromandus* breeding in Woolly-necked Stork *Ciconia episcopus* nests,' 1–5, for this interesting study.
176. From Suresh C. Sharma in a personal communication with the author.
177. First reported by Rakesh Ahlawat, and recorded in several eBird checklists, e.g. by Pankaj Gupta (10 February 2020), https://ebird.org/view/checklist/S64310306, and Pradyumna Majumdar (10 November 2020), https://ebird.org/view/checklist/S76066081.
178. These earlier records are from Frome, 'The Birds of Delhi and District,' 288; and Gaston, K.S. Gopi Sundar et al., 'Does the stork bring home the owl? Dusky Eagle-Owls *Bubo coromandus* breeding in Woolly-necked Stork *Ciconia episcopus* nests,' 123.
179. See Yasser Arafat et al., 'Oriental Pied Hornbill *Anthracocerus albirostris*, and Indian Pitta *Pitta brachyura* in Delhi, India,' *Indian BIRDS* 10, no. 5 (2015): 132–133; among subsequent reports, single females were recorded at Shahpur Jat, New Delhi, in early April 2017 and late February 2018; at Mangar Bani, Faridabad, in mid-March 2018; the Delhi Cantonment, in early April 2019; at JNU campus, South Delhi, in mid-September 2019. Finally, one female was recorded at Mangar Bani, Faridabad district in September–October 2021. Most, if not all, of these may well pertain to the same individual female bird.
180. From Nikhil Devasar in a personal communication with the author.
181. See Himanshu Sharma, eBird checklist, 3 April 2022, https://ebird.org/view/checklist/S106151060.
182. These records are from Jaswinder Waraich (2017, 112A); and eBird checklists by Udiyaman Shukla (01 August 2020), https://ebird.org/view/checklist/S66469325; and Kousheyo Bagchi (21 July 2021), https://ebird.org/view/checklist/S92086455, respectively.
183. From Y.M. Rai, 'The birds of Delhi and Meerut,' *Journal of the Bombay Natural History Society* 83, no. 1 (1986): 213; and Suresh C. Sharma in a personal communication with the author.
184. From Narender S. Khaira, ebird checklist, 11 June 2017, https://ebird.org/view/checklist/S37522528.
185. From Harvey et al., *Atlas of the Birds of Delhi and Haryana*, 318; and Mike Prince, eBird checklist, 30 April 2003, https://ebird.org/view/checklist/S19462903.
186. Citations from Hugh Whistler, *Popular handbook of Indian birds* (London: Gurney and Jackson, 1949), 382–384; and Ali and Ripley, *Handbook of the birds of India and Pakistan together with those of Nepal, Sikkim, Bhutan and Ceylon. Divers to Hawks*, 344–46, respectively.
187. See photograph by P. Kundu, ebird checklist, 14 May 2018, https://ebird.org/view/checklist/S45913122.
188. Records of Large Cuckooshrikes from Delhi gardens

and the New Delhi Ridge fall mainly in the months July–March: thus, from the 1960s (Ganguli 1975); in August 1971 and March 1973 (Gaston 1978a); in April 1979 and a pair in end July 1996 (Vyas 1996, 2002); in February 1992, November–December 2002 and winter 2003–04, from the Sultanpur National Park (Harvey et al. 2006); in July 2008 (a pair), September 2012 and January 2016, from the Asola Wildlife Sanctuary; in mid-October from Okhla; and on 1 March 2023 at Rewari. They become progressively more frequent towards the east of the area, from Hastinapur Wildlife Sanctuary but still as a non-breeding visitor, recorded between December and March.

189. For these records, see Harvey et al., *Atlas of the Birds of Delhi and Haryana*, 323 and eBird checklists by Tanuj Seth (23 December 2021), https://ebird.org/view/checklist/S99680617; and Pritpal Panjeta (31 December 2021), https://ebird.org/view/checklist/S99928731, respectively.
190. From Y.M. Rai, 'The Birds of Delhi and Meerut,' 214.
191. Ganguli (1975) mentions just three records of Black-hooded Orioles from Delhi parks, in March 1951, 1967 and 1969–1970 winters; there are recent observations, supported with photographs, at Okhla in November 2008 and March 2009, all immatures (Devasar and Suvarna 2018); at Surajpur, GautamBuddha Nagar district in March 2019; at the Buddha Jayanti Park on the New Delhi Ridge, where two adults remained from end January 2020 to mid March 2020; at Noida in October 2020 and again in October 2022, both records of an immature; at the Asola-Bhatti Wildlife Sanctuary, south Delhi, in November 2020; at the Surajpur wetlands, U.P., in the third week of January and mid-February 2022; at the Yamuna *khadar* in north Delhi in late February 2022; at Bhondsi in June 2022; at Badkhal, Faridabad, in February 2023; and at Damdama, Gurugram, in early March 2023.
192. From, respectively, Basil-Edwardes, 'A contribution to the ornithology of Delhi. Part 1,' 263–64; Ganguli, *A Guide to the Birds of the Delhi Area*, 215; and Vyas, 'Checklist of the birds of the Delhi region: an update,' 231.
193. From Richard Grimmett et al., *Birds of the Indian Subcontinent* (London: Christopher Helm, A & C Black, 1998), 586–87; Prasad Ganpule, 'Red-backed, Brown, Isabelline and Red-tailed Shrike in Gujarat,' *Flamingo* 15, no. 3 (2017): 1–7.
194. See Harvey et al., *Atlas of the Birds of Delhi and Haryana*, 191.
195. See Puja Sharma (20 December 2015), eBird checklist https://ebird.org/view/checklist/S61084490.
196. From Bill Harvey in a personal communication e-mail dated 23 August 2018.
197. From Ganguli, *A Guide to the Birds of the Delhi Area*, 196.
198. See Ganpule, 'Notes on the Great Grey Shrike (Laniidae: *Lanius excubitor*) complex in northwestern India: Variation, identification, and status,' *Indian BIRDS* 11, no. 1 (2016): 1–10.
199. From Frome, 'The Birds of Delhi and District,' 278; Vyas (2002), 329; the Bhindawas record is by S. C. Sharma.
200. These two records are from Ganguli, *A Guide to the Birds of the Delhi Area*, 253–54 and Vyas, 'Checklist of the birds of the Delhi region: an update,' 233.
201. See Ali and Ripley, 1973b, 195–97.
202. Extract from Hugh Whistler, *Popular handbook of Indian birds,* 257. The Singing Bushlark *Mirafra cantillans* has been recently lumped with Australasian Bushlark *M. javanica* on genetic evidence, despite variations in plumage. This lump has been accepted as *M. javanica*, widespread from Africa to Australia, with 20 subspecies in two groups (*cantillans* group with four races from Africa to India, and *javanica* group with 16 races from Southeast Asia to Australia).
203. From Harvey et al., *Atlas of the Birds of Delhi and Haryana*, 338.
204. See Martin Kelsey, 'Sight record of Horned Lark *Eremophila alpestris* near Delhi,' *Journal of the Bombay Natural History Society* 101, no. 2 (2004): 321, for this extraordinary record.
205. See Alström et al., 'Multilocus phylogeny of the avian family Alaudidae (larks) reveals complex morphological evolution, non-monophyletic genera and hidden species diversity,' *Molecular Phylogenetics and Evolution* 69, no. 3 (2013): 1043–56; Martin Stervander et al., 'Multiple instances of paraphyletic species and cryptic taxa revealed by mitochondrial and nuclear RAD data for *Calandrella* larks (Aves: Alaudidae),' *Molecular Phylogenetics and Evolution* 102 (2016): 233–45.
206. From Harvey et al., *Atlas of the Birds of Delhi and Haryana*, 339.
207. See Hemant Kirola, eBird checklist, 31 December 2016, https://ebird.org/view/checklist/S33304425; and Nitu Sethi, '*Asian/Lesser Short-toed Lark-probably the 1st record from India*,' in Google Groups, 11 January 2017, https://groups.google.com/forum/#!msg/delhibirdpix/WKLXIipzquo/8wVMLSXuAQAJ;context-place=forum/delhibirdpix.
208. These citations are from Christian (2019, 81–84) and Ganpule (2019, 97–111). The 'Lesser Short-toed Lark complex' is a widely distributed group spread across the drier biotopes of the Mediterranean and Central Asia, and recent studies (2020) indicate that it is better treated as four species – the Mediterranean Short-toed Lark *Alaudala rufescens* (with three races); the Turkestan Short-toed Lark *Alaudala heinei* (with three races in Central Asia, *heinei, aharonii, persica); the* Asian Short-toed Lark *Alaudala cheleensis* (with four races in Kazakhstan east to Mongolia, *cheleensis, tuvinica, leucophaea, kukunoorensis)*; and the Sand Lark *Alaudala raytal* of the Subcontinent (which is also part of this complex).
209. See Kavi Nanda, 'A Sykes's Nightjar *Caprimulgus*

mahrattensis, and a summer record of Eurasian Skylark *Alauda arvensis* from Delhi-NCR Region,' 27.

210. See Alström et et al., 'Morphology vocalizations, and mitochondrial DNA suggest that the Graceful Prinia is two species,' *Ornithology* 138, no 2. (2021) for details of the split.
211. From Horace Gundy Alexander, 'The Birds of Delhi and District,' *Journal of the Bombay Natural History Society* 48, no. 2 (1949): 370; Ganguli, *A Guide to the Birds of the Delhi Area*, 230–231.
212. See Gaston, 'The seasonal occurrence of birds on the New Delhi Ridge,' 123–24; Sudhir Vyas, 'Four additions to the Delhi list and other interesting records,' *Newsletter for Birdwatchers* 19, no. 11 (1979): 2–5.
213. Y.M. Rai, 'The Birds of Delhi and Meerut,' 213.
214. Pankaj Gupta in a personal communication with the author.
215. Harvey et al., *Atlas of the Birds of Delhi and Haryana*, 247. In the *Handbook*, Vol. 8 (1973a, 27–29), Ali and Ripley mention circumstantial evidence of their nesting in Punjab and the Kumaon *terai* in the past.
216. See Donahue, 'Notes on a collection of Indian birds, mostly from Delhi,' 425–26, for the first record of the Bristled Grassbird from Delhi; and Hutson, *The Birds About Delhi*, 7, for the quote.
217. See Vyas, 'Checklist of the birds of the Delhi region: an update,' 232, and Y.M. Rai, 'The Birds of Delhi and Meerut,' 213.
218. See Rasmussen and Anderton, *Birds of South Asia: the Ripley Guide Vols I and II*, 309–10 and Donahue, 'Notes on a collection of Indian birds, mostly from Delhi,' 421.
219. From Maitreya Sukumar, in a personal communication, e-mail dated 12 May 2022.
220. See Tom J. Roberts, *The Birds of Pakistan. Passeriformes: Pittas to Buntings, Vol 2*. (Karachi: OUP, 1992), 32–34.
221. From Douglas Ball, eBird checklist, 31 January 2020, https://ebird.org/view/checklist/S63956801 [Accessed 30 June 2022].
222. From 'The Birds of Delhi and District,' 279, and Ganguli, *A Guide to the Birds of the Delhi Area*, 216.
223. See Sankar et al., 'Birds of Sariska Tiger Reserve, Rajasthan, India,' *Forktail* 8 (February, 1993); Shahabuddin et al., 'Annotated checklist of the birds of Sariska Tiger Reserve, Rajasthan, India,' *Indian BIRDS* 2, no. 3 (2006): 71–76.
224. These historical records are from Ganguli, *A Guide to the Birds of the Delhi Area*, 239, and Gaston, 'The seasonal occurrence of birds on the New Delhi Ridge,' 124.
225. The closely related Dusky Warblers *Phylloscopus fuscatus* have also been claimed from the Delhi area but the evidence is not compelling. They are East Asian breeders, and winter in India (mostly towards the east) and in Southeast Asia. Like Smoky Warblers *P. fuligiventer*, they also tend to wander west- and southwards of their usual winter range on occasion.
226. Validated records of Smoky Warblers from the area include single birds at Mandnaka Jheel, Palwal, Faridabad district, on 15 February 2015; at the Basai marshes, Gurugram district, where one spent the 2014–2015 winter, December to early May; in GautamBuddha Nagar district on 3 December 2017; at Bohar village, Rohtak, on 24 April 2019 and Mandothi wetlands, Jhajjar district (where one spent the winter from late November 2020 to end-January 2021), both in Haryana; at Surajpur lake on 20 November 2022, and Dhanauri on 5 February 2023, both sites in GautamBuddha Nagar district.
227. From Harvey et al., *Atlas of the Birds of Delhi and Haryana*, 334.
228. From Majumdar et al., 'Green-crowned Warbler *Phylloscopus burkii*, a new species for Haryana, and Delhi NCR,' *Indian BIRDS* 17, no. 5 (2021): 157–58
229. Records of Whistler's Warblers from the area include: Hastinapur in March 1979 (Rai 1986, 219); the Sultanpur National Park in winter 2009–2010; Bhindawas Sanctuary in late December 2010; the Sultanpur National Park neighbourhood in end-October 2014 and in January–February 2016; and from woodland areas on the Southern Ridge (Mangar Bani, Faridabad and Sanjay Van) in end-February 2016, late January 2017 and late November 2020 (for details, see eBird checklists).
230. Records of Large-billed Leaf Warbler: from the New Delhi Ridge, including birds mist-netted in September in 1971 and 1972 (Gaston 1978a), on 9 September 2000 (Clive Harris, in Harvey et al. 2006) and 8 September 2011; from the Southern Ridge at Mangar Bani, Faridabad district (which appears to be a favoured site for this warbler), on 8 September 2019, 20 October 2019, 25 September 2021 and a report of several birds calling freely on 5 October 2021, these latter documented in eBird checklists.
231. Nikhil Devasar in a personal communication with the author.
232. These records are from Gaston (1978a, 124); Harvey et al. (2006, 337); and Douglas Ball, eBird checklist, 7 October 2018, https://ebird.org/view/checklist/S49036568 [Accessed 30 June 2022].
233. A recent taxonomic reassessment has reorganized the 13 races formerly included in 'Oriental White-eye' into five species, *viz.* Indian White-eye *Z. palpebrosus* with seven subspecies, with the remaining subspecies regrouped as part of four other species of East and Southeast Asian white-eyes. Only the first occurs in India.
234. In MacDonald, *Birds in my Indian Garden* (London: Jonathan Cape, 1960) Chapter 10.
235. See Gaston (1978c).
236. From Anthony Gaston, 'Notes on the Striated Babbler *Turdoides earlei* (Blyth) near Delhi,' *Journal of the Bombay Natural History Society* 75, no. 1 (1978b): 219–220.
237. These references are from Ganguli (1975, 255); Gaston (1978a, 124); and Harvey et al. (2006, 331) respectively.

238. From Vyas, 'Checklist of the birds of the Delhi region: an update,' 233.
239. From Y.M. Rai, 'The Birds of Delhi and Meerut,' 213.
240. From Pankaj Gupta and M.S. Srinivas, 'Snapshot sightings: Scaly Thrush from Faridabad, Haryana,' *Indian BIRDS* 13, no. 2 (2017): 56A.
241. From Harvey et al., *Atlas of the Birds of Delhi and Haryana*, 326. This is one of just four records in India outside its Himalayan range: Lahore, Punjab in January 1903; Sultanpur, Haryana in February 2002; Pachmarhi, Madhya Pradesh in February 2003; and Kachchh, Gujarat in December 2010 (Mishra 2015, 161).
242. Chhabra et al., 'Song Thrush *Turdus philomelos*: A first record for Delhi NCR,' *Indian BIRDS* 13, no. 2 (2017): 55.
243. See Ali and Ripley, 1972b, 145–148; and Rasmussen and Anderton, *Birds of South Asia: the Ripley Guide, Vol II* (2012), 373.
244. See S.C. Sharma, photograph in Birds of Haryana Facebook Group, 24 October 2016, https://www.facebook.com/photo.php? fbid=10210182579499213&set=gm.115151854 4923756&type=3&theater&ifg=1.
245. From Gaston (1978a), 123, and eBird checklists, Sohail Madan (17 April 2018), https://ebird.org/view/checklist/S44658538 [Accessed 30 June 2022] and Arvind Yadav, 10 October 2022, https://ebird.org/checklist/S120004871 [Accessed 10 October 2022].
246. From G. Shahabuddin and Amita Baviskar, 'Birdlife: Heirs of the ecological mosaic,' in *First Garden of the Republic: Nature in the President's Estate*, ed. Amita Baviskar (New Delhi: Publications Division, Government of India, 2016).
247. From Gaurav Kumar, eBird checklist, 1 October 2022, https://ebird.org/india/checklist/S119747803 [Accessed 6 October 2022].
248. From S. C. Sharma in an e-mail dated 29 October 2018.
249. These records are from Frome, 'The Birds of Delhi and District,' 280; and Vyas, 'Checklist of the birds of the Delhi region: an update,' 233.
250. See Luo et al., 'Deep phylogeographic divergence of a migratory passerine in Sino-Himalayan and Siberian forests: The Red-flanked Bluetail *Tarsiger cyanurus* complex,' *Ecology and Evolution*, 4, no. 7 (2014).
251. References for these records are in Pritpal Panjeta, eBird checklists, 21 November 2020 and 07 March 2021, https://ebird.org/view/checklist/S76541465 and https://ebird.org/view/checklist/S82920945; and Kavi Nanda, eBird checklist, 22 November 2020, https://ebird.org/view/checklist/S76530795.
252. From Ganguli, *A Guide to the Birds of the Delhi Area*, 223–224; and Gaston, 'The seasonal occurrence of birds on the New Delhi Ridge,' 123.
253. See Svensson (2005), 538-541.
254. See Sanjay Sharma, photograph in Google Groups [Delhibirdpix], 14 September 2015, https://groups.google.com/forum/#!searchin/delhibirdpix/kashmir$20flycatcher%7Csort:date/delhibirdpix/lWv-7sC5VPI/jZxG6PxjAQAJ [Accessed 30 June 2022].
255. Recorded by S.C. Sharma, in Harvey et al., *Atlas of the Birds of Delhi and Haryana*, 329.
256. Recorded by S.C. Sharma, in Harvey et al., *Atlas of the Birds of Delhi and Haryana*, 329.
257. From Harvey et al., *Atlas of the Birds of Delhi and Haryana*, 328.
258. See Asad Rahmani, 'Status and distribution of White-browed Bushchat *Saxicola macrorhyncha* in India,' *Forktail* 12 (August 1997): 61–77; and Suresh Sharma and P.S. Sangwan, 'Stoliczka's Bushchat *Saxicola macrorhyncha* in Hissar District, Haryana,' *Indian BIRDS* 1, no. 1 (2005): 6–7.
259. From Harvey et al., *Atlas of the Birds of Delhi and Haryana*, 215.
260. From Sunil Kumar, 'Snapshot sightings: Stoliczka's Bushchat from Delhi,' *Indian BIRDS* 12, nos. 2 and 3 (2016): 88A.
261. See Vyas, 'Checklist of the birds of the Delhi region: an update,' 233; and Rai (1986), 13.
262. From E. Sridharan and S. Bikhchandani, 'The Redtailed Wheatear *Oenanthe xanthoprymna* in the Delhi area,' *Journal of the Bombay Natural History Society* 78, no. 1 (1981): 170; Mohit Mehta and Piyush Dogra, 'Snapshot sightings: Red-tailed Wheatear in Sultanpur, Delhi NCR,' *Indian BIRDS* 14, no. 3 (2018b): 96A; Harvey et al., *Atlas of the Birds of Delhi and Haryana*, 329.
263. From Y.M. Rai, 'The Birds of Delhi and Meerut,' 213–14, and R. Bhargava, *Status of Finn's Weaver in India: Past & Present* (Mumbai: Bombay Natural History Society, 2017).
264. From Harvey et al., *Atlas of the Birds of Delhi and Haryana*, 342.
265. See Menon et al. (2021), 174–178.
266. See Gaston and Mackrell, 'Green Munia *Estrilda formosa* at Delhi, and other interesting records for 1978,' 144–145.
267. Ali and Ripley, (1974), 106.
268. From Restall (1997).
269. These historical records are from Benthall (1949, 368), and Hutson (1954, 68).
270. See Bill Harvey and Suresh Sharma, 'The initial colonisation of the Yamuna flood plain by the Sind Sparrow *Passer pyrrhonotus*,' *Journal of the Bombay Natural History Society* 99, no. 1 (2002): 35–43.
271. See Harvey et al., *Atlas of the Birds of Delhi and Haryana*, 275.
272. These records are from Ganguli (1975), 262; Arora (2017), 26–27; and eBird checklists, Narender Khaira 28 February 2020, https://ebird.org/view/checklist/S65233901, and Mohit Mehta (2 October 2022), https://ebird.org/checklist/S119827224 [Accessed 6 October 2022].
273. These records are from Jackson (1965, 304–05); S.C. Sharma (2005, 70–71); and Anand Arya, eBird checklist, 12 April 2009, https://ebird.org/view/checklist/S96356941 [Accessed 30 June 2022].
274. From Y.M. Rai, 'The Birds of Delhi and Meerut,' 213.

275. From Robson (1998), 44, and Pritpal Panjeta, eBird checklist of 3 April 2016, https://ebird.org/view/checklist/S28894943, respectively.
276. From Shivani Sharma, eBird checklist, 13 March 2022, https://ebird.org/view/checklist/S104724989.
277. See, e.g., Harvey et al., *Atlas of the Birds of Delhi and Haryana*, 296.
278. See Harvey et al., *Atlas of the Birds of Delhi and Haryana*, 343.
279. From Gaston, 'The seasonal occurrence of birds on the New Delhi Ridge,' 124. Gaston himself seems to have realized the discrepancy between his and other observers' assessments and noted that 'the status accorded to the species in the checklist is puzzling, and it may have become more abundant recently.'
280. See Gaston, 'The seasonal occurrence of birds on the New Delhi Ridge,' 124.
281. Ganguli, *A Guide to the Birds of the Delhi Area*, 279–80.
282. Harvey et al., *Atlas of the Birds of Delhi and Haryana*, 343.
283. Bill Harvey, in forwarded e-mail dated 14 June 2018.
284. E.S. Lewis, 'Bewick's Swan (*Cygnus bewickii* Yarrell) near Delhi,' *Journal of the Bombay Natural History Society* 40, no. 2 (1938), is the original published record.
285. Praveen et al., 'Notes on Indian rarities–2: Waterfowl, diving waterbirds, gulls and terns,' (2014), examines this record in his notes on Indian rarities, in *Indian BIRDS*.
286. Detailed in E.C.S Baker, *The Game-Birds of India, Burma and Ceylon: Ducks and their allies (swans, geese and ducks)* (London: Bombay Natural History Society, 1921), 303–304; also Ganguli, *A Guide to the Birds of the Delhi Area*, 58.
287. Detailed notes (with original records) in Blewitt (1875), Chill (1887) and Praveen (n.d.).
288. See Ganguli, *A Guide to the Birds of the Delhi Area*, 139.
289. From H.P.W. Hutson, *The Birds About Delhi, Together with a Complete List of Birds Observed in Delhi and the Surrounding Country* (Delhi: The Delhi Bird Watching Society, 1954), 154.
290. From Hume and Marshall (1881), Vol III, 11.
291. From A.O. Hume and C.H.T. Marshall, *The Game Birds of India, Burmah, and Ceylon* (Calcutta: Published by the authors, 1879), Vol I, 7.
292. See Ganguli, *A Guide to the Birds of the Delhi Area*, 41.
293. Ganguli (1975, 27), refers to a historical observation of three parties of five birds each of all three pelican species at Sultanpur, enabling direct comparison.
294. See Ganguli, *A Guide to the Birds of the Delhi Area*, 70.
295. From Frome, 'The Birds of Delhi and District,' 290; Ganguli, *A Guide to the Birds of the Delhi Area*, 77.
296. From Frome, 'The Birds of Delhi and District,' 281.
297. From Frome, 'The Birds of Delhi and District,' 287.
298. See Ganguli, *A Guide to the Birds of the Delhi Area*, 218.
299. From Anthony Gaston, 'Black Bulbuls *Hypsipetes madagascariensis* (P.L.S. Müller) in Delhi,' *Journal of the Bombay Natural History Society* 69, no. 3 (1973a) and 'The seasonal occurrence of birds on the New Delhi Ridge'.
300. See Ganguli, *A Guide to the Birds of the Delhi Area*, 239.
301. See N.S. Tyabji, 'Birds of New Delhi area,' *Newsletter for Birdwatchers* 2, no. 2 (1962), for a first-hand account of this record.
302. See Hutson, *The Birds About Delhi*, 16, and Ganguli, *A Guide to the Birds of the Delhi Area*, 256.
303. From Hutson, *The Birds About Delhi*, 13
304. From Hume, 'Notes,' *Stray Feathers* 7, nos. 3, 4 and 5 (1878); see also Ganguli, *A Guide to the Birds of the Delhi Area*, 276.
305. See Ganguli, *A Guide to the Birds of the Delhi Area*, 30 and Jackson, 'Some new bird records for Delhi,' *Journal of the Bombay Natural History Society* 65, no. 3 (1969), 782.
306. See Ali and Ripley (1968 and 1978), 53.
307. From Ganguli, *A Guide to the Birds of the Delhi Area*, 115.
308. In Harvey et al., *Atlas of the Birds of Delhi and Haryana*, 313.
309. See Ganguli, *A Guide to the Birds of the Delhi Area*, 65.
310. Hutson, *The Birds About Delhi*, 128.
311. Hans Peeters in an email dated 1 June 2016.
312. These reports are, respectively, in Ganguli (1975, 63); Gaston (1978a, 122); and Harvey et al. (2006, 317).
313. By D.K. Singal, blog report, 21 March 2015, http://indianbirdsphotography.blogspot.in/2015/03/delhibirdpix-upland-buzzard_21.html.
314. From Y.M. Rai, 'The Birds of Delhi and Meerut,' 213.
315. Ganguli, *A Guide to the Birds of the Delhi Area*, 163, and *The Birds About Delhi*.
316. From L. Khacher, 'Birding with Sálim Ali,' *Hornbill* (4) (1995).
317. Devasar and Suvarna, *Birds About Delhi: A Field Guide* (New Delhi: Dorling Kindersley, 2018), 187, accept this record.
318. From Y.M. Rai, 'The Birds of Delhi and Meerut,' 213.
319. Ganguli, *A Guide to the Birds of the Delhi Area*, 198.
320. From Rasmussen and Anderton, *Birds of South Asia: the Ripley Guide*, Vols. I and II, 487.
321. Ganguli, *A Guide to the Birds of the Delhi Area*, 231.
322. Devasar and Suvarna, *Birds About Delhi: A Field Guide*, 280, accept this species on the Delhi list, and categorize it as a passage migrant.
323. Ganguli, *A Guide to the Birds of the Delhi Area*, 267.

References

Abdulali, Humayun, and Jamshed D. Panday. *Checklist of the Birds of Delhi, Agra and Bharatpur with Notes on their Status in the Neighbourhood*. Bombay: Published by Humayun Abdulali, 1978.

Abhinav, C., and Devinder Singh Dhadwal. "Falcated Duck *Mareca falcata*: An addition to the avifauna of Himachal Pradesh." *Indian BIRDS* 13, no. 6 (2017): 154–56.

Abhinav, C., and Piyush Dogra. "Rufous-tailed Rock Thrush *Monticola saxatilis* at Pong Lake, Himachal Pradesh, and its status in India." *Indian BIRDS* 17, no 5. (2021): 135–38

Ahlawat, Rakesh. "Horned Grebe *Podiceps auritus* at Dighal, Haryana, India." *Indian BIRDS* 14, no. 4 (2018): 121.

Ahlawat, Rakesh, and Anshul Jain. "Snapshot sightings: Spotted Crake at Dighal, Haryana." *Indian BIRDS* 9, no. 4 (2014): 112A.

Ahlawat, Rakesh, Chetna Sharma, and Sonu Dalal. "White-naped Tit *Machlolophus nuchalis* in Haryana." *Indian BIRDS* 14, no. 4 (2018): 125.

Ahlawat, Rakesh, and Sanjay Tiwari. "Snapshot sightings: Eurasian Bittern at Dighal, Haryana." *Indian BIRDS* 9, no. 3 (2014): 84.

Alexander, Horace Gundy. "The Birds of Delhi and District". *Journal of the Bombay Natural History Society* 48, no. 2 (1949): 370–372.

Alexander, Horace Gundy. "Possible occurrence of the Black Tern [*Chlidonias niger* (L.)] near Delhi". *Journal of the Bombay Natural History Society* 49, no. 1 (1950): 120–121.

Alexander, Horace Gundy. "Rednecked Phalarope near Delhi." *Journal of the Bombay Natural History Society* 51, no. 2 (1953): 507–508.

Alexander, Horace Gundy. "Return to Delhi." *Newsletter for Birdwatchers* 4, no. 1 (1964): 1–3.

Alexander, Horace Gundy (1974). *Seventy Years of Birdwatching*. 1st ed. Berkhamsted: T. & A. D. Poyser.

Ali, Sálim, and Sidney Dillon Ripley. *Handbook of the birds of India and Pakistan together with those of Nepal, Sikkim, Bhutan and Ceylon. Divers to Hawks*. 1st ed. Vol. 1 of 10 vols.Bombay: Oxford University Press (OUP), 1968.

Ali, Sálim, and Sidney Dillon Ripley. *Handbook of the birds of India and Pakistan together with those of Nepal, Sikkim, Bhutan and Ceylon. Megapodes to Crab Plover*. 1st ed. Vol. 2 of 10 vols. Bombay: OUP, 1969a.

Ali, Sálim, and Sidney Dillon Ripley. *Handbook of the birds of India and Pakistan together with those of Nepal, Sikkim, Bhutan and Ceylon. Stone Curlews to Owls*. 1st ed. Vol. 3 of 10 vols. Bombay: OUP, 1969b.

Ali, Sálim, and Sidney Dillon Ripley. *Handbook of the birds of India and Pakistan together with those of Nepal, Sikkim, Bhutan and Ceylon. Frogmouths to Pittas*. 1st ed. Vol. 4 of 10 vols. Bombay: OUP, 1970.

Ali, Sálim, and Sidney Dillon Ripley. *Handbook of the birds of India and Pakistan together with those of Nepal, Sikkim, Bhutan and Ceylon. Cuckoo-shrikes to Babaxes*. 1st ed. Vol. 6 of 10 vols. Bombay: OUP, 1971.

Ali, Sálim, and Sidney Dillon Ripley. *Handbook of the birds of India and Pakistan together with those of Nepal, Sikkim, Bhutan and Ceylon. Larks to the Grey Hypocolius*. 1st ed. Vol. 5 of 10 vols. Bombay: OUP, 1972a.

Ali, Sálim, and Sidney Dillon Ripley. *Handbook of the birds of India and Pakistan together with those of Nepal, Sikkim, Bhutan and Ceylon. Laughing Thrushes to the Mangrove Whistler*. 1st ed. Vol. 7 of 10 vols. Bombay: OUP, 1972b.

Ali, Sálim, and Sidney Dillon Ripley. *Handbook of the birds of India and Pakistan together with those of Bangladesh, Nepal, Sikkim, Bhutan and Sri Lanka. Warblers to Redstarts*. 1st ed. Vol. 8 of 10 vols. Bombay: OUP, 1973a.

Ali, Sálim, and Sidney Dillon Ripley. *Handbook of the birds of India and Pakistan together with those of Bangladesh, Nepal, Sikkim, Bhutan and Sri Lanka. Robins to Wagtails*. 1st ed. Vol. 9 of 10 vols. Bombay: OUP, 1973b.

Ali, Sálim, and Sidney Dillon Ripley. *Handbook of the birds of India and Pakistan together with those of Bangladesh, Nepal, Sikkim, Bhutan and Sri Lanka. Flowerpeckers to Buntings*. 1st ed. Vol. 10 of 10 vols Bombay: OUP, 1974.

Ali, Sálim, and Sidney Dillon Ripley. *Handbook of the birds of India and Pakistan together with those of Bangladesh, Nepal, Bhutan and Sri Lanka*. 2nd ed. Vols. 1–10. Bombay: OUP, 1978–1999.

Alström, Per. "Common Gull *Larus canus* Linnaeus recorded in India." *Journal of the Bombay Natural History Society* 90, no. 3 (1994): 509–10.

Alström, Per, Keith N. Barnes, Urban Olsson, F. Keith Barker, Paulette Bloomer, Aleem Ahmed Khan, Masood Ahmed Qureshi, Alban Guillaumet, Pierre-André Crochet, Peter G. Ryan. "Multilocus phylogeny of the avian family Alaudidae (larks) reveals complex morphological evolution, non-monophyletic genera and hidden species diversity." *Molecular Phylogenetics and Evolution* 69, no. 3 (2013): 1043–1056.

Alström, Per, Pamela C. Rasmussen, Canwei Xia, Lijun Zhang, Chengyi Liu, Jesper Magnusson, Arya Shafaeipour, Urban Olsson. "Morphology, vocalizations, and mitochondrial DNA suggest that the Graceful Prinia is two species." *Ornithology* 138, no 2. (2021). https://doi.org/10.1093/ornithology/ukab014.

Anonymous. *Gazetteer of the Delhi District 1883–84*. Calcutta: Punjab Government, 1884.

Anonymous. Birds of Delhi and District, Field Checklist. Delhi: Delhi Birdwatching Society, 1967.

Appleton, Graham. "Following Sociable Lapwings." *Wadertales* (blog) 3 January 2021. https://wadertales.wordpress.com/2021/01/03/following-sociable-lapwings/.

Arafat, Yasser, M. Shah Hussain and Aisha Sultana. "Oriental Pied Hornbill *Anthracocerus albirostris*, and Indian Pitta *Pitta brachyura* in Delhi, India." *Indian BIRDS* 10, no. 5 (2015): 132–133.

Arora, Rohit. "First definitive documentation of Forest Wagtail *Dendronanthus indicus* from Okhla Bird Sanctuary, Delhi NCR." *Indian BIRDS* 13, no. 1 (2017): 26–27.

Badhwar, Fateh Chand. "Obituary. Lt. Gen. Sir Harold Williams (1897–1971)." *Journal of the Bombay Natural History Society* 69, no. 2 (1972): 395–397.

Baker, E.C.S. *The Game-Birds of India, Burma and Ceylon: Ducks and their allies (swans, geese and ducks)*. 2nd ed. Vol. 1 of 3 vols. London: Bombay Natural History Society, 1921.

Banerjee, Purbasha, and Vibhu Prakash. "Record Of Grey-Headed Fish-Eagle *Ichthyophaga ichthyaetus* from Bhindawas Wildlife Sanctuary, Haryana, India." *Journal of the Bombay Natural History Society* 114 (2017): 42.

Basil-Edwardes, Sydney. "A contribution to the ornithology of Delhi. Part 1." *Journal of the Bombay Natural History Society* 31, no. 2 (1926a): 261–273.

Basil-Edwardes, Sydney. "A contribution to the ornithology of Delhi. Part 2." *Journal of the Bombay Natural History Society* 31, no. 3 (1926b): 567–578.

Baviskar, Amita, ed. *First Garden of the Republic: Nature in the President's Estate*. Delhi: Publications Division, Government of India, 2016.

Benthall, Edward Charles. "The Birds of Delhi and District". *Journal of the Bombay Natural History Society* 48, no. 2 (1949): 368–370.

Bhargava, Rajat. *Status of Finn's Weaver in India: Past & Present*. Mumbai: Bombay Natural History Society, 2017.

Bhatt, Nirav, and Prasad Ganpule. "The identification of the Red-naped Shaheen *Falco peregrinus babylonicus*, its separation from *F. p. calidus* in the field, and its status and distribution in northwestern India." *Indian BIRDS* 13, no. 4 (2017): 85–92.

Billerman, Shawn M., Brooke K. Keeney, Paul G. Rodewald, and Tom S. Schulenberg, eds. *Birds of the World*. Ithaca, NY: Cornell Laboratory of Ornithology, 2022. Accessed 30 June 2022. https://birdsoftheworld.org/bow/home.

BirdLife International. "The flyways concept can help coordinate global efforts to conserve migratory birds." 2010. http://datazone.birdlife.org/sowb/casestudy/the-flyways-concept-can-help-coordinate-global-efforts-to-conserve-migratory-birds.

BirdLife International. *East Asia/East Africa. BirdLife Flyway Factsheet 6*. Cambridge: Birdlife International, 2010a.

BirdLife International. *Central Asia/South Asia. BirdLife Flyway Factsheet 7*. Cambridge: Birdlife International, 2010b.

BirdLife International. *East Asia/Australasia. BirdLife Flyway Factsheet 8*. Cambridge: Birdlife International, 2010c.

BirdLife International. Species factsheet: *Haliaeetus leucoryphus*. 2022. http://www.birdlife.org.

Blewitt, Robert. F. Letters to the Editor. ['On the evening of this 29th instant, settled in cultivated fields . . .']. *Stray Feathers* 3 (1, 2 and 3): 268. [See also: Praveen, J., (n.d.) https://ebird.org/view/checklist/S42116096.]

Bombay Natural History Society (BNHS). "Announcements: The Delhi Bird-watching Society." *Journal of the Bombay Natural History Society* 49, no. 3 (1950): 595.

Chandran, Abhinand, Pamela C. Rasmussen, Shah Jahan and Praveen J. "The Pallid Scops Owl *Otus brucei* in southwestern India, with notes on its identification." *Indian BIRDS* 12, nos. 2 and 3 (2016): 56–63.

Chhabra, Sheila, Hemant Kirola, and Nitu Sethi. "Song Thrush *Turdus philomelos*: A first record for Delhi NCR." *Indian BIRDS* 13, no. 2 (2017): 55.

Chill, W.N. "Letters to the Editor." *Stray Feathers* 10, no. 5 (1887): 427.

Christian, Catherene. "Status of Asian/Lesser Short-toed Larks *Alaudala cheleensis/rufescens* in India." *Indian BIRDS* 15, no. 3 (2019): 81–84.

Cicero, Carla, Nicholas A. Mason, Rosa Alicia Jiménez, Daniel R. Wait, Cynthia Y. Wang-Claypool, Rauri C.K. Bowie. "Integrative taxonomy and geographic sampling underlie successful species delimitation." *Ornithology* 138, no. 2 (2021): 1–15. URL: https://doi.org/10.1093/ornithology/ukab009.

Clements, J. F., T. S. Schulenberg, M. J. Iliff, T. A. Fredericks, J. A. Gerbracht, D. Lepage, S. M. Billerman, B. L. Sullivan, and C. L. Wood. 2022. The eBird/Clements checklist of Birds of the World: v2022. Downloaded from https://www.birds.cornell.edu/clementschecklist/download/.

Cuthbert, Richard, Rhys E. Green, Sachin Ranade, S. S. Saravanan, Deborah J. Pain, Vibhu Prakash, and Andrew A. Cunningham. "Rapid population declines of Egyptian Vulture *Neophron percnopterus* and Red-headed Vulture *Sarcogyps calvus* in India." *Animal Conservation* 9, no. 3 (2006): 349–354. https://doi.org/10.1111/j.1469-1795.2006.00041.x .

Das, Devojit., Richard J. Cuthbert, Ram D. Jakati, and Vibhu Prakash . "Diclofenac is toxic to the Himalayan Vulture *Gyps himalayensis*." *Bird Conservation International* 21, no. 1 (2011): 72–75. URL: https://doi.org/10.1017/S0959270910000171.

del Hoyo, Josep, and Nigel J. Collar. *HBW and BirdLife International Illustrated Checklist of the Birds of the World. Volume 1: Non-passerines*. Barcelona: Lynx Edicions, 2014.

del Hoyo, Josep, and Nigel J. Collar. *HBW and BirdLife International illustrated checklist of the birds of the world. Volume 2: Passerines*. Barcelona: Lynx Edicions, 2014.

Delhi Development Authority (DDA). *Landscape and Environmental Planning Department – Biodiversity Park*. 2021. URL: https://dda.gov.in/biodiversity-park.

Delhi Biodiversity Foundation DDA. *DDA Biodiversity Parks of Delhi*. 2021. URL: http://www.delhibiodiversityparks.org/.

Devasar, Nikhil, and Rajneesh Suvarna. *Birds About Delhi: A Field Guide*. New Delhi: Dorling Kindersley Publishing Private Limited, 2018.

Donahue, Julian P. "Notes on a collection of Indian birds, mostly from Delhi." *Journal of the Bombay Natural History Society* 64, no. 3 (1967): 410–429. Donahue, Julian P., and Usha Ganguli.

Donahue, Julian P., and Usha Ganguli. "Notes on a colony of the Whiskered Tern [*Chlidonias hybrida* (Pallas)] in Delhi, with comments on its breeding status in India." *Journal of the Bombay Natural History Society* 62 no. 2 (1965): 254–258.

Dookia, Sumit. "First record of Pied Tit *Parus nuchalis* in Thar Desert of Rajasthan." *Indian BIRDS* 3, no. 3 (2007): 112–113.

eBird. *eBird: An online database of bird distribution and abundance [web application]*. Ithaca, New York: Cornell Lab of Ornithology, 2022. URL: http://www.ebird.org.

Editors. "The Birds of Delhi and District". *Journal of the Bombay Natural History Society* 48, no. 4. (1950): 811–812.

Frome, Norman Fredrick. "The Birds of Delhi and district." *Journal of the Bombay Natural History Society* 47, no. 2. (1947): 277–300 and 47, no. 4 (1948): 751–3.

Galligan, Toby H., Tatsuya Amano, Vibhu Prakash, Mandar Kulkarni, Rohan Shringarpure, Nikita Prakash, Sachin Ranade, Rhys E. Green, and Richard J. Cuthbert. "Have population declines in Egyptian Vulture and Red-headed Vulture in India slowed since the 2006 ban on veterinary diclofenac?" *Bird Conservation International* 24, no. 3 (2014): 272–281. https://doi.org/10.1017/S0959270913000580.

Galushin, Vladimir M. "A huge urban population of birds of prey in Delhi, India (Preliminary note)." *Ibis* 113, no. 4 (1971): 522.

Ganguli, Usha. "The Large Sand Plover *Charadrius leschenaultii* in Delhi." *Newsletter for Birdwatchers* 5, no. 8 (1965): 5–6.

Ganguli, Usha. *A Guide to the Birds of the Delhi Area*. New Delhi: Indian Council of Agricultural Research, 1975.

Ganpule, Prasad, and Nirav Bhatt. "Steppe Buzzard *Buteo buteo vulpinus* in the Little Rann of Kachchh, and its distribution in the Saurashtra region of Gujarat, India." *Indian BIRDS* 9, no. 2 (2014): 41–45.

Ganpule, Prasad. "Roosting behaviour of Franklin's Nightjar *Caprimulgus affinis*." *Indian BIRDS* 6, nos. 4 and 5 (2010): 92–94.

Ganpule, Prasad. "Notes on the Great Grey Shrike (Laniidae: *Lanius excubitor*) complex in northwestern India: Variation, identification, and status." *Indian BIRDS* 11, no. 1 (2016): 1–10.

Ganpule, Prasad. "Red-backed, Brown, Isabelline and Red-tailed Shrike in Gujarat." *Flamingo* 15, no. 3 (2017): 1–7.

Ganpule, Prasad. "Field identification of Sand Lark *Alaudala raytal* and Lesser/Asian Short-toed Lark *Alaudala rufescens /cheleensis*: An unacknowledged pitfall." *Indian BIRDS* 15, no. 4 (2019): 97–111.

Gaston, Anthony, J. "Black Bulbuls *Hypsipetes madagascariensis* (P.L.S. Müller) in Delhi." *Journal of the Bombay Natural History Society* 69, no. 3 (1973a): 651–652.

Gaston, Anthony, J. "Ortolan Bunting *Emberiza hortulana* Linn. near Delhi." *Journal of the Bombay Natural History Society* 69, no. 3 (1973b): 654–655.

Gaston, Anthony, J. "The seasonal occurrence of birds on the New Delhi Ridge." *Journal of the Bombay Natural History Society* 75, no. 1 (1978a): 115–128.

Gaston, Anthony, J. "Notes on the Striated Babbler *Turdoides earlei* (Blyth) near Delhi." *Journal of the Bombay Natural History Society* 75, no. 1 (1978b): 219–220.

Gaston, Anthony, J. "Ecology of the Common Babbler *Turdoides caudatus*." *Ibis* 120 (4): 415–432.

Gaston, Anthony, J. "Some comments on the 'revival' of Sultanpur Lake." *Oriental Bird Club Bulletin* 20 (1994): 49–50.

Gaston, Anthony. J., and J. Mackrell. "Green Munia *Estrilda formosa* at Delhi, and other interesting records for 1978." *Journal of the Bombay Natural History Society* 77, no. 1 (1980): 144–145.

Gill, F., D. Donsker, and P. Rasmussen, eds. IOC World Bird List (v13.1), 2023. https://doi.org/10.14344/IOC.ML.13.1.

Grewal, Bikram. "Bristled Grassbird *Chaetornis striatus* at Okhla, Delhi." *Oriental Bird Club Bulletin* 24, (1996): 43–44.

Grimmett, Richard, Carol Inskipp, and Tim Inskipp. *Birds of the Indian Subcontinent*. 1st ed. London: Christopher Helm, A & C Black, 1998.

Gupta, Pankaj, and M.S. Srinivas. "Snapshot sightings: Scaly Thrush from Faridabad, Haryana." *Indian BIRDS* 13, no. 2 (2017): 56A.

Harris, Clive. Checklist of the Birds of Yamuna River (Okhla to Jaitpur village). Privately published, 2001.

Harvey, Bill, and Suresh C. Sharma. "The initial colonisation of the Yamuna flood plain by the Sind Sparrow *Passer pyrrhonotus*." *Journal of the Bombay Natural History Society* 99, no. 1 (2002): 35–43.

Harvey, Bill. *Checklist of the Birds of the Delhi Region*. Compiled by Delhibird and the Northern India Bird Network. Privately published, 2003.

Harvey, Bill, Nikhil Devasar, and Bikram Grewal. *Atlas of the Birds of Delhi and Haryana*. 1st ed. New Delhi: Rupa and Co, 2006.

Hume, Allan Octavian. "Notes." *Stray Feathers* 7, nos. 3, 4 and 5 (1878): 451–465.

Hume, Allan Octavian. "Notes." *Stray Feathers* 8, nos. 2–5 (1879): 406–413.

Hume, Allan Octavian. "Notes." *Stray Feathers* 10, nos. 1–3 (1887): 1–174 and no. 5 (1887): 329–434.

Hume, Allan Octavian. *The Nests and Eggs of Indian Birds*

Vol 1. 2nd ed. London: R.H. Porter, 1889.

Hume, Allan Octavian. *The Nests and Eggs of Indian Birds Vol II*. 2nd ed. London: R.H. Porter, 1890a.

Hume, Allan Octavian. *The Nests and Eggs of Indian Birds Vol III*. 2nd ed. London: R.H. Porter, 1890b.

Hume, Allan Octavian, and Charles Henry Tilson Marshall. *The Game Birds of India, Burmah, and Ceylon*. 1st ed. Vol. I of 3 vols. Calcutta: Published by the authors, 1879.

Hume, Allan Octavian and Charles Henry Tilson Marshall. *The Game Birds of India, Burmah, and Ceylon*. 1st ed. Vol. III of 3 vols. Calcutta: Published by the authors, 1881.

Hutson, Henry Porter Wolseley. *The Birds About Delhi, Together With a Complete List of Birds Observed in Delhi and the Surrounding Country*. Delhi: The Delhi Bird Watching Society, 1954.

Jackson, Peter F.R. "Whiteheaded Yellow Wagtail [*Motacilla flava leucocephala* (Przevalski)] near Delhi." *Journal of the Bombay Natural History Society* 62, no. 2 (1965): 304–305.

Jackson, Peter. "Some new bird records for Delhi." *Journal of the Bombay Natural History Society* 65, no. 3 (1968): 780–782.

Kalpavriksh. *What's That Bird? A Guide to Birdwatching, With Special Reference to Delhi*. New Delhi: Kalpavriksh, 1991.

Kazmierczak, Krys. *A Field Guide to the Birds of India, Sri Lanka, Pakistan, Nepal, Bhutan, Bangladesh and the Maldives*. 1st ed. New Delhi: Om Book Service, 2000.

Kelsey, Martin. "Sight record of Horned Lark *Eremophila alpestris* near Delhi." *Journal of the Bombay Natural History Society* 101, no. 2 (2004): 321.

Khan, Mohd. Shahnawaz, Zarreen Syed Aftab, Asghar Nawab, Orus Ilyas and Affifullah Khan. Composition and conservation status of avian species at Hastinapur Wildlife Sanctuary, Uttar Pradesh, India. *Journal of Threatened Taxa* 5 (2013): 4714–4721.

Khera, Neerja., V. Mehta, and B. C. Sabata. "Interrelationship of birds and habitat features in urban greenspaces in Delhi, India." *Urban Forestry & Urban Greening* 8, no. 3 (2009): 187–196.

Koparde, Pankaj, and Pierre Yesou. "Probable hybrids Little Egret x Indian Reef Heron in India and Sri Lanka." *Dutch Birding* 39 (2017): 238–246.

Krishen, Pradip. *Trees of Delhi: A Field Guide*. 1st ed. New Delhi: Dorling Kindersley (India) Pvt. Ltd, 2006.

Kumar, Sunil. "Snapshot sightings: Stoliczka's Bushchat from Delhi." *Indian BIRDS* 12, nos. 2 and 3 (2016): 88A.

Lambert, James. "Snapshot sightings: Little Gull at Okhla, Uttar Pradesh." *Indian BIRDS* 10, no. 1 (2015): 28A.

Lees, Alexander, and J. Gilroy. *Vagrancy in Birds*. Princeton: Princeton University Press, 2022.

Lewis, E.S. "Bewick's Swan (*Cygnus bewickii* Yarrell) near Delhi." *Journal of the Bombay Natural History Society* 40, no. 2 (1938): 333.

Luo, Site, Yuchun Wu, Qing Chang, Yang Liu, Xiaojun Yang, Zhengwang Zhang, Min Zhang, Qiang Zhang, and Fasheng Zou. "Deep phylogeographic divergence of a migratory passerine in Sino-Himalayan and Siberian forests: The Red-flanked Bluetail *Tarsiger cyanurus* complex." *Ecology and Evolution* 4, no. 7 (2014). 977–86. https://doi.org/10.1002/ece3.967.

MacDonald, Malcolm. *Birds in my Indian Garden*. London: Jonathan Cape, 1960.

Majumdar, Pradyumna Krishna, Rakesh Ahlawat, Sarita Majumdar, and Sarika Sharma. "Green-crowned Warbler *Phylloscopus burkii*, a new species for Haryana, and Delhi NCR." *Indian BIRDS* 17, no. 5 (2021): 157–58.

Maheshwari, Jai Kishan. *The Flora of Delhi*. Delhi: Council of Scientific and Industrial Research (CSIR), 1963.

McCarthy, Eugene M. *Handbook of Avian Hybrids of the World*. New York: OUP, 2006.

Mehta, Mohit, and Piyush Dogra. "Snapshot sightings: Jungle Bush Quail from Faridabad, Haryana." *Indian BIRDS* 13, no. 5 (2017): 140A.

Manral, Upma, Angshuman Raha, Ridhima Solanki, Syed Ainul Hussain, Mattozbiyil Mani Babu, Dhananjai Mohan, Gopi G. Veeraswami, K. Sivakumar, and Gautam Talukdar. "Plant species of Okhla Bird Sanctuary: A wetland of Upper Gangetic Plains, India." *Check List* 9 no. 2 (2013): 263–274.

Mehta, Mohit, and Piyush Dogra. "Snapshot sightings: Whimbrel at Najafgarh Drain, Haryana." *Indian BIRDS* 14, no. 2 (2018a): 64A.

Mehta, Mohit, and Piyush Dogra. "Snapshot sightings: Red-tailed Wheatear in Sultanpur, Delhi NCR." *Indian BIRDS* 14, no. 3 (2018b): 96A.

Menon, Tarun, Ashwin Viswanathan, Manjunath Desai, Prabhakar T. P., K. Dhanapal, Rajiv Ramaswamy, and Mohit Aggarwal. "Is the Black-breasted Weaver *Ploceus benghalensis* polytypic?" *Indian BIRDS* 17, no. 6 (2021): 174–178.

Menon, Vivek, Tara Gandhi, Mohit Aggarwal, and Rajesh Thadani. "Slenderbilled Gull *Larus genei* Brème in New Delhi." *Journal of the Bombay Natural History Society* 92, no. 3 (1995): 419.

Mishra, Veer Vaibhav. "A Mistle Thrush *Turdus viscivorus* from Banni Grasslands, Gujarat, India." *Indian BIRDS* 10, no. 6 (2015): 161.

Nanda, Kavi. "Tickell's Blue Flycatcher *Cyornis tickelliae*, and Spotted Flycatcher *Muscicapa striata* from the Delhi-NCR region." *Indian BIRDS* 12, no. 6 (2017): 175.

Nanda, Kavi. "A Sykes's Nightjar *Caprimulgus mahrattensis*, and a summer record of Eurasian Skylark *Alauda arvensis* from Delhi-NCR Region." *Indian BIRDS* 15, no. 1 (2018): 27.

Naoroji, Rishad. *Birds of Prey of the Indian Subcontinent*. New Delhi: Om Books International, 2006.

"Notes on a colony of the Whiskered Tern [*Chlidonias hybrida* (Pallas)] in Delhi, with comments on its breeding status in India." *Journal of the Bombay Natural History Society* 62, no. 2 (1965): 254–258.

Olson, David. M., Eric Dinerstein, Eric D. Wikramanayake, Neil D. Burgess, George V. N. Powell, Emma C. Underwood, et al. "Terrestrial Ecoregions of the

World: A New Map of Life on Earth: A new global map of terrestrial ecoregions provides an innovative tool for conserving biodiversity." *BioScience*, Vol. 51, no.11 (2001): 933–38. https://doi.org/10.1641/0006-3568(2001)051[0933:TEOTWA]2.0.CO;2.

Olsson, Urban., Per Alström, Lars Svensson, Mansour Aliabadian, and Per Sundberg. "The *Lanius excubitor (Aves, Passeriformes)* conundrum—taxonomic dilemma when molecular and non-molecular data tell different stories." *Molecular Phylogenetics and Evolution* 55, no. 2 (2010): 347–57.

Pittie, Aasheesh. Bibliography of South Asian Ornithology: 2018. Accessed 30 June 2022. http://www.southasiaornith.in.

Prakash, Vibhu., Mohan Chandra Bishwakarma, Anand Chaudhary, Richard Cuthbert, Ruchi Dave, Mandar Kulkarni, Sashi Kumar, Khadananda Paudel, Sachin Ranade, Rohan Shringarpure, Rhys E. Green. "The population decline of *Gyps* vultures in India and Nepal has slowed since veterinary use of Diclofenac was banned." *PloS ONE* 7, no. 11 (2012): 1–10. https://doi.org/10.1371/journal.pone.0049118.

Prakash, Vibhu., Toby H. Galligan, Soumya S. Chakraborty, Ruchi Dave, Mandar D. Kulkarni, Nikita Prakash, Rohan N. Shringarpure, Sachin P. Ranade, and Rhys E. Green. "Recent changes in populations of Critically Endangered Gyps vultures in India." *Bird Conservation International*, 29, no. 1 (2017). 1–16. https://doi.org/10.1017/S0959270917000545.

Prakash, Vibhu., Rhys E. Green, Debbie Pain, Sachin P. Ranade, S. Saravanan, Nikita Prakash, R. Venkitachalam, Richard J. Cuthbert, Asad Rafi Rahmani, and Andrew A. Cunningham. "Recent changes in populations of resident *Gyps* vultures in India." *Journal of the Bombay Natural History Society* 104, no. 2 (2007): 127–133.

Praveen, J. "Pin-tailed Sandgrouse." URL: https://ebird.org/view/checklist/S42116096 (n.d.) [See also: Blewitt 1875, above.]

Praveen, J., Rajah Jayapal, and Aasheesh Pittie. "Notes on Indian rarities-2: Waterfowl, diving waterbirds, gulls and terns." *Indian BIRDS* 9 nos. 5 and 6 (2014): 113–136.

Praveen, J., Rajah Jayapal, and Aasheesh Pittie. "A Checklist of the Birds of India." *Indian BIRDS* 11, nos. 5 and 6 (2016): 113–172A.

Praveen, J., and R. Jayapal. Checklist of the birds of India (v.7.0) (2023). http://www.indianbirds.in/india/ [Date of publication : 28 February 2023].

Rahmani, Asad R. "Status of the Blacknecked Stork, *Ephippiorhyncus asiaticus*, in the Indian subcontinent.' *Forktail* 5 (December, 1989): 99–110.

Rahmani, Asad R. "Status and distribution of White-browed Bushchat *Saxicola macrorhyncha* in India." *Forktail* 12 (1996): 61–77.

Rahmani, Asad R., Zafar-ul Islam, and Raju M. Kasambe. *Important Bird and Biodiversity Areas in India: Priority sites for conservation*. Revised and updated ed. Mumbai: Bombay Natural History Society, Indian Bird Conservation Network, Royal Society for the Protection of Birds and BirdLife International, 2016.

Rai, Yado Mohan. "The Birds of Delhi and Meerut." *Journal of the Bombay Natural History Society* 83, no. 1 (1986): 212–214.

Rasmussen, Pamela C. and John C Anderton. *Birds of South Asia: the Ripley Guide: Field Guide Vols I and II*. 1st ed. Washington, D.C. and Barcelona: Smithsonian Institution and Lynx Edicions, 2005.

Rasmussen, Pamela C. and John C. Anderton. *Birds of South Asia: the Ripley Guide Vols I and II*. 2nd ed. Washington, D.C. and Barcelona: Smithsonian Institution and Lynx Edicions, 2012.

Restall, Robin. *Munias and Mannikins*. New Haven: Yale University Press, 1997.

Roberts, Tom J. *The birds of Pakistan: Regional Studies and Non-Passeriformes Vol 1*. 1st ed. Karachi: OUP, 1991.

Roberts, Tom J. *The Birds of Pakistan. Passeriformes: Pittas to Buntings Vol 2*. 1st ed. Karachi: OUP, 1992.

Robson, Craig. "From the field: India." *Oriental Bird Club Bulletin* 28: 44.

Robson, Craig. From the field: Sri Lanka. *Oriental Bird Club Bulletin* 31 (2000): 52.

Robson, Craig. From the Field: India. *BirdingASIA* 3 (2005): 78–79.

Sangha, Harkirat Singh. *Waders of the Indian Subcontinent*. Jaipur: Privately published, 2021.

Sangha, Harkirat Singh, Gobind Sagar Bhardwaj, and Surat Singh Poonia. "Huge concentrations of the White-eyed Buzzard *Butastur teesa* at Tal Chhapar Wildlife Sanctuary, Churu district, Rajasthan." *Indian BIRDS* 7, no. 5 (2011): 139–40.

Sangster, George, Kim Manzon Cancino, and Robert O. Hutchinson. "Taxonomic revision of the Savanna Nightjar (*Caprimulgus affinis*) complex based on vocalizations reveals three species." *Avian Research* 12, no. 54 (2021). https://doi.org/10.1186/s40657-021-00288-z.

Sankar, Kalyanasundaram, Dhananjai Mohan, and Sanjeeva Pandey. "Birds of Sariska Tiger Reserve, Rajasthan, India." *Forktail* 8 (February 1993): 133–41.

Sant, Niranjan, Rahul Prabhukhanolkar, Vidhyadhar Shelke, and Shridhar Shelke. "Breeding ecology of the Indian Spotted Eagle *Clanga hastata* around Belgaum, India." *Indian BIRDS* 16, no. 6 (2020): 176–84.

Schweizer, Manuel and Raffael Ayé. "Identification of the Pale Sand Martin *Riparia diluta* in Central Asia." *Alula* 13, no. 4 (2007): 152–58.

Shahabuddin, Ghazala, and Amita Baviskar. "Birdlife: Heirs of the ecological mosaic." In *First Garden of the Republic: Nature in the President's Estate*. Edited by Amita Baviskar, 160–197. New Delhi: Publications Division, Government of India, 2016.

Shahabuddin, Ghazala, Raman Kumar, and Ashok Verma. "Annotated checklist of the birds of Sariska Tiger Reserve, Rajasthan, India." *Indian BIRDS* 2, no. 3 (2006): 71–76 (with one table).

Sharma, Manoj, Harkirat Singh Sangha, Sharad Sridhar, and C. Abhinav "Long-billed Dowitcher *Limnodromus scolopaceus* at Sultanpur National Park, Haryana, India." *Indian BIRDS* 8, no. 4 (2013): 101–03.

Sharma, Suresh C. "White-headed Yellow Wagtail *Motacilla flava leucocephala* (Przevalski) near Delhi, India." *Indian BIRDS* 1, no. 3 (2005): 70–71.

Sharma, Suresh C., and Jaideep Chanda. "Indian Spotted Eagle *Aquila hastata* nesting in Sonepat, Haryana, India." *Indian BIRDS* 6, no. 1 (2010): 18.

Sharma, Suresh C., and P.S. Sangwan. "Stoliczka's Bushchat *Saxicola macrorhyncha* in Hissar District, Haryana." *Indian BIRDS* 1, no. 1 (2005): 6–7.

Singal, Dinesh K. Upland Buzzard. URL: http://indianbirdsphotography.blogspot.in/2015/03/delhibirdpix-upland-buzzard_21.html (2015) [Accessed: 30 June 2022].

Sinha, G.N. ed. *An Introduction to the Delhi Ridge*. New Delhi: Department of Forests and Wildlife, Government of NCT of Delhi, 2014.

Sondhi, Sanjay, Ashish Kothari, Balwant Singh Negi, Bhupinder Singh, Deep Chandra Joshi, Naveen Upadhyay, Puran Singh Pilkhwal, and Virender Singh. "First record of the Pompadour ('Ashy-headed') Green Pigeon *Treron pompadora conoveri/phayrei* from Uttarakhand, India" *Indian BIRDS* 11, no. 1 (2016): 21–22.

Sridharan, E., and S. Bikhchandani. 'The Redtailed Wheatear *Oenanthe xanthoprymna* in the Delhi area.' *Journal of the Bombay Natural History Society* 78 no. 1 (1981): 170.

Stervander, Martin, Per Alström, Urban Olsson, Ulf Ottosson, Bengt Hansson, and Staffan Benscha. "Multiple instances of paraphyletic species and cryptic taxa revealed by mitochondrial and nuclear RAD data for *Calandrella* larks (Aves: Alaudidae)." *Molecular Phylogenetics and Evolution* 102 (2016): 233–45.

Sultana, Aisha. "An updated checklist of birds of Sariska Tiger Reserve, Rajasthan, India." *Journal of Threatened Taxa* 5, no. 13 (2013): 4791–4804.

Gopi Sundar, K.S., Rakesh Ahlawat, Devender Singh Dalal, and Swati Kittur. "Does the stork bring home the owl? Dusky Eagle-Owls *Bubo coromandus* breeding in Woolly-necked Stork *Ciconia episcopus* nests." *BioTROPICA* 54, no. 2 (2022): 1–5. https://doi.org/10.1111/btp.13086.

Svensson, Lars, *et al*. "Species limits in the Red-breasted Flycatcher." *British Birds* 98 (2005): 538–541.

Tak, P.C. and J.P. Sati, 'Aves.' In *Fauna of Delhi*. Edited by Director, Zoological Survey of India: 699–821. Calcutta: Zoological Survey of India, 1997.

Tobias, Joseph A., Nathalie Seddon, Claire N. Spottiswoode, John D. Pilgrim, Lincoln D C Fishpool, and Nigel J. Collar "Quantitative criteria for species delimitation." *Ibis* 152, no. 4 (2010): 724–46.

Trivedi, Rajni, and B. M. Parasharya. "Probable breeding of Little Bittern *Ixobrychus minutus* at Nalsarovar Bird Sanctuary, Gujarat, western India, with notes on identification of juveniles." *Indian BIRDS* 15, no. 1 (2019): 17–20.

Turaga, Janaki. "Oriental Scops Owl *Otus sunia* sighted in Delhi after nearly a century." *Indian BIRDS* 9, no. 3 (2014): 76–77.

Turaga, Janaki. "Birds and trees in an urban context: An ecosystem paradigm for Vasant Vihar, New Delhi, India." *Indian BIRDS* 10, nos. 3 and 4 (2015): 85–93.

Tyabji, Nadir S. "Birds of Najafgarh." *Newsletter for Birdwatchers* 1, no. 11 (1961): 4–7.

Tyabji, Nadir S. "Birds of New Delhi area." *Newsletter for Birdwatchers* 2, no. 2 (1962): 7.

Urfi, Abdul Jamil. "Wetlands of ornithological significance in the Delhi region." *Oriental Bird Club Bulletin* 22 (1995): 38–41.

Urfi, Abdul Jamil. "The significance of Delhi Zoo for wild waterbirds, with special reference to the Painted Stork *Mycteria leucocephala*." *Forktail* 12 (1996): 87–97.

Urfi, Abdul Jamil. "The birds of Okhla barrage bird sanctuary, Delhi, India." *Forktail* 19 (2003): 39–50.

Urfi, Abdul Jamil. *The Painted Stork: Ecology and Conservation*. New York: Springer-Verlag, 2011.

Urfi, Abdul Jamil, Thangarasu Meganathan, and Abdul Kalam. "Nesting ecology of the Painted Stork *Mycteria leucocephala* at Sultanpur National Park, Haryana, India." *Forktail* 23 (2007): 150–153.

Urfi, Abdul Jamil, and S.C. Sharma. "Bird conservation at some lesser known wetlands around Delhi." *Newsletter for Birdwatchers* 32, nos. 5-6 (1992): 2–5.

Vittery, A. "The birds of Pakistan: supplementary observations from the northern Punjab and hills." *Forktail*. 9 (1994): 143–47.

Vyas, Sudhir. "Four additions to the Delhi list and other interesting records." *Newsletter for Birdwatchers* 19, no. 11 (1979): 2–5.

Vyas, Sudhir. "Checklist of the birds of the Delhi region: an update." *Journal of the Bombay Natural History Society* 93, no. 2 (1996): 219–37.

Vyas, Sudhir. "Some interesting bird records from the Delhi area." *Journal of the Bombay Natural History Society* 99, no. 2 (2002): 325–30.

Vyas, Sudhir. *The Birds of the Delhi Area: An Annotated Checklist. Indian BIRDS Monograph* 1. Hyderabad: New Ornis Foundation, 2019. http://www.indianbirds.in/pdfs/IB_Mono1_Vyas_DelhiBirds1.pdf.

Waraich, Jaswinder. "Snapshot sightings: Chestnut-headed Bee-eater from Okhla, Delhi NCR." *Indian BIRDS* 13, no. 4 (2017): 112A.

Whistler, Hugh. *Popular handbook of Indian birds*. Revised and Enlarged 4th ed. London: Gurney and Jackson, 1949.

Whistler, Hugh. "Some notes on the Genus *Caprimulgus* (Nightjars) in the Punjab ['With a note on the nightjars of Sind by Dr. C. B. Ticehurst']." *Journal of the Bombay Natural History Society* 27, no. 2 (1920): 363–70.

Winkler, David W., Shawn M. Billerman, and Irby J. Lovette. Hawks, Eagles, and Kites (*Accipitridae)*, version 1.0. In *Birds of the World*. Edited by Billerman et al. Ithaca, New York. Cornell Lab of Ornithology, 2022.

Worfolk, Tim. "Identification of Red-backed, Isabelline and Brown Shrikes." *Dutch Birding* 22, no. 6 (2000): 323–62.

Index

[Page numbers in **bold** indicate a main profile.]

Common Names

Scientific Names

A Note on the Author and Photographer

A career diplomat, now retired, **Sudhir Vyas** has pursued his passion for birdwatching for over 50 years around Delhi, elsewhere in India, and overseas as well, including in some of India's neighbouring countries that he had the opportunity to serve in. He knows the birds of Delhi well, and has authored several articles and studies on the ornithology of the Delhi area. He's also the author of a monograph on the subject published in the scholarly journal *Indian Birds*.

A wildlife and bird photographer who has documented 377 avian species in Haryana and over a thousand internationally, **Amit Sharma**'s work has featured in books such as the *Birds of Haryana and 100 Best Birdwatching Sites in India*, as well as in leading national newspapers. He has photographed some of the rarest birds seen in the Delhi region. Internationally, his work on documenting Ethiopian wildlife has featured in a wildlife film broadcast on Ethiopia's national TV network. An adventure enthusiast at heart, he also enjoys trekking and has climbed three 6,000-metre peaks in India.

About Indian Pitta

Indian Pitta is India's first dedicated book imprint for bird lovers, conservationists, and policy makers. Our books about birds and natural history go beyond field/identification guides, to explore the bigger mosaic of habitats, ecosystems and human interactions that touch the lives of birds. Successful conservation programmes, troubling environmental challenges, personal exploration of a landscape, deep dives into the ecology of a species, the quest for a rare bird and the sheer joy of birding – these are some of the ideas that you can expect to explore within the pages of our books.

Upcoming Titles in 2023

***Women in the Wild* edited by Anita Mani**

How many Indian wildlife biologists can you name? How many Indian women wildlife biologists can you name?

Yet, there are several, and their lives and work have been extraordinary.

Written by natural historians, *Women in the Wild* will not only explore the details of their work – and their impact – but also what it took for them to get there. Inspiring and eye-opening, it will showcase the untold stories of some of India's most brilliant women wildlife scientists.

***Birder on the Road* by Shashank Dalvi**

Wildlife biologist Shashank Dalvi's *Birder on the Road* recounts his year-long trip across India to watch and record 1,128 species of birds. As much about India as its extraordinarily rich and diverse bird life, this book will be a travel and wildlife classic.